BROOKS/COLE
CENGAGE Learning

ASTRO 2010–2011 Edition
Michael Seeds
Dana Backman

Publisher: Mary Finch

Acquisitions Editor: Kilean Kennedy

Developmental Editor: Laura Rush,
B-books, Ltd.

Product Development Manager,
4LTR Press: Steven E. Joos

Project Manager, 4LTR Press:
Clara Goosman

Editorial Assistant: Laura Bowen

Brand Executive Marketing Manager,
4LTR Press: Robin Lucas

Marketing Director: Patrick Leow

Marketing Manager: Nicole Mollica

Marketing Coordinator: Kevin Carroll

Marketing Communications Manager:
Belinda Krohmer

Production Director: Amy McGuire,
B-books, Ltd.

Content Project Manager: Jill Clark

Media Editor: Rebecca Berardy-Schwartz

Manufacturing Coordinator:
Miranda Klapper

Production Service: B-books, Ltd.

Senior Art Director: Cate Rickard Barr

Cover Design: Irene Morris

Cover Image Credit: NASA and The Hubble
Heritage Team (STScI)

Photography Manager: Deanna Ettinger

Photo Researchers: Laura Rush, B-books,
Ltd., and Charlotte Goldman

Cover Image: A brilliant white core is
encircled by thick dust lanes in this spiral
galaxy, seen edge-on. The galaxy is
50,000 light-years across and 28 million
light years from Earth.

For product information and technology assistance, contact us at
Cengage Learning Academic Resource Center, 1-800-423-0563

For permission to use material from this text or product,
submit all requests online at **www.cengage.com/permissions**
Further permissions questions can be emailed to
permissionrequest@cengage.com

Library of Congress Control Number: 2009943794

SE ISBN-13:978-0-538-73804-0
SE ISBN-10: 0-538-73804-9

Brooks/Cole Cengage Learning
20 Channel Center Street
Boston, MA 02210
USA

Cengage Learning products are represented in Canada by
Nelson Education, Ltd.

For your course and learning solutions, visit **academic.cengage.com**
Purchase any of our products at your local college store or at our
preferred online store **www.CengageBrain.com**

Printed in the United States of America

3 4 5 6 7 13 12

astro Brief Contents

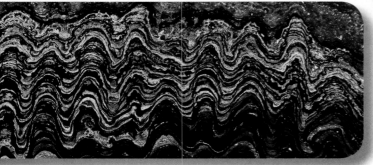

astro
Contents

Contents

Contents

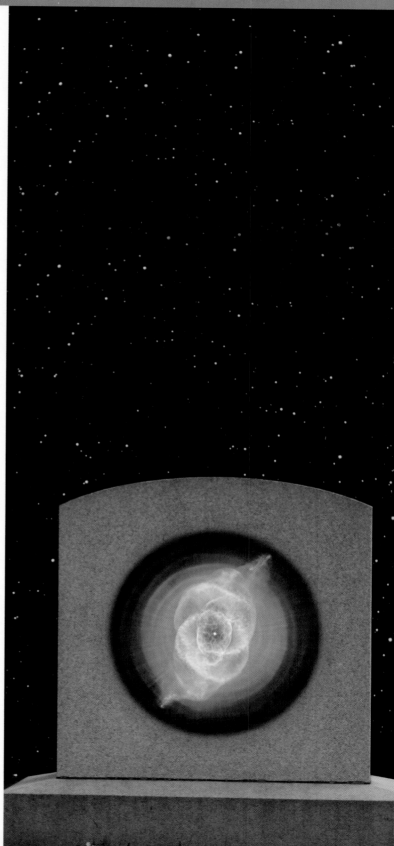

astro Contents

astro
Contents

Contents

The Scale of
the Cosmos: From Solar System to Galaxy to Universe

> The longest journey begins with a single step.
> —Lao Tse

You are about to go on a voyage to the limits of the known universe, traveling outward, away from your home on Earth, past the moon and the sun and the other planets of our solar system, past the stars you see in the night sky, and beyond billions more stars that can be seen only with the aid of telescopes. You will visit the most distant galaxies—great globes and whirlpools of stars—and continue on, carried only by experience and imagination, seeking to understand the structure of the universe. Astronomy is more than the study of planets, stars, and galaxies—it is the study of the whole universe in which you live. Although humanity is confined to a small planet circling an average star, the study of astronomy can take you beyond these boundaries and help you not only see *where* you are but also understand *what* you are.

getting started

You are already an expert in astronomy. You have enjoyed sunsets and moonrises, admired the stars, and may know a few constellations. You have probably read about Mars rovers and the Hubble Space Telescope. That is more than most Earthlings know about astronomy. Still, you owe it to yourself to understand where you are. You should know what it means to live on a planet that whirls around a star sailing through one galaxy in a universe full of galaxies.

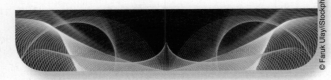

Your imagination is the key to discovery; it will be your scientific space-and-time machine transporting you across the universe and into the past and future. Go back in time to watch the formation of the sun and Earth, the birth of the first stars, and ultimately the creation of the universe. Then, rush into the future to see what will happen when the sun dies and Earth withers.

Although you will discover a beginning to the universe, you will not find an edge or an end in space. No matter how far you voyage, you will not run into a wall. In a later chapter you will discover evidence that the universe may be infinite; that is, it may extend in all directions without limit.

Astronomy will introduce you to sizes, distances, and times far beyond your usual experience on Earth. Your task in this chapter is to grasp the meaning of these unfamiliar sizes, distances, and times. The solution lies in a single word: scale. In this chapter, you will compare objects of different sizes in order to comprehend the scale of the universe.

looking ahead

It is easy to learn a few facts, but it is the relationships among facts that are important. This chapter will give you the sense of scale that you need to understand where you are in the universe. The remaining chapters in this book will fill in the details, give evidence, describe theories, and illustrate the wonderful intricacy and beauty of the universe. That journey begins here.

So distant that light has taken 2500 years to reach Earth, the Veil Nebula was produced by the explosion of a star 15,000 years ago.

astro

Figure 1.1

16 meters

Figure 1.2

You should begin with something familiar, like the size of yourself and your surroundings. Figure 1.1 shows a region about 16 meters (52 feet) across occupied by a human being, a sidewalk, and a few trees—all objects with sizes you can relate to. Each picture in the following sequence shows you a frame or **field of view** within the universe that is 100 times wider than the preceding picture.

In Figure 1.2, your field of view has increased in size by a factor of 100, and you can now see an area about 1.6 kilometers (1 mile) in diameter. The 16-meter area of Figure 1.1 is quite small from this view. Now you can see an entire college campus and the surrounding streets and houses. This is still the world you know and can relate to the scale of your body.

Figure 1.3 has a span of 160 kilometers (100 miles). Take a look at the picture and notice the red color. This is an infrared photograph, in which healthy green leaves and crops show up as red. Your eyes are sensitive to only a narrow range of colors. As you explore the universe, you will need to learn to use a wide range of "colors," from X rays to radio waves, revealing sights invisible to your unaided eyes.

The college campus is now invisible, and the patches of gray are towns and cities, including Wilmington, Delaware, visible in the lower right corner. At this scale, you can see natural features of Earth's surface: the Allegheny Mountains of southern Pennsylvania crossing the picture at the upper left and the Susquehanna River flowing southeast into

Field of view: The area visible in an image. Usually given as the diameter of the region.

Chesapeake Bay. What look like white bumps are actually puffs of clouds. Mountains and valleys are only temporary features on Earth that are slowly but constantly changing. As you explore the universe, you will come to see that it is also always evolving.

In the next step of the journey, you will see the entire planet Earth (Figure 1.4), which is about 13,000 kilometers (8,000 miles) in diameter. This picture shows most of the daylight side of the planet. However, the blurriness at the extreme right is the sunset line. The rotation of Earth on its axis each 24 hours carries you eastward, and as you

Figure 1.3

cross the sunset line into darkness you say that the sun has set. At the scale of this figure, the atmosphere on which your life depends is thinner than a strand of thread.

Enlarge your field of view again by a factor of 100, and you see a region 1,600,000 kilometers (1 million miles) wide (Figure 1.5). Earth is the small blue dot in the center, and the moon, with a diameter only about one-fourth that of Earth, is an even smaller dot along its orbit. If you've had a high-mileage car, it may have traveled the equivalent of a trip to the moon, which has an average distance from Earth of 380,000 kilometers (240,000 miles). These numbers are so large that it is inconvenient to write them out. Astronomy is the science of big numbers, and you will use numbers much larger than these to describe the universe.

Rather than writing out numbers such as those in the previous paragraph, it is more convenient to write them in **scientific notation**. This is nothing more than a simple way to write numbers without writing a great many zeros. For example, in scientific notation, you can write 380,000 as 3.8×10^5.

When you once again enlarge your view by a factor of 100 (Figure 1.6), Earth, its moon, and the moon's orbit all lie in the small red box at lower left. This figure has a diameter of 1.6×10^8 kilometers. Now you see the sun and two other planets that are part of our solar system. The **solar system** consists of the sun, its family of planets, and some smaller bodies, such as moons, asteroids, and comets.

Like Earth, Venus and Mercury are **planets**, small, nonluminous bodies that shine by reflecting sunlight. Venus is about the size of Earth, and Mercury is a bit larger

than Earth's moon. In this figure they are both too small to be seen as anything but tiny dots. The sun is a **star**, a self-luminous ball of hot gas that generates its own energy. The sun is about 110 times larger in diameter than Earth (inset), but it too is nothing more than a dot in this view. Earth orbits the sun once a year.

Another way astronomers deal with large numbers is to define new units. The average distance from Earth to the sun is called the **Astronomical Unit (AU)**, a distance of 1.5×10^8 kilometers (93 million miles). Using that unit you can then say, for example, that the average distance from Venus to the sun is about 0.7 AU.

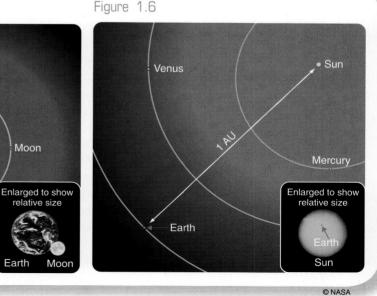

Each of the pictures in the sequence of images shows you a field of view within the universe 100 times wider than the preceding picture.

Figure 1.4

Figure 1.5

Figure 1.6

Earth → Moon

Enlarged to show relative size

Earth   Moon

Venus

Sun

1 AU

Mercury

Earth

Enlarged to show relative size

Earth

Sun

© NASA

Figure 1.7　　　　　　　　Figure 1.8 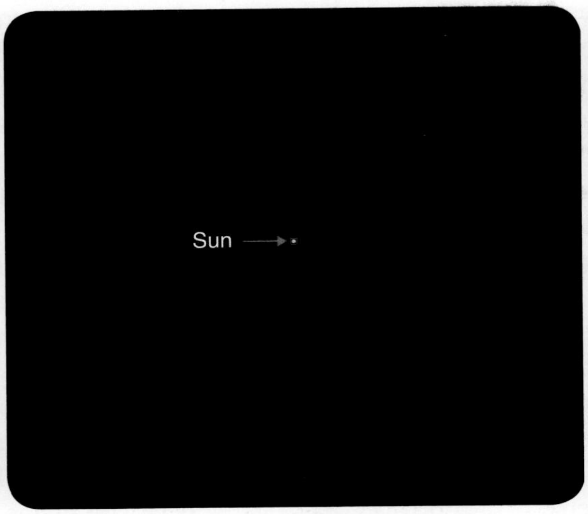　　　　　　　　Figure 1.9

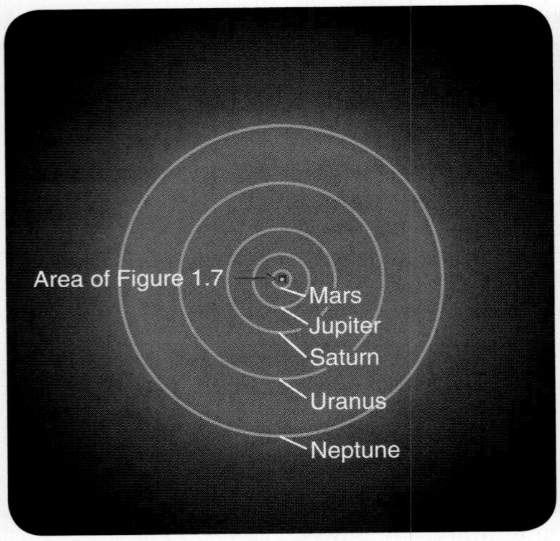

Your first field of view was only about 16 meters (52 feet) in width. After just six steps, each enlarging by a factor of 100, you now see the entire solar system, all the major plants, and their slightly elliptical orbits (Figure 1.7). Your view now is 1 trillion (10^{12}) times wider than in the first figure. The details of the previous figure are lost in the red square at the center of this figure. You see only the brighter, more widely separated objects as you back away. The sun, Mercury, Venus, and Earth are so close together that you cannot separate them at this scale.

Mars, the next outward planet, lies only 1.5 AU from the sun. In contrast, Jupiter, Saturn, Uranus, and Neptune are so far from the sun that they are easy to find in this figure. Light from the sun reaches Earth in only 8 minutes, but it takes over 4 hours to reach Neptune. Pluto orbits mostly outside Neptune's orbit, but it is no longer considered a major planet.

When you again enlarge your view by a factor of 100, the solar system becomes invisibly small (Figure 1.8). The sun is only a point of light, and all the planets and their slightly elliptical orbits are now crowded into the small red square at the center. The planets are too small and reflect too little light to be visible so near the brilliance of the sun.

Nor are any stars visible except for the sun. The sun is a fairly typical star, a bit larger than average, and it is located in a fairly normal neighborhood in the universe. Although there are many billions of stars like the sun, none is close enough to be visible in this figure. The stars are separated by average distances about 30 times larger than this view, which has a

Light year (ly): Unit of distance equal to the distance light travels in 1 year.

diameter of 11,000 AU. It is difficult to grasp the isolation of the stars. If the sun were represented by a golf ball in New York City, the nearest star would be another golf ball in Chicago.

In Figure 1.9, your view has expanded to a diameter a bit over 1 million AU. The sun is at the center, and you see a few of the nearest stars. These stars are so distant that it is not reasonable to give their distances in AU. Astronomers have defined a new larger unit of distance, the light-year. One **light-year (ly)** is the distance that light travels in 1 year, roughly 10^{13} km or 63,000 AU. The diameter of your view in Figure 1.9 is 17 ly. The nearest star to the sun, Proxima Centauri, is 4.2 ly from Earth. In other words, light from Proxima Centauri takes 4.2 years to reach Earth.

Although these stars are roughly the same size as the sun, they are so far away that you cannot see them as anything but points of light. Even with the largest telescopes on Earth, you still see only points of light when you look at stars, and any planets that might circle those stars are much too small and faint to be visible. In Figure 1.9 the sizes of the dots represent not the sizes of the stars but their brightness. This is the custom in astronomical diagrams, and it is also how starlight is recorded. Bright stars make larger spots in a photograph or electronic picture than faint stars. The size of a star image in a photograph tells you not how big the star really is but only how bright it looks.

In Figure 1.10, you expand your field of view by another factor of 100, and the sun and its neighboring stars vanish into the background of thousands of stars. This figure is 1700 ly in diameter. Of course, no one has ever journeyed thousands of light-years from Earth to look back and photograph the sun's neighborhood, so this is a representative picture from Earth of a part of the sky that can be used as a reasonable simulation. The sun is faint enough that it would not be easily located in a picture at this scale.

Some things that are invisible in this figure are actually critically important. You do not see the thin gas that fills the spaces between the stars. Although those clouds of gas are thinner than the best vacuum produced in laboratories on Earth, it is those clouds that give birth to new stars. The sun formed from such a cloud about 5 billion years ago. You will see more star formation in the next figure.

If you expand your view again by a factor of 100, you see our galaxy (Figure 1.11). A **galaxy** is a great cloud of stars, gas, and dust bound together by the combined gravity of all the matter. In the night sky, you see our galaxy from the inside as a great, cloudy band of stars ringing the sky as the **Milky Way**, and our galaxy is called the **Milky Way Galaxy**. Of course, no one has photographed our galaxy, so this figure actually shows a galaxy that astronomers have evidence is similar to our own. Our sun would be invisible in such a picture, but if you could see it, you would find it about two-thirds of the way from the center to the edge. Our galaxy contains over 100 billion stars, and, like many others, has graceful **spiral arms** winding outward through the disk. You will discover in a later chapter that stars are born in great clouds of gas and dust as they pass through the spiral arms.

The visible disk of our galaxy is roughly 80,000 ly in diameter. Only a century ago, astronomers thought it was the entire universe—an island universe of stars in an otherwise empty vastness. Now the Milky Way Galaxy is known to be not unique; it is a typical galaxy in many respects, although larger than most. In fact, ours is only one of many billions of galaxies scattered throughout the universe.

As you expand your field of view by another factor of 100, our galaxy appears as a tiny luminous speck surrounded by other specks (Figure 1.12). This figure includes a region 17 million ly in diameter, and each of the dots represents a galaxy. Notice that our galaxy is part of a cluster of a few dozen galaxies. You will find that galaxies are commonly grouped together in clusters. Some of these galaxies have beautiful spiral patterns like our own galaxy, but others do not. One of the questions you will investigate in a later chapter is what produces these differences among the galaxies.

Figure 1.13 represents a view with a diameter of 1.7 billion light years by combining observations with

Galaxy: A large system of stars, star clusters, gas, dust, and nebulae orbiting a common center of mass.

Milky Way: The hazy band of light that circles our sky, produced by the glow of our galaxy.

Milky Way Galaxy: The spiral galaxy containing our sun, visible in the night sky as the Milky Way.

Spiral arms: Long spiral pattern of bright stars, star clusters, gas, and dust. Spiral arms extend from the center to the edge of the disk of spiral galaxies.

Figure 1.10

Figure 1.11

Figure 1.12

Milky Way Galaxy →

Figure 1.13

Detail from galaxy map from M. Seldner, B. L. Siebers, E. J. Groth, and P. J. E. Peebles, Astronomical Journal 82 (1977).

theoretical calculations. The figure shows clusters of galaxies connected in a vast network. Clusters are grouped into **superclusters**—clusters of clusters—and the superclusters are linked to form long filaments and walls outlining voids that seem nearly empty of galaxies. These filaments and walls appear to be the largest structures in the universe.

Were you to expand your view frame one more time, you would probably see a uniform fog of filaments and voids. When you puzzle over the origin of these structures, you are at the frontier of human knowledge. The sequence of figures ends here because it has reached the limits of the best telescopes. Humanity's view does not extend as far as the region that would be covered by a figure 100 times larger than Figure 1.13.

A problem in studying astronomy is keeping a proper sense of scale. Remember that each of the billions of galaxies contains billions of stars. Many of those stars probably have families of planets like our solar system, and on some of those billions of planets liquid-water oceans and protective atmospheres may have sheltered

Superclusters: A cluster of galaxy clusters.

the spark of life. It is possible that some other planets are inhabited by intelligent creatures who share your curiosity, wonder at the scale of the cosmos, and are looking back at you when you gaze into the heavens.

RECAP

Now that you've taken a "cosmic zoom," you should understand how objects you know relate to the largest things in the universe. Now you should be able to answer a few essential questions:

1. Where is Earth in relation to the sun, the planets, the stars, and the galaxies?

2. How do astronomers describe distances?

3. Which of these objects are big relative to the others, and which are small?

4. Are there other worlds like Earth?

Astronomy is important because it helps us understand what we are. We human beings live on the surface of planet Earth as it orbits the star we call the sun. What are we? How did we and our planet and our star come to be here, and what does the future hold for us? Astronomy helps us answer those questions.

Notice that the question is not, "Who are we?" If you want to know who we are, you may want to talk to a minister, priest, or rabbi. "What are we?" is a fundamentally different question. Astronomy helps to answer the question "What are we?" by placing us in the physical universe. It helps us locate ourselves on a small planet in a whirling complex universe filled with stars and galaxies. But astronomy also locates us in the physical processes that govern the universe. Gravity and atoms work together to make stars, light the universe, generate energy, and create the chemical elements in our bodies. Astronomy locates us in that cosmic cycle of time.

Although astronomers use physics and mathematics to study specific problems in astronomy, they do enjoy the beauty of the universe. Many astronomers chose their profession because of the beauty of the night sky and the drama of photographs of galaxies and glowing clouds of gas in space. But astronomers understand that ideas can also be beautiful. The elegance of the processes that form stars, galaxies, and planets, and the harmonious purity of the mathematics of astronomy assure that we are all part of something very large and very beautiful.

Astronomy is a journey that leads to self-knowledge. As you learn about the lives of the stars, the birth of Earth and the death of the sun, you learn about what we are. Astronomy enriches our lives and gives us perspective on what it means to be here on Earth.

User's Guide to
the Sky: Patterns and Cycles

The night sky is the rest of the universe as seen from our planet. When you look up at the stars, you look out through a layer of air only about 100 kilometers deep. Beyond that, space is nearly empty, with the planets of our solar system several AU away and the far more distant stars scattered many light-years apart. You can begin your understanding of the natural laws that govern the universe by carefully noting what the universe looks like and how it behaves. As you read this chapter, keep in mind that you live on a planet, a moving platform. Earth rotates on its axis once a day, so from our viewpoint sky objects appear to rotate around us each day. For example, the sun rises in the east and sets in the west, and also so do the stars. The sun, the moon, planets, stars, and galaxies all have an apparent daily motion that is not real but is caused by a real motion of Earth.

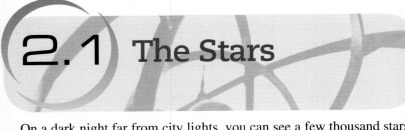

2.1 The Stars

On a dark night far from city lights, you can see a few thousand stars in the sky. Your observations can be summarized by naming individual stars and groups of stars and by specifying their relative brightness.

Constellations

All around the world, ancient cultures celebrated heroes, gods, and mythical beasts by naming groups of stars called **constellations**. You should not be surprised that the star patterns do not look like the creatures they are named after any more than Columbus, Ohio, looks like Christopher Columbus.

Constellation: One of the stellar patterns identified by name, usually of mythological gods, people, animals, or objects. Also, the region of the sky containing that star pattern.

The constellations named within western culture originated in ancient civilizations of Mesopotamia, Babylon, Egypt, and Greece beginning as much as 5,000 years ago. Of those ancient constellations, 48 are still in use. In those former times, a constellation was simply a loose grouping of bright stars, and many of the fainter stars were not included in any constellation.

looking back

As you learn about astronomy, you learn about yourself. The "cosmic zoom" in the previous chapter showed you and Earth in relation to other objects in the universe. Even in that quick preview, as you learned about stars and galaxies, you were also learning about your own place in the universe.

looking ahead

The remaining chapters of this book will fill in details and help you understand objects in the sky. Appearances can be deceiving; although it seems as if you and Earth don't move and are at the center of the universe, Chapter 3 will show how humans finally came to understand that we live on a moving planet. In fact, you will discover that modern science was born as the result of people trying to understand fully the appearance of the sky and its cycles.

astro

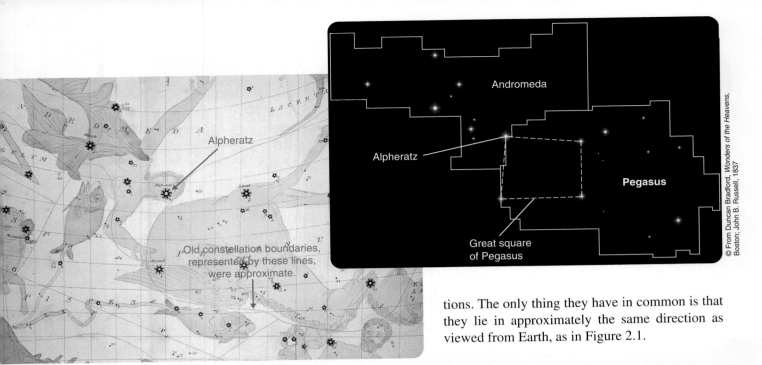

Alpheratz

Old constellation boundaries, represented by these lines, were approximate.

Andromeda

Alpheratz

Pegasus

Great square of Pegasus

tions. The only thing they have in common is that they lie in approximately the same direction as viewed from Earth, as in Figure 2.1.

The Names of the Stars

In addition to naming groups of stars, ancient astronomers named the brighter stars, and modern astronomers still use many of those names. The names of the constellations are in Latin or Greek, the languages of science in Medieval and Renaissance Europe. Most individual star names derive from ancient Arabic, much altered over centuries. For example, the name of Betelgeuse, the bright red star in Orion, comes from the Arabic phrase *yad aljawza,* meaning "armpit of Jawza [Orion]." Aldebaran, the bright red eye of Taurus the bull, comes from the Arabic *al-dabar an,* meaning "the follower."

Another way to identify stars is to assign Greek letters to the bright stars in a constellation in approximate order of brightness. Thus the brightest star is usually designated alpha (α), the second brightest beta (β), and so on. For many constellations, the letters follow the order of brightness, but some constellations, like Orion, are exceptions. A Greek-letter star name also includes the possessive form of the constellation name; for example, the brightest star in the constellation Canis Major is alpha Canis Majoris. This name identifies the star and the constellation and gives a clue to the relative brightness of the star. Compare this with the ancient individual name for that star, Sirius, which tells you nothing about its location or brightness.

The Brightness of Stars

Astronomers measure the brightness of stars using the **magnitude scale**. The ancient astronomers divided the stars into six brightness groups. The brightest were called

Regions of the southern sky not visible to ancient astronomers living at northern latitudes also were not identified with any constellations. Constellation boundaries, when they were defined at all, were only approximate, so a star like Alpheratz could be thought of both as part of Pegasus and also part of Andromeda.

In recent centuries, astronomers have added 40 modern constellations to fill gaps, and in 1928 the International Astronomical Union (IAU) established a total of 88 official constellations with clearly defined permanent boundaries that together cover the entire sky. Thus a constellation now represents not a group of stars but a section of the sky—a viewing direction—and any star within the region belongs only to that one constellation.

In addition to the 88 official constellations, the sky contains a number of less formally defined groupings called **asterisms**. The Big Dipper, for example, is an asterism you probably recognize that is part of the constellation Ursa Major (the Great Bear). Another asterism is the Great Square of Pegasus that includes three stars from Pegasus and the previously mentioned star Alpheratz, now considered to be part of Andromeda only.

Although constellations and asterisms are named as if they were real groupings, most are made up of stars that are not physically associated with one another. Some stars may be many times farther away than others in the same constellation and moving through space in different direc-

Asterism: A named grouping of stars that is not one of the recognized constellations.

Magnitude scale: The astronomical brightness scale. The larger the number, the fainter the star.

Figure 2.1

The stars you see in the Big Dipper are not at the same distance from Earth. You see the stars in a group in the sky because they lie in the same general direction as seen from Earth, not because they are all actually near each other. The sizes of the star dots in the star chart represent the apparent brightness of the stars.

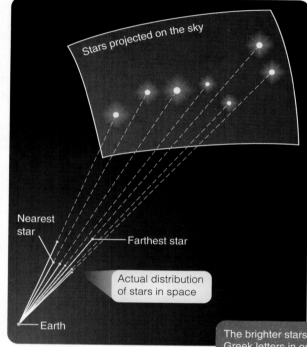

how the stars look to human eyes observing from Earth.

Brightness is quite subjective, depending on both the physiology of eyes and the psychology of perception. To be scientifically accurate you should refer to **flux**—a measure of the light energy from a star that hits one square meter in one second. With modern scientific instruments astronomers can measure the flux of starlight with high precision and then use a simple mathematical relationship that relates light flux to apparent visual magnitude. Instead of saying that the star known by the charming name Chort (Theta Leonis) is about third magnitude, you can say its magnitude is 3.34. Thus, precise modern measurements of the brightness of stars are still connected to observations of apparent visual magnitude that go back to the time of Hipparchus.

Limitations of the apparent visual magnitude system have motivated astronomers to supplement it in various ways: (1) Some stars are so bright the magnitude scale must extend into negative numbers, as demonstrated in

> **Apparent visual magnitude (m_v):** A measure of the brightness of a star as seen by human eyes on Earth.
>
> **Flux:** A measure of the flow of energy out of a surface. Usually applied to light.

first-magnitude stars. The scale continued downward to sixth-magnitude stars, the faintest visible to the human eye. Thus, the *larger* the magnitude number, the *fainter* the star. This makes sense if you think of bright stars as first-class stars and the faintest stars visible as sixth-class stars.

The Greek astronomer Hipparchus (190–120 BCE) is believed to have compiled the first star catalog, and there is evidence he used the magnitude system in that catalog. About 300 years later (around 140 CE) the Egyptian-Greek astronomer Claudius Ptolemy definitely used the magnitude system in his own catalog, and successive generations of astronomers have continued to use the system. Star brightnesses expressed in this system are known as **apparent visual magnitudes (m_v)**, describing

The brighter stars in a constellation are usually given Greek letters in order of decreasing brightness.

Orion

α Orionis is also known as Betelgeuse.

β Orionis is also known as Rigel.

In Orion β is brighter than α, and κ is brighter than η. Fainter stars do not have Greek letters or names, but if they are located inside the constellation boundaries, they are part of the constellation.

Figure 2.2. On this scale, Sirius, the brightest star in the sky, has a magnitude of −1.47. (2) With a telescope you can find stars much fainter than the limit for your unaided eyes. Thus the magnitude system has also been extended to numbers larger than sixth magnitude to include faint stars (Figure 2.2). (3) Although some stars emit large amounts of infrared or ultraviolet light, those types of radiation (discussed further in Chapters 4 and 5) are invisible to human eyes. The subscript "V" in m_V is a reminder that you are counting only light that is visible. Other magnitude systems have been invented to express the brightness of invisible light arriving at Earth from the stars. (4) An apparent magnitude tells only how bright the star is as seen from Earth but doesn't tell anything about a star's true power output because the star's distance is not included. You can describe the true power output of stars with another magnitude system that will be described in Chapter 6.

2.2 The Sky and Its Motions

The sky above you seems to be a great blue dome in the daytime and a sparkling ceiling at night. Learning to understand the sky requires that you first recall the perspectives of people who observed the sky thousands of years ago.

The Celestial Sphere

Ancient astronomers believed the sky was a great sphere surrounding Earth with the stars stuck on the inside like thumbtacks in a ceiling. Modern astronomers know that the stars are scattered through space at different distances, but it is still convenient and useful in some contexts for you to think of the sky as a great sphere enclosing Earth with stars all at one distance. The **celestial sphere** is an example of a **scientific model**, a common feature of scientific thought. You can use the celestial sphere as a convenient model of the sky. You will learn about many scientific models in the chapters that follow.

As you study pages 16–17, notice three important points:

1. Sky objects appear to rotate westward around Earth each day, but that is a consequence of the eastward rotation of Earth. This produces day and night.

2. What you can see of the sky depends on where you are on Earth. For example, Australians see many constellations and asterisms invisible from North America, but they never see the Big Dipper.

3. Astronomers measure distances across the sky as angles expressed in units of degrees and subdivisions of degrees called arc minutes and arc seconds.

Precession

In addition to the daily motion of the sky, Earth's rotation adds a second motion to the sky that can be detected only over centuries. More than 2000 years ago, Hipparchus compared a few of his star positions with those made by other astronomers nearly two centuries before him and realized that the celestial poles and equator were

Figure 2.2

The scale of apparent visual magnitudes extends into negative numbers to represent the brightest objects and to positive numbers larger than 6 to represent objects fainter than the human eye can see.

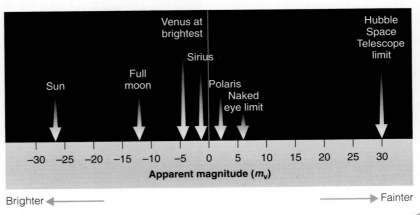

scientific models

Frameworks for Thinking about Nature

A scientific model is a carefully devised mental conception of how something works, a framework that helps scientists think about some aspect of nature. For example, astronomers use the celestial sphere as a way to think about the motions of the sky, sun, moon, and stars.

Some models are imprecise—the psychologist's model of how the human mind processes visual information into images, for instance. But other models are so specific that they can be expressed as a set of mathematical equations, like those used to describe how gas falls into a black hole. You could use metal and plastic to build a celestial globe, but the importance of a model is its use as a mental conception more than its physical presence.

A model is not meant to be a statement of truth. The celestial sphere is not real; you know the stars are scattered through space at various distances. Nevertheless, you can imagine a celestial sphere and use it to help you think about the sky. A scientific model does not have to be true to be useful. Chemists, for example, think about the atoms in molecules

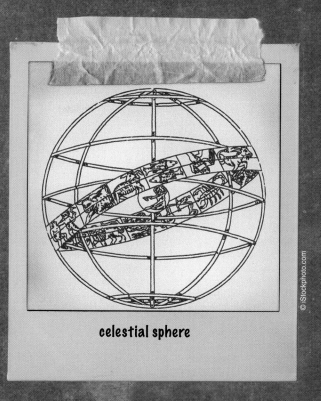
celestial sphere

by visualizing them as spheres joined together by sticks. This model of a molecule is not really correct, but it is a helpful way to think about molecules; it gives chemists a framework within which to organize their ideas.

Because scientific models are not meant to be totally correct, you must always remember the assumptions on which they are based. If you begin to think a model is true, it can mislead you instead of helping you. The celestial sphere, for instance, can help you think about the sky, but you must remember that the universe is much larger and more complex than this ancient scientific model of the heavens.

slowly moving relative to the stars. Later astronomers understood that this apparent motion is caused by a special motion of Earth called **precession**.

If you have ever played with a toy top or gyroscope, you may recall that the axis of such a rapidly spinning object sweeps around relatively slowly in a circle. Look at Figure 2.3 on page 18 and think about how that top moved. The weight of the top tends to make it tip, and this combines with its rapid rotation to make its axis sweep around slowly in precession motion.

Comparing Figure 2.3b to the motion of the top in Figure 2.3a, you can see that Earth spins like a giant top, but it does not spin upright relative to its orbit around the sun. You can say either that Earth's axis is tipped 23.5 degrees from vertical or that Earth's equator is tipped 23.5 degrees relative to its orbit; the two statements are equivalent. Earth's large mass and

> **Precession:** The slow change in orientation of the Earth's axis of rotation. One cycle takes nearly 26,000 years.

The Sky around Us

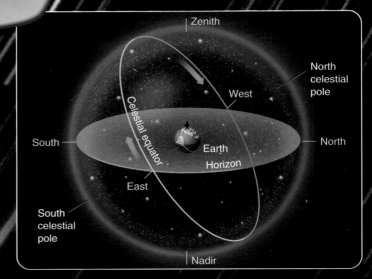

1 The eastward rotation of Earth causes the sun, moon, and stars to move westward in the sky as if the **celestial sphere** were rotating westward around Earth. From any location on Earth you see only half of the celestial sphere, the half above the **horizon**. The **zenith** marks the top of the sky above your head, and the **nadir** marks the bottom of the sky directly under your feet. The drawing at right shows the view for an observer in North America. An observer in South America would have a dramatically different horizon, zenith, and nadir.

The apparent pivot points are the **north celestial pole** and the **south celestial pole** located directly above Earth's north and south poles. Halfway between the celestial poles lies the **celestial equator**. Earth's rotation defines the directions you use every day. The **north point** and **south point** are the points on the horizon closest to the celestial poles. The **east point** and the **west point** lie halfway between the north and south points. The celestial equator always touches the horizon at the east and west points.

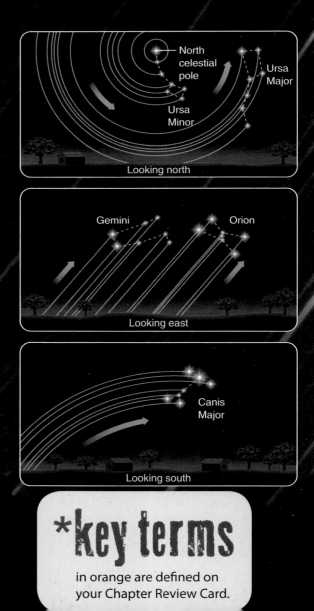

© AURA/NOAO/NSF

1a This time exposure of about 30 minutes shows stars as streaks, called star trails, rising behind an observatory dome. The camera was facing northeast to take this photo. The motion you see in the sky depends on which direction you look, as shown at right. Looking north, you see the star Polaris, the North Star, located near the north celestial pole. As the sky appears to rotate westward, Polaris hardly moves, but other stars circle the celestial pole. Looking south from a location in North America, you can see stars circling the south celestial pole, which is invisible below the southern horizon.

*key terms

in orange are defined on your Chapter Review Card.

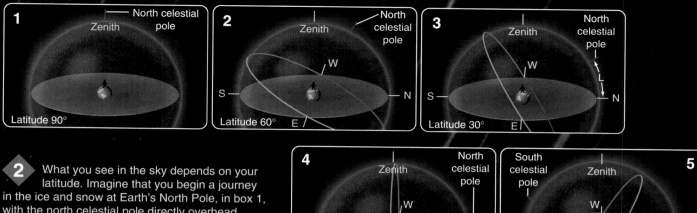

1 — Latitude 90° — Zenith, North celestial pole

2 — Latitude 60° — Zenith, North celestial pole, W, S, N, E

3 — Latitude 30° — Zenith, North celestial pole, W, L, S, N, E

4 — Latitude 0° — Zenith, North celestial pole, W, S, N, E

5 — Latitude −30° — South celestial pole, Zenith, W, S, N, E

2 What you see in the sky depends on your latitude. Imagine that you begin a journey in the ice and snow at Earth's North Pole, in box 1, with the north celestial pole directly overhead. As you walk southward, the celestial pole moves toward the horizon, and you can see further into the southern sky. The angular distance from the horizon to the north celestial pole always equals your latitude—the basis for celestial navigation. As you cross Earth's equator, the celestial equator would pass through your zenith, and the north celestial pole would sink below your northern horizon.

3 Astronomers might say, "The star was only 2 degrees from the moon." Of course, the stars are much farther away than the moon, but when you think of the celestial sphere, you can measure distances *on the sky* as **angular distances** in degrees, minutes of arc, and seconds of arc. An **arc minute** is 1/60th of a degree, and an **arc second** is 1/60th of a minute of arc. Then the **angular diameter** of an object is the angular distance from one edge to the other. The sun and moon are each about half a degree in diameter, and the bowl of the Big Dipper is about 10° wide.

3a **Circumpolar constellations** are those that never rise or set. From mid-northern latitudes, as shown at left, you see a number of familiar constellations circling Polaris and never dipping below the horizon. As the sky rotates, the pointer stars at the front of the Big Dipper always point toward Polaris. Circumpolar constellations near the south celestial pole never rise as seen from mid-northern latitudes. From a high latitude such as Norway, you would have more circumpolar constellations, and from Quito, Ecuador, located on Earth's equator, you would have no circumpolar constellations at all.

A few circumpolar constellations

Cassiopeia

Cepheus

Perseus

Rotation of sky

Rotation of sky

Polaris

Ursa Minor

Ursa Major

Angular distance

Astronomers measure distance across the sky as angles.

rapid rotation keep its axis of rotation pointed toward a spot near the star Polaris (alpha Ursa Minoris), and its axis direction would not move if Earth were a perfect sphere. However, Earth has a slight bulge around its middle, and the gravity of the sun and moon both pull on this bulge, tending to twist Earth's axis upright relative to its orbit. The combination of these forces and Earth's rotation causes Earth's axis to precess in a slow circular sweep, taking about 26,000 years for one cycle.

Because the celestial poles and equator are defined by Earth's rotational axis, precession moves these reference marks. Figure 2.3c shows the apparent path followed by the north celestial pole over thousands of years. You would notice no change at all from night to night or year to year, but precise measurements reveal their slow apparent motion. Over centuries, precession has dramatic effects. Egyptian records show that 4,800 years ago the north celestial pole was near the star Thuban (alpha Draconis). The pole is now approaching Polaris and will be closest to it in about the year 2100. In another 12,000 years the pole will have moved to the apparent vicinity of the very bright star Vega (alpha Lyrae).

2.3 The Cycle of the Sun

The English language defines **rotation** as the turning of a body on its axis. **Revolution** means the motion of a body around a point outside the body. Earth rotates on its axis, and that produces day and night. Earth also revolves around the sun, and that produces the yearly cycle.

The Annual Motion of the Sun

Even in the daytime the sky is actually filled with stars, but the glare of sunlight fills our atmosphere with scattered light, and you can see only the brilliant blue sky. If the sun were fainter and you could see the stars in the daytime, you would notice that the sun appears to be moving slowly eastward relative to the background of the distant stars, as demonstrated in Figure 2.4. This apparent motion is caused by the real orbital motion of Earth around the sun. In January you would see the sun in front of the constellation Sagittarius. By March you would see the sun in front of Aquarius. Note that your

Figure 2.3

Precession. (a) A spinning top precesses in a slow circular motion around the perpendicular to the floor because its weight tends to make it fall over. (b) Earth precesses around the perpendicular to its orbit because the gravity of the sun and moon tend to twist it "upright." (c) Precession causes the north celestial pole to drift among the stars, completing a circle in 26,000 years.

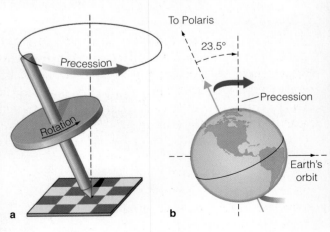

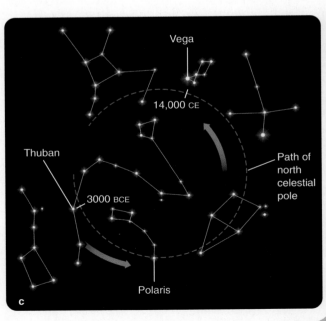

*Figure 2.4

The motion of Earth around the sun makes the sun appear to move against the background of the stars. The circular orbit of Earth is thus projected on the sky as the ecliptic, the circular path of the sun during the year as seen from Earth. If you could see the stars in the daytime, you would notice the sun crossing in front of the distant constellations as Earth moves along its orbit.

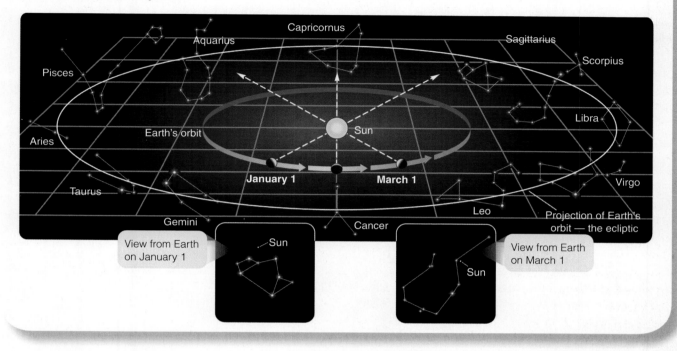

angle of view in Figure 2.4 makes the Earth's orbit seem very elliptical when it is really almost a perfect circle.

Through the year, the sun moves eastward among the stars following a line called the **ecliptic**, the apparent path of the sun among the stars. Recall the concept of the celestial sphere; if the sky were a great screen, the ecliptic would be the shadow cast by Earth's orbit. In other words, you can call the ecliptic the projection of Earth's orbit on the celestial sphere. Earth circles the sun in 365.26 days, and consequently the sun appears to go once around the sky in the same period. You don't notice this motion because you cannot see the stars in the daytime, but the apparent motion of the sun caused by a real motion of Earth has an important consequence that you do notice—the seasons.

The Seasons

The seasons are caused by the revolution of Earth around the sun combined with a simple fact you have already encountered: Earth's equator is tipped 23.5 degrees relative to its orbit. As you study pages 20–21, notice two important principles:

1. The seasons are *not* caused by variation in the distance between Earth and the sun. Earth's orbit is nearly circular, so it is always about the same distance from the sun.

2. The seasons *are* caused by changes in the amount of solar energy that Earth's northern and southern hemispheres receive at different times of the year, resulting from the tip of the Earth's equator and axis relative to its orbit.

The seasons are so important as a cycle of growth and harvest that cultures around the world have attached great significance to the ecliptic. It marks the center line of the **zodiac** ("circle of animals"), and the motion of the sun, moon, and the five visible planets (Mercury, Venus, Mars, Jupiter, and Saturn) are the basis of the ancient superstition of astrology. The signs of the zodiac are no longer important in astronomy. You can look for the planets along the ecliptic appearing like bright stars. Mars looks quite orange in color. Because Venus and

Ecliptic: The apparent path of the sun around the sky.

Zodiac: A band centered on the ecliptic and encircling the sky.

The Cycle of the Seasons

1 You can use the celestial sphere to help you think about the seasons. The celestial equator is the projection of Earth's equator on the sky, and the ecliptic is the projection of Earth's orbit on the sky. Because Earth is tipped in its orbit, the ecliptic and equator are inclined to each other by 23.5° as shown at right. As the sun moves eastward around the sky, it spends half the year in the southern half of the sky and half of the year in the northern half. That causes the seasons.

The sun crosses the celestial equator going northward at the point called the **vernal equinox**. The sun is at its farthest north at the point called the **summer solstice**. It crosses the celestial equator going southward at the **autumnal equinox** and reaches its most southern point at the **winter solstice**. These labels refer to seasons in the northern hemisphere.

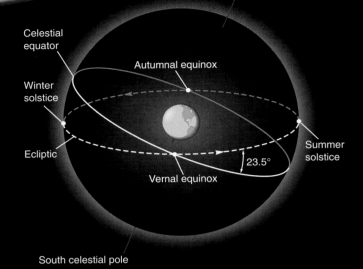

1a The seasons are defined by the dates when the sun crosses these four points, as shown in the table at the right. *Equinox* comes from the word for "equal"; the day of an equinox has equal amounts of daylight and darkness. *Solstice* comes from the words meaning "sun" and "stationary." *Vernal* comes from the word for "green." The "green" equinox marks the beginning of spring in the northern hemisphere.

Event	Date*	Season
Vernal equinox	March 20	Spring begins
Summer solstice	June 22	Summer begins
Autumnal equinox	September 22	Autumn begins
Winter solstice	December 22	Winter begins

* Give or take a day due to leap year and other factors.

1b On the day of the summer solstice in late June, Earth's northern hemisphere is inclined toward the sun, and sunlight shines almost straight down at northern latitudes. At southern latitudes, sunlight strikes the ground at an angle and spreads out. North America has warm weather, and South America has cool weather.

Earth's axis of rotation points toward Polaris, and, like a top, the spinning Earth holds its axis fixed as it orbits the sun. On one side of the sun, Earth's northern hemisphere leans toward the sun; on the other side of its orbit, it leans away. However, the direction of the axis of rotation does not change.

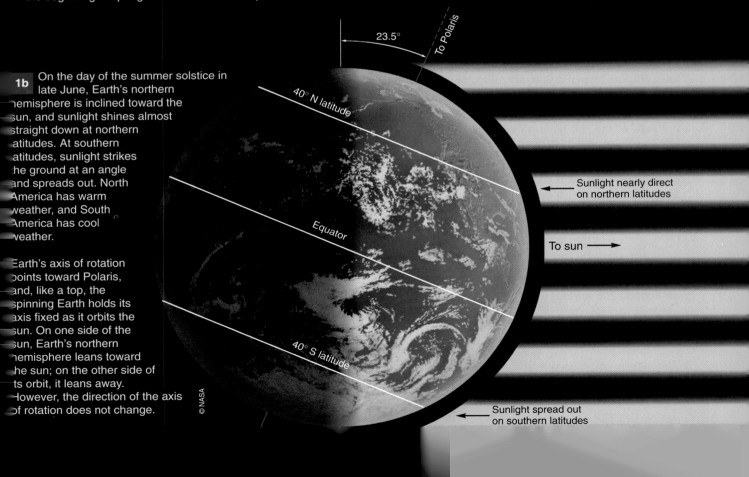

© NASA

Summer solstice light

1c Light striking the ground at a steep angle spreads out less than light striking the ground at a shallow angle. Light from the summer-solstice sun strikes northern latitudes from nearly overhead and is concentrated.

Winter solstice light

Light from the winter-solstice sun strikes northern latitudes at a much steeper angle and spreads out. The same amount of energy is spread over a larger area, so the ground receives less energy from the winter sun.

2 The two causes of the seasons are shown at right for someone in the northern hemisphere. First, the noon summer sun is higher in the sky and the winter sun is lower, as shown by the longer winter shadows. Thus winter sunlight is more spread out. Second, the summer sun rises in the northeast and sets in the northwest, spending more than 12 hours in the sky. The winter sun rises in the southeast and sets in the southwest, spending less than 12 hours in the sky. Both of these effects mean that northern latitudes receive more energy from the summer sun, and summer days are warmer than winter days.

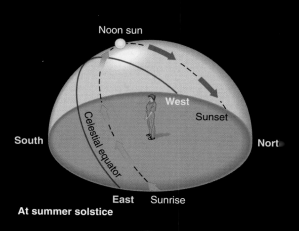

At summer solstice

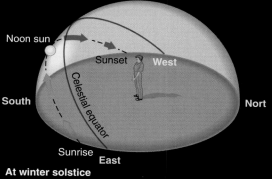

At winter solstice

23.5° — To Polaris

Sunlight spread out on northern latitudes

40° N latitude

← To sun

Equator

Sunlight nearly direct on southern latitudes

40° S latitude

© NASA

1d On the day of the winter solstice in late December, Earth's northern hemisphere is inclined away from the sun, and sunlight strikes the ground at an angle and spreads out. At southern latitudes, sunlight shines almost straight down and does not spread out. North America has cool weather and South America has warm weather.

Earth's orbit is only very slightly elliptical. On about January 3, Earth is at **perihelion**, its closest point to the sun, when it is only 1.7 percent closer than average. On about July 5, Earth is at **aphelion**, its most distant point from the sun, when it is only 1.7 percent farther than average. This small variation does not significantly affect the seasons.

Figure 2.5
Mercury and Venus follow orbits that keep them near the sun, and they are visible only soon after sunset or just before sunrise. Venus takes 584 days to move from morning sky to evening sky and back again, but Mercury zips around in only 116 days.

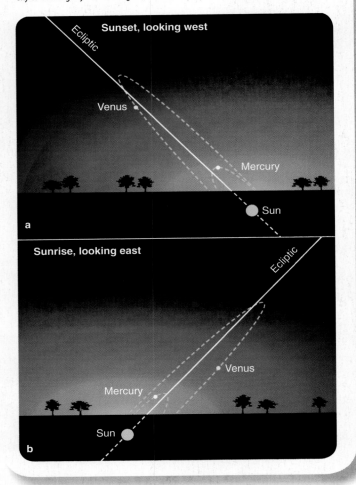

Months later it will switch back to being an evening star again.

2.4 The Cycles of the Moon

The moon orbits eastward around Earth once a month. Starting this evening, look for the moon in the sky. If it is a cloudy night or if the moon is in the wrong part of its orbit, you may not see it; but keep trying on successive evenings, and within a week or two you will see the moon. Then, watch for the moon on following evenings, and you will see it move along its orbit around Earth and cycling through its phases as it has done for billions of years.

The Motion of the Moon

If you watch the moon night after night, you will notice two things about its motion. *First,* you will see it moving relative to the background of stars; *second,* you will notice that the markings on its face don't change. These two observations will help you understand the motion of the moon and the origin of the moon's phases.

The moon moves rapidly among the constellations. If you watch the moon for just an hour, you can see it move eastward against the background of stars by slightly more than its own apparent diameter. Each night when you look at the moon, you will see it is roughly half the width of a zodiac constellation (about 13 degrees) to the east of its location the night before. This movement is the result of the motion of the moon along its orbit around Earth.

The Cycle of Moon Phases

The changing shape of the illuminated part of the moon as it orbits Earth is one of the most easily observed phenomena in astronomy. You have surely seen the full moon rising dramatically or a thin crescent moon hanging in the evening sky. Study "The Phases of the Moon" on pages 24–25 and notice three important points:

1. The moon always keeps the same side facing Earth, and you never see the far side of the moon. "The man in the moon" (some cultures see "the rabbit in

Mercury orbit inside Earth's orbit, they never get far from the sun and are visible in the west after sunset or in the east before sunrise, as you can see in Figure 2.5. Venus can be very bright, but Mercury is difficult to see near the horizon. By tradition, any planet in the sunset sky is called an **evening star**, and any planet in the dawn sky is called a **morning star**. Perhaps the most beautiful is Venus, which can become as bright as magnitude –4.7. As Venus moves around its orbit, it can dominate the western sky each evening for many weeks, but eventually its orbit appears to carry it back toward the sun as seen from Earth, and it is lost in the haze near the horizon. A few weeks later you can see Venus reappear in the dawn sky as a brilliant morning star.

Evening star: Any planet visible in the sky just after sunset.

Morning star: Any planet visible in the sky just before sunrise

astrology and pseudoscience

Misusing the Rules of Science

Astronomers have a low opinion of astrology because it pretends to be a science. It is a pseudoscience, from the prefix *pseudo,* meaning "false."

A pseudoscience is a set of beliefs that appear to be based on scientific ideas but that fail to obey the most basic rules of science. For example, some years ago a fad arose in which people placed objects under pyramids made of paper, plastic, and so on. The claim was that a pyramidal shape would focus cosmic forces on anything inside to preserve fruit and sharpen razor blades, among other things. Experiments revealed that a cover in any shape had the same preservation and "sharpening" effect observed in the pyramids. Nevertheless, supporters of the theory declined to abandon or revise their claims.

Pseudoscience appeals to our needs and desires, which often creates claims that are self-fulfilling. For example, some people bought pyramidal tents to put over their beds to improve their rest. Many people slept more soundly because they wanted and expected the claim to be true.

Astrology is a pseudoscience. Over the centuries, astrology has been tested repeatedly, and no correlation between the movement of the heavens and human personalities or fates has been found. But it survives, and its supporters disregard any evidence that it doesn't work. Like all pseudosciences, astrology is not open to revision in the face of contradictory evidence. Furthermore, astrology fulfills our human need to believe that there is order and meaning to our lives. It may comfort you to believe that your sweetheart has rejected you because of the motions of the sun, moon, and planets along the zodiac rather than to admit that you behaved badly on your last date. Comfort aside, astrology is a poor basis for life decisions.

Human nature and human needs probably ensure that pseudoscientific beliefs will continue to plague humans like emotional viruses propagating from person to person. If you recognize them for what they are, you can more easily guide your life by rational principles and not assign the stars blame for your failures and credit for your successes.

---

the moon" instead) is produced by familiar features on the moon's near side.

2. The changing shape of the moon as it passes through its cycle of phases is produced by sunlight illuminating different parts of the side of the moon you can see. You always see the same side of the moon looking down on you, but the shifting shadows make the "man in the moon" change moods as the moon cycles through its phases (see also Figure 2.6 on page 26).

3. The orbital period of the moon around the Earth is not the same as the length of a moon phase cycle.

2.5 Eclipses

Eclipses are due to a seemingly complicated combination of apparent motions of the sun and moon, yet they are actually easy to predict once all the cycles are understood. Eclipses are also among the most spectacular of nature's sights you might witness.

The Phases of the Moon

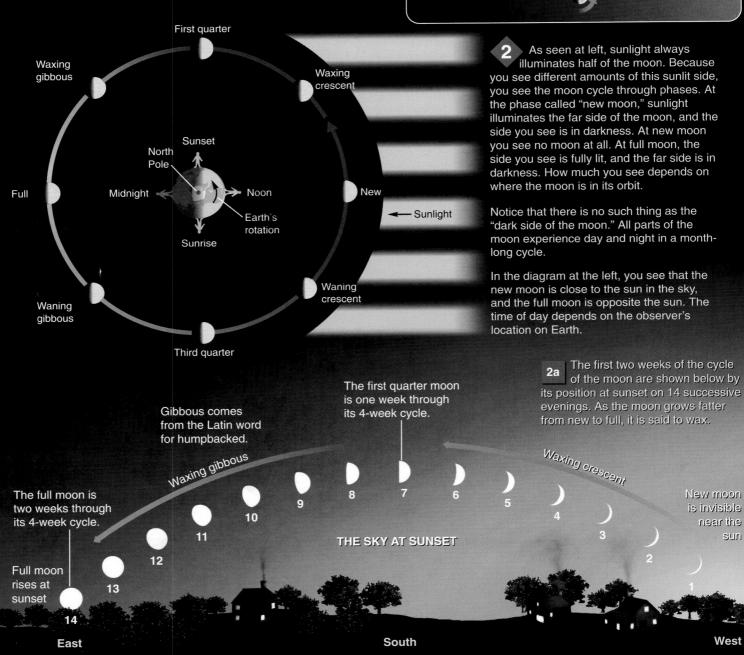

1 As the moon orbits Earth, it rotates to keep the same side facing Earth as shown at right. Consequently you always see the same features on the moon, and you never see the far side of the moon. A mountain on the moon that points at Earth will always point at Earth as the moon revolves and rotates.

(Not to scale)

First quarter

Waxing gibbous

Waxing crescent

Sunset

North Pole

Midnight — Noon

Earth's rotation

New

Full

Sunrise

Sunlight

Waning crescent

Waning gibbous

Third quarter

2 As seen at left, sunlight always illuminates half of the moon. Because you see different amounts of this sunlit side, you see the moon cycle through phases. At the phase called "new moon," sunlight illuminates the far side of the moon, and the side you see is in darkness. At new moon you see no moon at all. At full moon, the side you see is fully lit, and the far side is in darkness. How much you see depends on where the moon is in its orbit.

Notice that there is no such thing as the "dark side of the moon." All parts of the moon experience day and night in a month-long cycle.

In the diagram at the left, you see that the new moon is close to the sun in the sky, and the full moon is opposite the sun. The time of day depends on the observer's location on Earth.

2a The first two weeks of the cycle of the moon are shown below by its position at sunset on 14 successive evenings. As the moon grows fatter from new to full, it is said to wax.

The first quarter moon is one week through its 4-week cycle.

Gibbous comes from the Latin word for humpbacked.

Waxing gibbous

Waxing crescent

The full moon is two weeks through its 4-week cycle.

Full moon rises at sunset

New moon is invisible near the sun

THE SKY AT SUNSET

14 13 12 11 10 9 8 7 6 5 4 3 2 1

East

South

West

3 The moon orbits eastward around Earth in 27.32 days. This is how long the moon takes to circle the sky once and return to the same position among the stars.

A complete cycle of lunar phases takes 29.53 days.

Although you think of the lunar cycle as being about 4 weeks long, it is actually 1.53 days longer than 4 weeks. The calendar divides the year into 30-day periods called months (literally "moonths") in recognition of the 29.53-day synodic cycle of the moon.

* To think about the changing phases of the moon, imagine facing the southern sky, which is where people living in the Northern Hemisphere find the ecliptic. The moon crosses from West to East night by night, following the ecliptic.

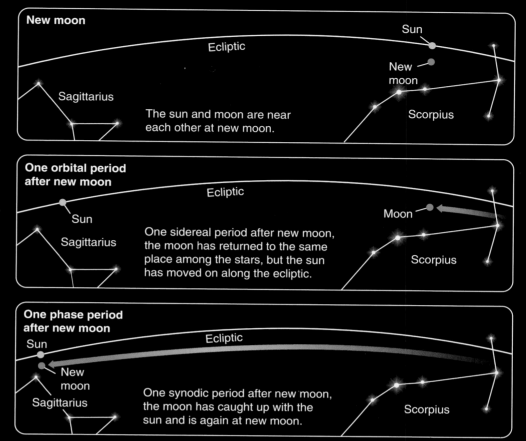

New moon

Ecliptic

Sun

New moon

Sagittarius

Scorpius

The sun and moon are near each other at new moon.

One orbital period after new moon

Ecliptic

Sun

Sagittarius

Moon

Scorpius

One sidereal period after new moon, the moon has returned to the same place among the stars, but the sun has moved on along the ecliptic.

One phase period after new moon

Sun

Ecliptic

New moon

Sagittarius

Scorpius

One synodic period after new moon, the moon has caught up with the sun and is again at new moon.

You can use the diagram on the opposite page to determine when the moon rises and sets at different phases.

TIMES OF MOONRISE AND MOONSET

Phase	Moonrise	Moonset
New	Dawn	Sunset
First quarter	Noon	Midnight
Full	Sunset	Dawn
Third quarter	Midnight	Noon

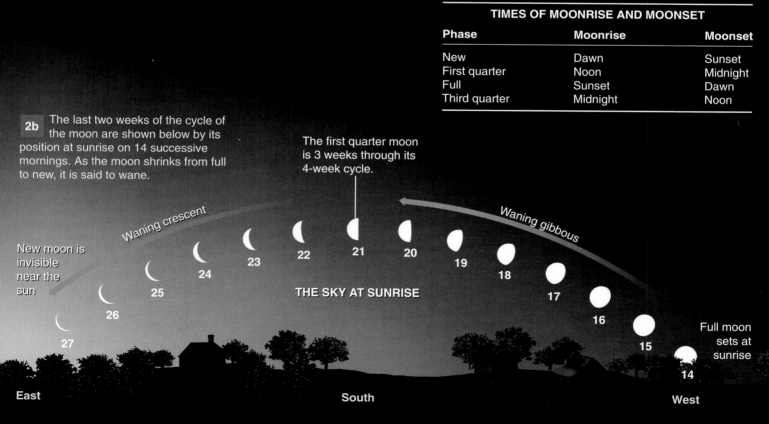

2b The last two weeks of the cycle of the moon are shown below by its position at sunrise on 14 successive mornings. As the moon shrinks from full to new, it is said to wane.

The first quarter moon is 3 weeks through its 4-week cycle.

Waning crescent

Waning gibbous

New moon is invisible near the sun

27   26   25   24   23   22   21   20   19   18   17   16   15   14

THE SKY AT SUNRISE

Full moon sets at sunrise

East   **South**   **West**

Figure 2.6

In this sequence of lunar phases, the moon shrinks from a thin crescent, to full, and back to a thin crescent. You see the same face of the moon, the same mountains, craters, and plains, but the changing direction of sunlight produces the lunar phases.

Image courtesy of Tim Hunter

Solar Eclipses

From planet Earth you can see a phenomenon that is not visible from most planets. It happens that the sun is 400 times larger than our moon and, on the average, 390 times farther away, so the sun and moon have nearly equal apparent diameters. Thus, the moon is just about the right size to cover the bright disk of the sun and cause a **solar eclipse**. In a solar eclipse, it is the sun that is being hidden (eclipsed) and the moon that is "in the way."

A shadow consists of two parts, as you can see in Figure 2.7. The **umbra** is the region of total shadow. For example, if you were in the umbra of the moon's shadow, you would see no portion of the sun. The umbra of the moon's shadow usually just barely reaches Earth's surface and covers a relatively small circular zone (Figure 2.7a). Standing in that umbral zone, you would be in total shadow, unable to see any part of the sun's surface. That is called a total eclipse, as seen in Figure 2.7b. If you moved into the **penumbra**, however, you would be in partial shadow but could also see part of the sun peeking around the edge of the moon. This is called a partial eclipse. Of course, if you are outside the penumbra, you see no eclipse at all.

Because of the orbital motion of the moon and the rotation of Earth, the moon's shadow sweeps rapidly across Earth in a long, narrow path of totality. If you want to see a total solar eclipse, you must be in the path of totality. When the umbra of the moon's shadow sweeps over you, you see one of the most dramatic sights in the sky—the totally eclipsed sun.

The eclipse begins as the moon slowly crosses in front of the sun. It takes about an hour for the moon to cover the solar disk, but as the last sliver of sun disappears, darkness arrives in a few seconds. Automatic streetlights come on, drivers turn on their cars' headlights, and birds go to roost. The sky usually becomes so dark you can even see the brighter stars. The darkness lasts only a few minutes because the umbra is never more than 270 km (170 miles) in diameter on the surface of Earth and sweeps across the landscape at over 1,600 km/hr (1,000 mph). On average, the period of totality lasts only 2 or 3 minutes and never more than 7.5 minutes. During totality you can see subtle features of the sun's atmosphere, such as red flamelike projections that are visible only during these moments when the brilliant disk of the sun is completely covered by the moon. (The sun's atmosphere will be discussed in more detail in Chapter 5.) As soon as part of the sun's disk reappears, the fainter features vanish in the glare, and the period of totality is over. The moon moves on in its orbit and in an hour the sun is completely visible again.

Sometimes when the moon crosses in front of the sun it is too small to fully cover the sun, and then you would witness an **annular eclipse**. That is a solar eclipse in which an annulus (meaning "ring") of the sun's disk is visible around the disk of the moon. The eclipse never becomes total; it never quite gets dark, and you can't see the faint features of the solar atmosphere. Annular eclipses occur because the moon follows a slightly elliptical orbit around Earth. If the moon is in the farther part of its orbit during totality, its apparent diameter will be less than the apparent diameter of the sun, and thus you see an annular eclipse. Furthermore, Earth's orbit is slightly elliptical,

Solar eclipse: The event that occurs when the moon passes directly between Earth and the sun, blocking your view of the sun.

Umbra: The region of a shadow that is totally shaded.

Penumbra: The portion of a shadow that is only partially shaded.

Annular eclipse: A solar eclipse in which the solar photosphere appears around the edge of the moon in a bright ring, or annulus. Features of the solar atmosphere cannot be seen during an annular eclipse.

so the Earth-to-sun distance varies slightly, and so does the apparent diameter of the solar disk, contributing to the effect of the moon's varying apparent size.

If you plan to observe a solar eclipse, remember that the sun is bright enough to burn your eyes and cause permanent damage if you look at it directly. This is true whether there is an eclipse or not. Solar eclipses can be misleading, tempting you to look at the sun in spite of its brilliance and thus risking your eyesight.

During the few minutes of totality, the brilliant disk of the sun is hidden, and it is safe to look at the eclipse, but the partial eclipse phases and annular eclipses can be dangerous. See Figure 2.8 for a safe way to observe the partially eclipsed sun. Table 2.1 will allow you to determine when some upcoming solar eclipses will be visible from your location.

Figure 2.7

(a) The umbral shadow of the moon sweeps over a narrow strip of Earth during a solar eclipse. (b) From a location inside the umbral shadow, you would see the moon cover the bright surface of the sun in a total solar eclipse. Note faint features of the sun's atmosphere around the edge of the moon's disk.

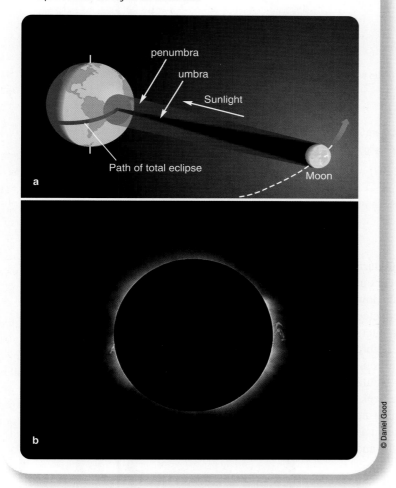

© Daniel Good

SOLAR ECLIPSE

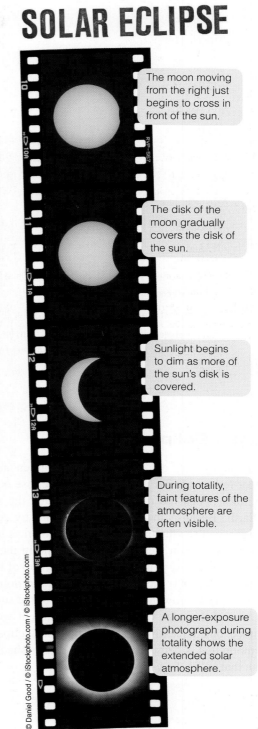

The moon moving from the right just begins to cross in front of the sun.

The disk of the moon gradually covers the disk of the sun.

Sunlight begins to dim as more of the sun's disk is covered.

During totality, faint features of the atmosphere are often visible.

A longer-exposure photograph during totality shows the extended solar atmosphere.

© Daniel Good / © iStockphoto.com

Table 2.1 **Total and Annular Eclipses of the Sun, 2010–2017***

Date	Total or Annular (T/A)	Time of Mid-eclipse** (GMT)	Maximum Length of Total or Annular Phase (Min:Sec)	Area of Visibility
2010 Jan. 15	A	7 ^{h}	11:10	Africa, Indian Ocean
2010 July 11	T	20 ^{h}	5:20	Pacific, S. America
2012 May 20	A	23 ^{h}	5:46	Japan, N. Pacific, W. USA
2012 Nov. 13	T	22 ^{h}	4:02	Australia, S. Pacific
2013 May 10	A	0 ^{h}	6:04	Australia, Pacific
2013 Nov. 3	AT	13 ^{h}	1:40	Atlantic, Africa
2015 March 20	T	10 ^{h}	2:47	N. Atlantic, Arctic
2016 March 9	T	2 ^{h}	4:10	Borneo, Pacific
2016 Sept. 1	A	9 ^{h}	3:06	Atlantic, Africa, Indian Ocean
2017 Feb. 26	A	15 ^{h}	1:22	S. Pacific, S. America, Africa
2017 Aug. 21	T	18 ^{h}	2:40	Pacific, USA, Atlantic

The next major total solar eclipse visible from the United States will occur on August 21, 2017, when the path of totality will cross the United States from Oregon to South Carolina.

*There are no total or partial solar eclipses in 2011 or 2014.

**Times are Greenwich Mean Time. Subtract 5 hours for Eastern Standard Time, 6 hours for Central Standard Time, 7 hours for Mountain Standard Time, and 8 hours for Pacific Standard Time. For Daylight Savings Time, add 1 hour to Standard Time.

^{h} hours.

Lunar Eclipses

Occasionally you can see the moon darken and turn copper-red in a **lunar eclipse**. A lunar eclipse occurs at full moon when the moon moves through the shadow of Earth. Because the moon shines only by reflected sunlight, you see the moon gradually darken as it enters the shadow. If you were on the moon and in the umbra of Earth's shadow, you would see no portion of the sun. If you moved into the penumbra, however, you would be in partial shadow and would see part of the sun peeking around the edge of Earth so the sunlight would be dimmed but not extinguished. In a lunar eclipse, it is the moon that is being hidden in the Earth's shadow and Earth that is "in the way" of the sunlight.

If the orbit of the moon carries it through the umbra of Earth's shadow, you see a total lunar eclipse, demonstrated

Lunar eclipse: The darkening of the moon when it moves through Earth's shadow.

Figure 2.8
A safe way to view the partial phases of a solar eclipse. Use a pinhole in a card to project an image of the sun on a second card. The greater the distance between the cards, the larger (and fainter) the image will be.

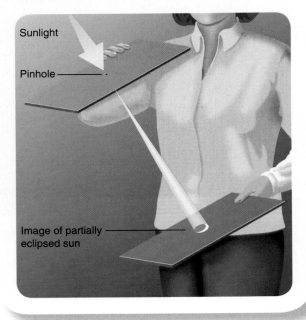

Sunlight

Pinhole

Image of partially eclipsed sun

in Figure 2.9a. As you watch the moon in the sky, it first moves into the penumbra and dims slightly; the deeper it moves into the penumbra, the more it dims. In about an hour, the moon reaches the umbra, and you see the umbral shadow darken part of the moon. It takes about an hour for the moon to enter the umbra completely and become totally eclipsed. The period of total eclipse may last as long as 1 hour 45 minutes, though the timing of the eclipse depends on where the moon crosses the shadow.

When the moon is totally eclipsed, it does not disappear completely. Although it receives no direct sunlight, the moon in the umbra does receive some sunlight refracted (bent) through Earth's atmosphere. If you were on the moon during totality, you would not see any part of the sun because it would be entirely hidden behind Earth. However, you would be able to see Earth's atmosphere illuminated from behind by the sun. The red glow from this ring consisting of all the Earth's simultaneous sunsets and sunrises illuminates the moon during totality and makes it glow coppery red, as shown in Figure 2.9b.

If the moon passes a bit too far north or south of the center of Earth's shadow, it may only partially enter the umbra, and you see a partial lunar eclipse. The part of the moon that remains outside the umbra in the penumbra receives some direct sunlight, and the glare is usually great enough to prevent your seeing the faint coppery glow of the part of the moon in the umbra.

Lunar eclipses always occur at full moon but not at every full moon. The moon's orbit is tipped about 5 degrees to the ecliptic, so most full moons cross the sky north or south of Earth's shadow and there is no lunar eclipse that month (see Figure 2.10 on the next page). For the same reason, solar eclipses always occur at new moon but not at every new moon. The orientation of the moon's orbit in space varies slowly and as a result solar and lunar eclipses repeat in a pattern called the **Saros cycle** lasting

Saros cycle: An 18-year, $11\frac{1}{3}$-day period after which the pattern of lunar and solar eclipses repeats.

Figure 2.9

A longer exposure was used to record the moon while it was totally eclipsed. The moon's path appears curved in the photo because of photographic effects.

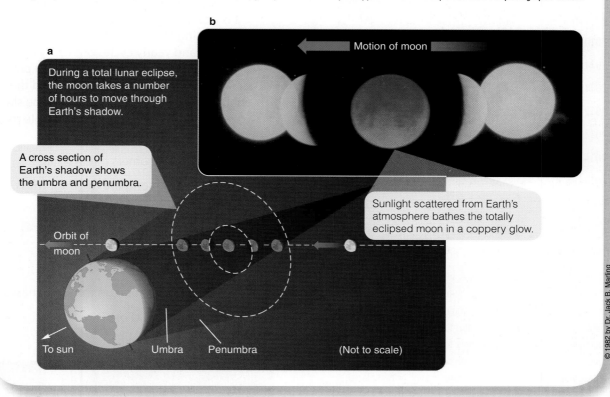

b

Motion of moon

a

During a total lunar eclipse, the moon takes a number of hours to move through Earth's shadow.

A cross section of Earth's shadow shows the umbra and penumbra.

Sunlight scattered from Earth's atmosphere bathes the totally eclipsed moon in a coppery glow.

Orbit of moon

To sun   Umbra   Penumbra   (Not to scale)

Figure 2.10

Umbral shadows of Earth and the moon. Because of the tilt of the moon's orbit relative to the ecliptic, it is easy for the shadows to miss their mark at full moon and at new moon and fail to produce eclipses. (The diameters of Earth and the moon are exaggerated by a factor of 2 for clarity.)

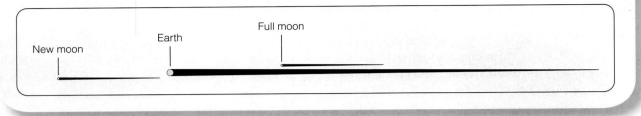

18 years and 11⅓ days. Ancient peoples who understood the Saros cycle could predict eclipses without understanding what the sun and moon really were.

Although there are usually no more than one or two lunar eclipses each year, it is not difficult to see one. You need only be on the dark side of Earth when the moon passes through Earth's shadow. That is, the eclipse must occur between sunset and sunrise at your location to be visible. Table 2.2 will allow you to determine when some upcoming lunar eclipses will be visible.

Table 2.2 **Total and Partial Eclipses of the Moon, 2010 to 2017***

Date	Time** of Mid-eclipse (GMT)	Length of Totality (Hr:Min)	Length of Eclipse† (Hr:Min)
2010 June 26	11:40	Partial	2:42
2010 Dec. 21	8:18	1:12	3:28
2011 June 15	20:13	1:40	3:38
2011 Dec. 10	14:33	0:50	3:32
2012 June 4	11:03	Partial	2:08
2013 April 25	20:10	Partial	0:28
2014 April 15	7:48	1:18	3:34
2014 Oct. 8	10:55	0:58	3:18
2015 April 4	12:02	Partial	3:28
2015 Sept. 28	2:48	1:12	3:30
2017 Aug. 7	18:22	Partial	1:54

*There are no total or partial lunar eclipses during 2016.
**Times are Greenwich Mean Time. Subtract 5 hours for Eastern Standard Time, 6 hours for Central Standard Time, 7 hours for Mountain Standard Time, and 8 hours for Pacific Standard Time. For Daylight Savings Time, add 1 hour to Standard Time. From your time zone, lunar eclipses that occur between sunset and sunrise will be visible, and those at midnight will be best placed.
†Does not include penumbral phase.

recap This chapter gave you a tour of the sky and vocabulary to describe it. This chapter also illustrates a powerful way to analyze certain kinds of problems. If a process repeats, you can understand it by searching for regular cycles. The daily, monthly, and yearly cycles of the sky have direct influences on your life and also produce dramatic sights. Now that you've read the chapter, you are more familiar with the sky's patterns and cycles and should be able to answer these questions:

1. How do astronomers refer to stars?

2. How can you compare the brightness of stars?

3. How does the sky appear to move as Earth rotates?

4. What causes the seasons?

5. Why does the moon go through phases?

6. What is a solar eclipse?

7. What is a lunar eclipse?

the earth and you

We are Earth creatures. We live on the exposed surface of a world spinning on its axis and revolving around the sun. Those motions produce the cycles of day and night and winter and summer, and we have evolved to live within those cycles. One theory holds that we sleep at night because dozing in the back of a cave (or in a comfortable bed) is safer than wandering around in the dark. The night is filled with predators, so sleeping may keep us safe. People who live and work in the Arctic or Antarctic, where the cycle of day and night does not occur, can suffer psychological problems from the lack of the daily cycle.

The cycle of the seasons controls the migration of game and the growth of crops, so cultures throughout history have followed the motions of the sun along the ecliptic with special reverence. The people who built Stonehenge were marking the summer solstice sunrise because it was a moment of power, order, and promise in the cycle of their lives.

The moon's cycles mark the passing days and divide our lives into weeks and months. In a Native American story, the mythical character Coyote gambles with the sun to see if the sun will continue to warm Earth, and the moon keeps score. The moon is a symbol of regularity, reliability, and dependability. When you notice the moon in the sky, remember that it is the scorekeeper counting out your days and months. Like the ticking of a cosmic clock, the passing weeks, months, and seasons mark the passage of time on Earth, but the cycle of the seasons is also affected by longer period changes in the motion of Earth such as precession. Ice ages come and go, and Earth's climate cycles in ways we do not entirely understand. If you don't feel quite as secure as you did when you started this chapter, then you are beginning to catch on. Astronomy tells us that Earth is a beautiful world, but it is also a complicated planet, spinning in a complicated universe, and there will probably be many more surprises for the humans who walk its surface.

The Origin
of Modern Astronomy

I f you are concerned about the environment, you owe a debt to the 16th century Polish astronomer, Nicolaus Copernicus. He proposed that Earth is not the center of the universe, but just one of the planets that circle the sun. His theory made Earth part of the cosmos and led to the modern understanding of humanity's place among all the creatures of Earth. As you read about Copernicus and his theory, you will see astronomers struggling with two related problems—the place of the Earth and the motion of the planets. That struggle led Galileo before the Inquisition and it prompted Isaac Newton to discover gravity. As you read about the birth of modern astronomy, notice that it is also the story of the invention of science as a way to know about the world we live in. Before Copernicus the world seemed filled with mysterious influences; after Newton, scientists understood that the world is described by natural laws that are open to human study. The mysteries of nature are mysteries because they are unknown, not because they are unknowable.

3.1 Astronomy before Copernicus

To understand why Copernicus's model was so important, you first need to backtrack to ancient Greece and meet the two great authorities of ancient astronomy, the brilliant philosopher Aristotle and a later follower of Aristotle's principles, Claudius Ptolemy (pronounced TAHL-eh-mee; the initial P is silent).

First, though, you should remember the point made in Chapter 1 that the terms *solar system, galaxy,* and *universe* have very different meanings. You know now that our solar system, consisting of Earth, the sun, Earth's sibling planets, their moons, and so on, is your very local neighborhood, much smaller than the Milky Way Galaxy, which in turn is tiny compared with the observable universe. However, from ancient times up through Copernicus's day, it was thought that the whole universe, everything that exists, did not extend much beyond the farthest planet of our solar system. Asking whether Earth or the sun is the center of the solar system was

looking back

The previous chapters gave you a view of the sky. You are now familiar with what you see there, and you also can imagine how the sun, moon, and planets move through space. Now you are ready to understand the first worldwide revolution in human thought: the realization that we live on a moving planet.

looking ahead

This is a book of science, and every chapter that follows will use the ways of thinking that were invented when Copernicus tried to repair the ancient model that put Earth at the center of the universe. As you think about planets, stars, and galaxies through the rest of this book, you will be using the intellectual methods that Galileo, Newton, and many others struggled to devise.

astro

then the same question as asking whether Earth or the sun is the center of the universe. As you read this chapter—but only this chapter—you can pretend to have the old-fashioned view in which "solar system" and "universe" meant much the same thing.

Aristotle's Universe

Philosophers of the ancient world attempted to deduce truth about the universe by reasoning from **first principles**. A first principle was something that seemed obviously true to everyone and supposedly needed no further examination. That may strike you as peculiar; modern thinkers tend to observe how things work and then from that evidence make principles and conclusions that can always be reexamined. Before the Renaissance, however, reasoning from evidence (what you might call "scientific thinking") was not widespread.

Study "The Ancient Universe" on pages 36–37 and notice three important ideas:

1. Ancient philosophers and astronomers accepted without question—as first principles—that heavenly objects must move on circular paths at constant speeds, and that the Earth is motionless at the

center of the universe. Although a few ancient writers mentioned the possibility that Earth might move, most of them did so in order to point out how that idea is "obviously" wrong.

2. As viewed by you from Earth, the planets seem to follow complicated paths in the sky, including episodes of "backward" motion that are difficult to explain in terms of motion on circular paths at constant speeds.

3. Finally, you can see how Ptolemy created an elaborate geometrical and mathematical model to explain details of the observed motions of the planets while assuming Earth is motionless at the center of the universe.

Aristotle lived in Greece from 384 to 322 BCE. He believed as a first principle that the heavens were perfect. Because the sphere and circle were considered the only perfect geometrical figures, Aristotle also believed that all motion in the perfect heavens must be caused by the rotation of spheres carrying objects around in uniform circular motion. Aristotle's writings became so famous that he was known throughout the Middle Ages as *The Philosopher*," and the geocentric universe of nested spheres that he devised dominated astronomy. His opinions on the nature of Earth and the sky were widely accepted for almost 2000 years.

Claudius Ptolemy, a mathematician who lived roughly 500 years after Aristotle, believed in the basic ideas of Aristotle's universe but was interested in practical rather than philosophical questions. For Ptolemy, first principles took second place to accuracy. He set about making an accurate mathematical description of the motions of the planets. Ptolemy weakened the first principles of Aristotle by moving Earth a little off-center in the model and inventing a way to slightly vary the planets' speeds. This made his model (published around 140 CE) a better match to the observed motions.

Aristotle's universe, as embodied in the mathematics of Ptolemy's model, dominated ancient astronomy. At first the Ptolemaic model predicted positions of the planets with fair accuracy; but as centuries passed, errors accumulated, and Islamic and later European astronomers had to update the model, adjusting the sizes and locations of the circles and changing the rates of motion.

THE EARTH IS THE CENTER OF THE UNIVERSE!

3.2 Nicolaus Copernicus

Nicolaus Copernicus (originally, Mikolaj Kopernik) was born in 1473 in what is now Poland. At the time of his birth—and throughout his life—astronomy was based on Ptolemy's model of Aristotle's universe. In spite of many revisions, the Ptolemaic model was still a poor predictor of planet positions, but because of the authority of Aristotle, it was the officially accepted model. Moreover, because in Aristotle's philosophy the most perfect region was in the heavens and the most imperfect region was at Earth's center, the classical **geocentric universe** model matched the commonly held Christian view of the geometry of heaven and hell. Anyone who criticized Aristotle's model of the universe was thereby also challenging belief in the locations of heaven and hell, risking at least criticism and perhaps a serious charge of heresy with a possible death penalty.

Copernicus's Model

Copernicus was associated with the Roman Catholic Church throughout his life.

NO WAY, MAN! IT'S THE SUN!

His uncle, by whom he was raised and educated, was a bishop. After studying medicine and Church law in some of the major universities in Europe, Copernicus became a Church employee, serving as secretary and personal physician to his powerful uncle for 15 years and moving into quarters in the cathedral after the uncle died. Because of this connection to the Church and his fear of persecution, he hesitated to publish his revolutionary ideas that challenged the Ptolemaic model and the geometry of heaven and hell.

> **Geocentric universe:** A model universe with Earth at the center, such as the Ptolemaic universe.
>
> **Heliocentric universe:** A model of the universe with the sun at the center, such as the Copernican universe.

What were these revolutionary ideas? Copernicus believed that the sun and not Earth was the center of the universe and that Earth rotated on its axis and revolved around the sun. Copernicus apparently began doubting Ptolemy's geocentric model during his college days. A **heliocentric universe** model (sun-centered, from the Greek word for sun, *helios*) had been discussed occasionally before Copernicus's time, but Copernicus was the first person to produce a detailed model with substantial justifying arguments. Sometime before 1514, Copernicus wrote a short pamphlet summarizing his model and distributed it in handwritten form while he worked on his book.

De Revolutionibus

Copernicus's book *De Revolutionibus Orbium Coelestium (On the Revolutions of Celestial Spheres)* was essentially finished by about 1530. He hesitated to publish, although other astronomers, and even church officials concerned about reform of the calendar, knew about his work, sought his advice, and looked forward to the book's publication. In 1542 Copernicus finally sent the manuscript for *De Revolutionibus* off to be printed. He died in 1543 before the printing was completed.

The most important idea in the book was that the sun was the center of the universe. That single innovation had an impressive consequence—the retrograde motion of the planets was immediately explained in a straightforward way without the epicycles that Ptolemy used.

In Copernicus's model, which you can see in Figure 3.1, Earth moves faster along its orbit than the planets that lie farther from the sun. Consequently, Earth periodically overtakes and passes these planets. Imagine that you are a runner on a track moving along an inside

The Ancient Universe

1 For 2000 years, the minds of astronomers were shackled by a pair of ideas. The Greek philosopher Plato argued that the heavens were perfect. Because the only perfect geometrical shape is a sphere, which carries a point on its surface around in a circle, and because the only perfect motion is uniform motion, Plato concluded that all motion in the heavens must be made up of combinations of circles turning at uniform rates. This idea was called **uniform circular motion**.

Plato's student Aristotle argued that Earth was imperfect and lay at the center of the universe. Such a model is known as a **geocentric universe**. His model contained 55 spheres turning at different rates and at different angles to carry the moon, Mercury, Venus, the sun, Mars, Jupiter, and Saturn across the sky.

Aristotle was known as the greatest philosopher in the ancient world, and for 2000 years his authority chained the minds of astronomers with uniform circular motion and geocentrism. See the model at right.

From *Cosmographica*
by Peter Apian (1539).

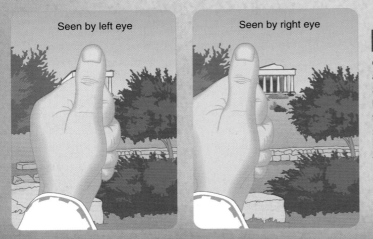

Seen by left eye

Seen by right eye

1a Ancient astronomers believed that Earth did not move because they saw no **parallax**, the apparent motion of an object because of the motion of the observer. To demonstrate parallax, close one eye and cover a distant object with your thumb held at arm's length. Switch eyes, and your thumb appears to shift position as shown at left. If Earth moves, ancient astronomers reasoned, you should see the sky from different locations at different times of the year, and you should see parallax distorting the shapes of the constellations. They saw no parallax, so they concluded Earth could not move. Actually, the parallax of the stars is too small to see with the unaided eye.

2 Planetary motion was a big problem for ancient astronomers. In fact, the word *planet* comes from the Greek word for "wanderer," referring to the eastward motion of the planets against the background of the fixed stars. The planets did not, however, move at a constant rate, and they could occasionally stop and move westward for a few months before resuming their eastward motion. This backward motion is called **retrograde motion**.

Every 2.14 years, Mars passes through a retrograde loop. Two successive loops are shown here. Each loop occurs further east along the ecliptic and has its own shape.

Gemini

March 10, 2010

Cancer

West

Leo

Dec. 18, 2009

Position of Mars
at 5 day intervals

April 17, 2012

Ecliptic

Regulus

2a Simple uniform circular motion centered on Earth could not explain retrograde motion, so ancient astronomers combined uniformly rotating circles much like gears in a machine to try to reproduce the motion of the planets.

3 Uniformly rotating circles were key elements of ancient astronomy. Claudius Ptolemy created a mathematical model of the Aristotelian universe in which the planet followed a small circle called the **epicycle** that slid around a larger circle called the **deferent**. By adjusting the size and rate of rotation of the circles, he could approximate the retrograde motion of a planet. See illustration at right.

To adjust the speed of the planet, Ptolemy supposed that Earth was slightly off center and that the center of the epicycle moved such that it appeared to move at a constant rate as seen from the point called the **equant**.

To further adjust his model, Ptolemy added small epicycles (not shown here) riding on top of larger epicycles, producing a highly complex model.

3a Ptolemy's great book *Mathematical Syntaxis* (c. 140 CE) contained the details of his model. Islamic astronomers preserved and studied the book through the Middle Ages, and they called it *Al Magisti* (The Greatest). When the book was found and translated from Arabic to Latin in the 12th century, it became known as *Almagest*.

3b The Ptolemaic model of the universe shown below was geocentric and based on uniform circular motion. Note that Mercury and Venus were treated differently from the rest of the planets. The centers of the epicycles of Mercury and Venus had to remain on the Earth–Sun line as the sun circled Earth through the year.

Equants and smaller epicycles are not shown here. Some versions contained nearly 100 epicycles as generations of astronomers tried to fine-tune the model to better reproduce the motion of the planets.

Notice that this modern illustration shows rings around Saturn and sunlight illuminating the globes of the planets, features that could not be known before the invention of the telescope.

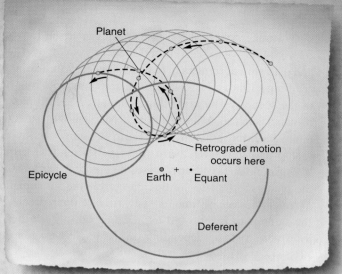

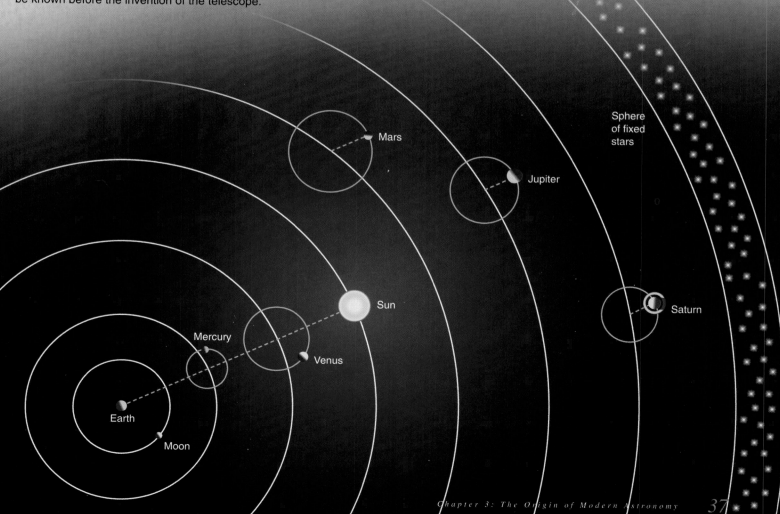

lane. Runners well ahead of you appear to be moving forward relative to background scenery. As you overtake and pass slower runners in outside lanes, they fall behind, seeming to move backward for a few moments relative to the scenery. The same thing happens as Earth passes a planet such as Mars. Although Mars moves steadily along its orbit, as seen from Earth it seems to slow to a stop and move westward (retrograde) relative to the background stars as Earth passes it (Figure 3.1). Because the planets' orbits do not lie in precisely the same plane, a planet does not resume its eastward motion in precisely the same path it followed earlier. Instead, it describes a loop with a shape depending on the angle between the two orbital planes.

Figure 3.1

Earth and Mars are shown at equal intervals to show that as Earth overtakes Mars (a–c), Mars appears to slow its eastward motion. As Earth passes Mars (d), Mars appears to move westward. As Earth draws ahead of Mars (e–g), Mars resumes its eastward motion against the background stars.

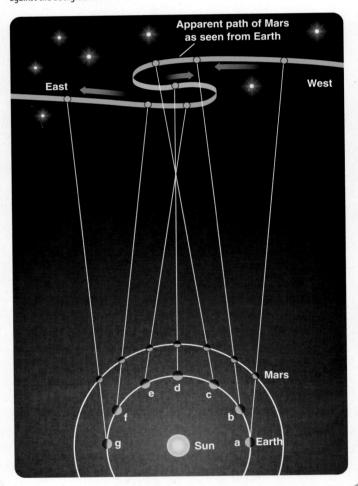

Copernicus's model was simple and straightforward compared with the multiple off-center circles of the Ptolemaic model. However, *De Revolutionibus* failed to immediately disprove the geocentric model for one critical reason—the Copernican model could not predict the positions of the planets any more accurately than the Ptolemaic model could.

Although Copernicus proposed a revolutionary idea in making the solar system heliocentric, he was a classically trained astronomer with great respect for the old concept of uniform circular motion. Copernicus objected to Ptolemy's schemes for moving Earth slightly off-center and varying the speeds of planet motions. That seemed arbitrary and ugly to Copernicus, so he returned to a strong but incorrect belief in uniform circular motion. Therefore, even though his model put the sun correctly at the center of the solar system, it could not accurately predict the positions of the planets as seen from Earth. Copernicus even had to reintroduce small epicycles to match minor variations in the motions of the sun, moon, and planets. Astronomers today recognize those variations as due to the planets' real motions in elliptical orbits (discussed in Section 3.3).

You should notice the difference between the Copernican model and the Copernican hypothesis. The Copernican *model* is inaccurate. It includes uniform circular motion and thus does not precisely describe the motions of the planets. But the Copernican *hypothesis* that the solar system is heliocentric is correct—the planets do circle the sun, not Earth. Why that hypothesis gradually won acceptance is a question historians still debate. There are probably a number of reasons, including the revolutionary spirit of the times, but the most important factor may be the simplicity of the idea. For one thing, placing the sun at the center of the universe produced a symmetry among the motions of the planets that is elegant, pleasing to the eye and mind (Figure 3.2). In the Ptolemaic model, Mercury and Venus had to be treated differently from the rest of the planets: Their epicycles had to remain centered on the Earth-sun line. In Copernicus's model, all of the planets were treated the same. They all followed orbits that circled the sun at the center.

Figure 3.2

(a) The Copernican universe, as reproduced in his book *De Revolutionibus*. Earth and all the known planets revolve in separate circular orbits about the sun (Sol) at the center. The outermost sphere carries the immobile stars of the celestial sphere. Notice the orbit of the moon around Earth (Terra). (b) The model is simple not only in the arrangement of the planets but also in their motions. Orbital velocities (blue arrows) decrease from that of Mercury, the fastest, to that of Saturn, the slowest.

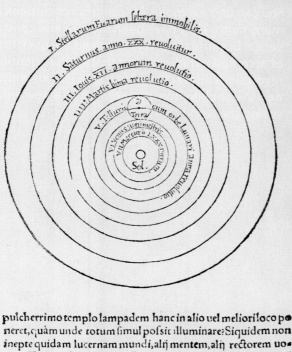

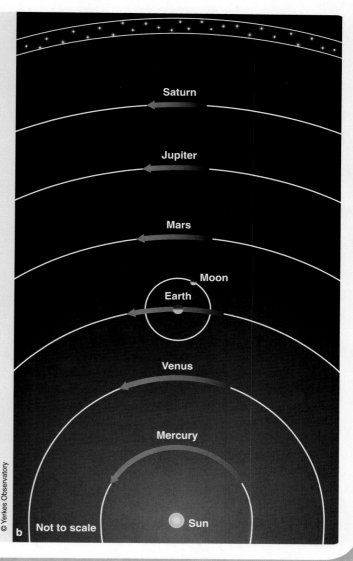

Although astronomers throughout Europe read and admired *De Revolutionibus* and found Copernicus's astronomical observations and mathematics to have great value, few astronomers believed, at first, that the sun actually was the center of the solar system and that Earth moved. How the Copernican hypothesis was gradually recognized as correct has been called the Copernican Revolution, because it was not just the adoption of a new idea but a total change in the way astronomers and the rest of humanity thought about the place of the Earth.

The most important consequence of the Copernican hypothesis was not what it said about the sun but what it said about Earth. By placing the sun at the center, Copernicus made Earth move along an orbit like the other planets. By making Earth a planet, Copernicus revolutionized humanity's view of its place in the universe and triggered a controversy that would eventually bring the astronomer Galileo Galilei before the Inquisition, a controversy over the nature of scientific and religious truths that continues even today.

The Copernican Revolution is often cited as the perfect example of a scientific revolution. Over a few decades, astronomers abandoned a way of thinking about the universe that was almost 2000 years old and adopted a new paradigm (pronounced para-dyme), or set of scientific ideas and assumptions. The pre-Copernicus geocentric paradigm survived for many centuries, until a new generation of astronomers overthrew the old paradigm and established a new, heliocentric paradigm.

A scientific paradigm is powerful because it shapes perceptions by determining which questions are important and what evidence is significant. Thus, it is often difficult to recognize how paradigms limit what you can understand. Though the geocentric paradigm contained problems that seem obvious to a modern mind, astronomers before Copernicus lived and worked inside that paradigm and had difficulty seeing those problems. Overthrowing an existing paradigm is not easy because you must learn to see nature in an entirely new way. Galileo, Kepler, and Newton viewed nature in a way that would have been almost incomprehensible to astronomers of earlier centuries.

You can find examples of scientific revolutions in many fields. They have been difficult and controversial because they have involved the overthrow of accepted paradigms, but that is why scientific revolutions are exciting. They give you an entirely new insight into how nature works—a new way of seeing the world.

Paradigm: A commonly accepted set of scientific ideas and assumptions.

©Carmen Martinez Banús/iStockphoto.com

3.3 Tycho Brahe, Johannes Kepler, and Planetary Motion

As astronomers struggled to understand the place of the Earth, they also faced the problem of planetary motion. How exactly do the planets move? That problem was solved by a nobleman who built a fabulous observatory and a poor commoner with a talent for mathematics.

Tycho Brahe

The Danish nobleman Tycho Brahe is remembered in part for wearing false noses to hide a dueling scar from his college days. He was reportedly very proud of his noble station, so his disfigurement probably did little to improve his lordly disposition.

In 1572, astronomers were startled to see a new star (now called Tycho's supernova) appear in the sky. Aristotle had argued that the heavens were perfect and therefore unchanging, so astronomers concluded that the new star had to be nearer than the moon. Tycho could not detect the new star's parallax (see p. 36), meaning it had

Tycho Brahe

Johannes Kepler

No one could have been more different from Tycho Brahe than was Johannes Kepler. He was born in 1571, the oldest of six children in a poor family living in what is now southwest Germany. His father, who fought as a mercenary soldier for whomever could afford his fees, eventually disappeared. Kepler's mother was apparently an unpleasant and unpopular woman. She was accused of witchcraft in her later years, and Kepler defended her (successfully) in a trial that dragged on for three years. Kepler himself had poor health, even as a child, so it is surprising that he did well in school, winning promotion to a Latin school and eventually a scholarship to the university at Tübingen, where he studied to become a Lutheran pastor.

While still a college student, Kepler had become a believer in the Copernican hypothesis. During his last year of study, Kepler accepted a teaching job in the town of Graz, in what is now Austria, which allowed him to continue his studies in mathematics and astronomy. By 1596, the same year Tycho arrived in Prague, Kepler had learned enough to publish a book called *The Forerunner of Dissertations on the Universe, Containing the Mystery of the Universe*. The book, like nearly all scientific works of that age, was written in Latin and is now known as *Mysterium Cosmographicum*.

to be far beyond the moon and that it was a change in the supposedly unchanging starry sphere.

When Tycho wrote a book about his discovery, the king of Denmark honored him with a generous income and the gift of an island, Hveen, where Tycho built a fabulous observatory. Tycho lived before the invention of the telescopes, so his observatory was equipped with wonderful instruments for measuring the positions of the sun, moon, and planets using the naked eye and peering along sight lines. For 20 years, Tycho and his assistants measured the positions of the stars and planets.

After the death of the Danish king, Tycho moved to Prague where he became the Imperial Mathematician to the Holy Roman Emperor Rudolph II. Tycho hired a few assistants including a German school teacher named Johannes Kepler. Just before Tycho died in 1601, he asked Rudolph II to make Kepler Imperial Mathematician. Thus the newcomer, Kepler, became Tycho's replacement (though at one-sixth Tycho's salary).

By modern standards, the book contains almost nothing of value. It begins with a long appreciation of Copernicus's model and then goes on to mystical speculation on the reason for the spacing of the planets' orbits. The second half has one virtue—as Kepler tried to understand planet orbits, he demonstrated that he was a talented mathematician and that he had become well versed in astronomy. He sent copies to Tycho and to Galileo, who both recognized Kepler's talent in spite of the mystical parts of the book.

Life was unsettled for Kepler in Graz because of the persecution of Protestants in that region, so when Tycho Brahe invited him to Prague in 1600, Kepler went eagerly, ready to work with the famous astronomer. Tycho's sudden death in 1601 left Kepler in a position to use Tycho's extensive records of observations to analyze the motions of the planets.

Kepler began by studying the motion of Mars, trying to deduce from the observations how the planet actually moved. By 1606, he had solved the mystery: The orbit of Mars is an ellipse, not a circle. Thus, he abandoned the ancient belief in the circular motion of the planets. But the mystery was even more complex. The planets do not move at uniform speeds along their elliptical orbits.

Kepler recognized that they move faster when close to the sun and slower when farther away. Thus Kepler abandoned both uniform motion and circular motion and thereby finally solved the problem of planetary motion. Later, he discovered that the period of each planet's orbit is related to that orbit's radius. Kepler published his results in 1609 and 1619 in books called, respectively, *Astronomia Nova (New Astronomy)* and *Harmonice Mundi (The Harmony of the World).*

Kepler's Three Laws of Planetary Motion

Although Kepler dabbled in the philosophical arguments of the day, he was a mathematician, and his triumph was the solution of the problem of the motion of the planets. The key to his solution was the **ellipse**. An ellipse is a figure drawn around two points, called the *foci,* in such a way that the distance from one focus to any point on the ellipse and back to the other focus equals a constant. This makes it easy to draw ellipses with two thumbtacks and a loop of string. Press the thumbtacks into a board, hook the string about the tacks, and place a pencil in the loop. As you can see in Figure 3.3a, if you keep the string taut as you move the pencil, it traces out an ellipse.

The geometry of an ellipse is described by two simple numbers. The **semi-major axis, *a*,** is half of the longest diameter (Figure 3.3b). The **eccentricity, *e*,** of an ellipse is half the distance between the foci divided by the semi-major axis. The eccentricity of an ellipse tells you its shape: If *e* is nearly equal to one, the ellipse is very elongated; if *e* is close to zero, the ellipse is more

Figure 3.3

The geometry of elliptical orbits. (a) Drawing an ellipse with two tacks and a loop of string. (b) The semi-major axis, a, is half of the longest diameter. (c) Kepler's second law is demonstrated by a planet that moves from A to B in one month and from A′ to B′ in the same amount of time. The two blue segments have the same area.

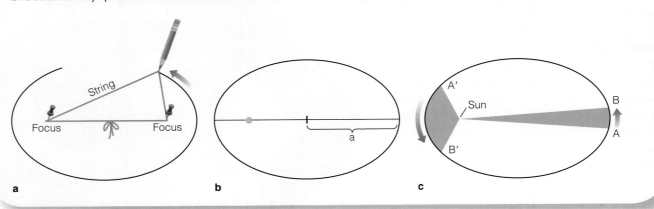

circular. To draw a circle with the string and tacks shown in Figure 3.3a, you would move the two thumbtacks together, which shows that a circle is the same as an ellipse with eccentricity equal to zero. As you move the thumbtacks farther apart, the ellipse becomes flatter, and the value of its eccentricity moves closer to 1.

Kepler used ellipses to describe the motion of the planets in three fundamental rules that have been tested and confirmed so many times that astronomers now refer to them as "natural laws". They are commonly called Kepler's laws of planetary motion, summarized in Table 3.1.

Kepler's first law states that the orbits of the planets around the sun are ellipses with the sun at one focus. Thanks to the precision of Tycho's observations and the sophistication of Kepler's mathematics, Kepler was able to recognize the elliptical shape of the orbits even though they are nearly circular. Of the planets known to Kepler, Mercury has the most elliptical orbit, which you can measure in Figure 3.4, but even it deviates only slightly from a circle.

Kepler's second law states that a line from the planet to the sun sweeps over equal areas in equal intervals of time. This means that when the planet is closer to the sun and the line connecting it to the sun is shorter, the planet moves more rapidly to sweep over the same area that is swept over when the planet is farther from the sun. Thus the planet in Figure 3.3c would move from point A' to point B' in one month, sweeping over the area shown. But when the planet is farther from the sun, one month's motion would be shorter, from A to B. The time that a planet takes to travel around the sun once is its orbital period, P, and its average distance from the sun equals the semi-major axis of its orbit, a. Kepler's third law tells us that these two quantities, orbital period and semi-major axis, are related: Orbital period squared is proportional to the semi-major axis cubed. For example, Jupiter's average distance from the sun (which equals the semi-major axis of its orbit) is 5.2 AU. The semi-major axis cubed would be about 140.6, so the period must be the square root of 140.6, roughly 11.8 years.

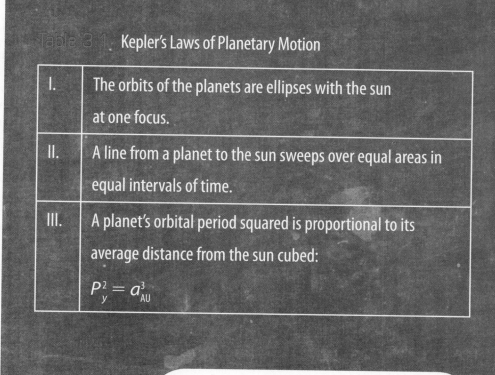

Table 3.1	Kepler's Laws of Planetary Motion
I.	The orbits of the planets are ellipses with the sun at one focus.
II.	A line from a planet to the sun sweeps over equal areas in equal intervals of time.
III.	A planet's orbital period squared is proportional to its average distance from the sun cubed: $$P_y^2 = a_{AU}^3$$

Figure 3.4

The orbits of the planets are nearly circular. You can measure the horizontal and vertical diameters of this orbit to detect its elliptical shape.

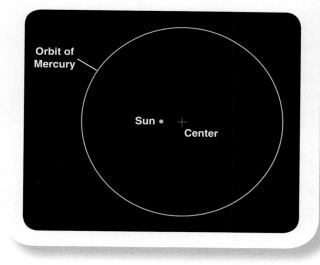

It is important to notice that Kepler's three laws are **empirical**. That is, they describe a phenomenon without explaining why it occurs. Kepler derived them from Tycho's extensive observations without referring to any first principles, fundamental assumptions, or theory. In fact, Kepler never knew what held the planets in their orbits or why they continued to move around the sun in the ways he discovered.

> **Empirical:** Description of a phenomenon without explaining why it occurs.

Hypothesis, Theory, and Law: Levels of Confidence

Even scientists misuse the words *hypothesis, theory,* and *law.* You must try to distinguish these terms from one another because they are key elements in science.

A **hypothesis** is a single assertion or conjecture that can be tested. It could be true or false. "All Texans love chili" is a hypothesis. To know whether it is true or false, you need to test it against reality by making observations or performing experiments. Copernicus asserted that the universe was heliocentric; his assertion was a hypothesis subject to testing.

In Scientific Models in Chapter 2, you saw that a model is a description of some natural phenomenon; it can't be right or wrong. A model is not a conjecture of truth but merely a convenient way to think about a natural phenomenon. Consequently, a model such as the celestial sphere is not a hypothesis. Copernicus used his hypothesis to build a model, but they are not the same thing.

A **theory** is a system of rules and principles that can be applied to a wide variety of circumstances. A theory may have begun as one or more hypotheses, but it has been tested, expanded, and generalized. Many textbooks refer to the "Copernican theory," but some historians argue that it was not complete and had not been tested enough to be a theory. It is probably better to call it the Copernican hypothesis.

A **natural law** is a theory that has been refined, tested, and confirmed so often that scientists have great confidence in it. Laws are the most fundamental principles of scientific knowledge. Kepler's laws are good examples.

Confidence is the key to understanding these terms. Scientists have more confidence in a theory than in a hypothesis and great confidence in a natural law. Nevertheless, scientists are not always consistent about these words. For example, Einstein's theory of relativity is much more accurate than Newton's laws of gravity and motion, but traditionally scientists refer to "Einstein's theories" and "Newton's laws." Darwin's theory of evolution has been tested many times, and scientists have great confidence in it, but no one refers to Darwin's law. These distinctions are subtle and sometimes depend more on custom than on levels of confidence.

3.4 Galileo Galilei

Hypothesis: A conjecture, subject to further tests, that accounts for a set of facts.

Theory: A system of assumptions and principles applicable to a wide range of phenomena that has been repeatedly verified.

Natural law: A theory that has been so well confirmed that it is almost universally accepted as correct.

Galileo Galilei was born in the Italian city of Pisa in 1564 and studied medicine at the university there. His true love, however, was mathematics, and he eventually became professor of mathematics at the university at Padua, where he remained for 18 years. During this time, Galileo seems to have adopted the Copernican model, although he admitted in a 1597 letter to Kepler that he did not support that model publicly, fearing criticism.

Most people know two "facts" about Galileo, and both are wrong. Galileo did not invent the telescope, and he was not condemned by the Inquisition for believing that Earth moved around the sun. As you learn about Galileo, you will discover that what was on trial were not just his opinions about the place of the Earth but also the methods of science itself.

Telescopic Observations

It was the telescope that drove Galileo to publicly defend the heliocentric model. Galileo did not invent the telescope. It was apparently invented around 1608 by lens makers in Holland. Galileo, hearing descriptions in the fall of 1609, was able to build working telescopes in his workshop. Galileo was also not the first person to look

Figure 3.5

(a) On the night of January 7, 1610, Galileo saw three small "stars" near the bright disk of Jupiter and sketched them in his notebook. On subsequent nights (except January 9, which was cloudy), he saw that the stars were actually four moons orbiting Jupiter. (b) This photograph, taken through a modern telescope, shows the overexposed disk of Jupiter and three of the four Galilean moons.

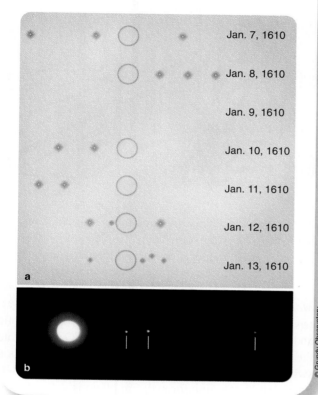

Jan. 7, 1610

Jan. 8, 1610

Jan. 9, 1610

Jan. 10, 1610

Jan. 11, 1610

Jan. 12, 1610

Jan. 13, 1610

a

b

© Grundy Observatory

at the sky through a telescope, but he was the first person to observe the sky carefully and apply his observations to the main theoretical problem of the day—the place of Earth.

What Galileo saw through his telescopes was so amazing he rushed a small book into print, *Sidereus Nuncius (The Starry Messenger)*. In that book he reported two major discoveries about the solar system. First, the moon was not perfect. It had mountains and valleys on its surface, and Galileo used the shadows to calculate the height of the mountains. Aristotle's philosophy held that the moon was perfect, but Galileo showed that it was not only imperfect but was even a world like Earth. Second, Galileo's telescope revealed four new "planets" circling Jupiter, planets that we know today as the Galilean moons of Jupiter, shown in Figure 3.5.

The moons of Jupiter supported the Copernican over the Ptolemaic model. Critics of Coper-

nicus had said Earth could not move because the moon would be left behind; but Jupiter moved and kept its satellites. Galileo's discovery suggested that Earth, too, could move and keep its moon. Also, Aristotle's philosophy included the belief that all heavenly motion was centered on Earth. Galileo showed that Jupiter's moons revolve around Jupiter, so there could be centers of motion other than Earth. Later, after the *Messenger* was published, Galileo noticed that Jupiter's innermost moon had the shortest orbital period and the moons further from Jupiter had proportionally longer periods. In this way, Jupiter's moons made up a harmonious system ruled by Jupiter, just as the planets in the Copernican universe were a harmonious system ruled by the sun. This similarity didn't constitute proof, but Galileo saw it as an indication that the solar system could be sun centered and not Earth centered.

In the years of further exploration with his telescope, Galileo made additional fundamental discoveries. When he observed Venus, Galileo saw that it was going through phases like those of the moon. In the Ptolemaic model, Venus moves around an epicycle centered on a line between Earth and the sun. If that were true, it would always be seen as a crescent, like the model in Figure 3.6 on the next page. But Galileo saw Venus go through a complete set of phases, including full and gibbous, which proved that it did indeed revolve around the sun (Figure 3.6b).

© Giorgio Malino/Photographer's Choice/Getty Images / © Steven Wynn/iStockphoto.com

Sidereus Nuncius was popular and made Galileo famous. In 1611, Galileo visited Rome and was treated with great respect. He had friendly discussions with the powerful Cardinal Barberini, but because he was outspoken, forceful, and sometimes tactless he offended other important people who questioned his telescopic discoveries. Some critics said he was wrong, and others said he was lying. Some refused to look through a telescope lest it mislead them, and others looked and claimed to see nothing (hardly surprising given the awkwardness of those first telescopes). When Galileo visited Rome again in 1616, Cardinal Bellarmine interviewed him privately and ordered him to cease public debate about models of the universe, an order Galileo appears to have mostly followed.

The Inquisition (formally named the Congregation of the Holy Office) banned books relevant to the Copernican hypothesis, although *De Revolutionibus* itself was only suspended pending revision because it was recognized as useful for its predictions of planet positions. Everyone who owned a copy of the book was required to cross out certain statements and add handwritten corrections stating that Earth's motion and the central location of the sun were only theories and not facts, a situation that you will recognize as recurring today in connection with textbooks discussing biological evolution.

Dialogo and Trial

In 1623 Galileo's friend Cardinal Barberini became pope, taking the name Urban VIII. Galileo went to Rome in an attempt to have the 1616 order to cease debate lifted. Although that attempt was unsuccessful, Galileo began to

© Yerkes Observatory

write a massive defense of Copernicus's model, completing it in 1629. After some delay, Galileo's book was approved by both the local censor in Florence and the head censor of the Vatican in Rome. It was printed in 1632.

Called *Dialogo Sopra i Due Massimi Sistemi del Mondo (Dialogue concerning the Two Chief World Systems),* it confronts the ancient astronomy of Aristotle and Ptolemy with the Copernican model. Galileo wrote the book as a debate among three friends. Salviati is a swift-tongued defender of Copernicus; Sagredo is intelligent but largely uninformed; Simplicio is a dim-witted defender of Ptolemy. The book was a clear defense of

Figure 3.6

(a) If Venus moved in an epicycle centered on the Earth-sun line, it would always appear as a crescent. (b) Galileo's telescope showed that Venus goes through a full set of phases, proving that it must orbit the sun.

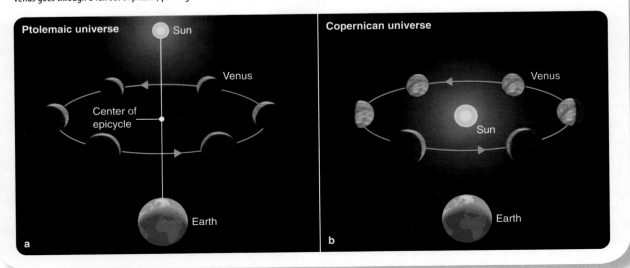

Copernicus, and, either intentionally or unintentionally, Galileo exposed the pope's authority to ridicule. Urban VIII was fond of arguing that, as God was omnipotent, God could construct the universe in any form while making it appear to humans to have a different form, and thus its true nature could not be deduced by mere observation. Galileo placed the pope's argument in the mouth of Simplicio. The pope took offense and ordered Galileo to face the Inquisition.

Galileo was interrogated by the Inquisition and threatened with torture. He must have thought of Giordano Bruno, a monk who was tried, condemned, and burned at the stake in Rome in 1600 for, among other offenses, supporting Copernicus. However, Galileo's trial did not center on his belief in Copernicus's model. *Dialogo* had been approved by two censors. Rather, the trial centered on a record of the meeting in 1616 between Galileo and Cardinal Bellarmine that included the statement that Galileo was "not to hold, teach, or defend in any way" the principles of Copernicus. Many historians believe that this document, which was signed neither by Galileo nor by Bellarmine nor by a legal secretary, was a forgery, or perhaps a draft that was never used. By this time Bellarmine was dead and could not testify about the meeting or the document.

The Inquisition condemned Galileo not primarily for heresy but for disobeying the orders given him in 1616. In 1633, at the age of 70, kneeling before the Inquisition, Galileo read a recantation admitting his errors. Tradition has it that as he rose he whispered, *"E pur si muove"* ("Still it moves"), referring to Earth. Although he was sentenced to life imprisonment, he was actually confined at his villa for the next 10 years, perhaps through the intervention of the pope. He died there in 1642, 99 years after the death of Copernicus.

Two Ways to Understand the World

Galileo was tried and condemned on a charge you might call a technicality. Why then is his trial so important that historians have studied it for almost four centuries? Why have some of the world's greatest authors, including Bertolt Brecht, written about Galileo's trial? Why in 1979 did Pope John Paul II create a commission to reexamine the case against Galileo?

To understand the trial, you must recognize that it was the result of a conflict between two ways of understanding the universe. Since the Middle Ages, European scholars had taught that the only path to true understanding was through religious faith. St. Augustine (354–430 CE)

wrote *"Credo ut intelligame,"* which can be translated as, "Believe in order to understand." Galileo and other scientists of the Renaissance, however, used their own observations to try to understand the universe; and, when their observations contradicted religious authorities, they assumed their observations of reality were correct. Galileo paraphrased Cardinal Baronius in saying, "The Bible tells us how to go to heaven, not how the heavens go." The significance of Galileo's trial is about the birth of modern science as a new way to understand the universe.

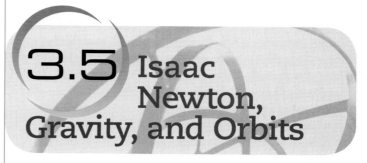

3.5 Isaac Newton, Gravity, and Orbits

The birth of modern astronomy and of modern science dates from the 144 years between the publication of Copernicus's book in 1543 and Newton's book in 1687. The Renaissance is commonly taken to be the period approximately between 1350 and 1600, so the 144 years of this story lie at the climax of the European reawakening of learning in all fields.

The problem of the place of Earth was resolved by the Copernican Revolution, but the problem of planetary motion was only partly solved by Kepler's laws. For the last 10 years of his life, Galileo studied the nature of motion, especially the accelerated motion of falling bodies. Although he made some important progress, he was not able to relate his discoveries about motion to the heavens. That final step was taken by Isaac Newton.

Isaac Newton

Galileo died in January 1642. Some 11 months later, on Christmas day 1642, Isaac Newton was born in the English village of Woolsthorpe. Newton's life represented the flowering of the seeds planted by the previous four astronomers in this story, Copernicus, Tycho, Kepler, and Galileo.

Newton was a quiet child from a farming family, but his work at school was so impressive that his uncle financed his education at Trinity College, where he studied mathematics and physics. In 1665, plague swept through England, and the colleges were closed. During 1665 and 1666, Newton spent his time back home in Woolsthorpe, thinking and studying. It was during these years that he

Mass: A measure of the amount of matter making up an object.

Weight: The force that gravity exerts on an object.

made most of his scientific discoveries. Among other things, he studied optics, developed three laws of motion, probed the nature of gravity, and invented calculus. The publication of his work in his book *Principia* in 1687 placed the fields of physics and astronomy on a new firm base.

It is beyond the scope of this book to analyze all of Newton's work, but his laws of motion and gravity had an important impact on the future of astronomy. From his study of the work of Galileo, Kepler, and others, Newton extracted three laws that relate the motion of a body to the forces acting on it (Table 3.2). These laws made it possible to predict exactly how a body would move if the forces were known.

When Newton thought carefully about motion, he realized that some force must pull the moon toward Earth's center. If there were no such force altering the moon's motion, it would continue moving in a straight line and leave Earth forever. It can circle Earth only if Earth attracts it. Newton's insight was to recognize that the force that holds the moon in its orbit is the same as the force that makes apples fall from trees.

Table 3.2 Newton's Three Laws of Motion

I.	A body continues at rest or in uniform motion in a straight line unless acted upon by some force.
II.	A body's change of motion is proportional to the force acting on it, and is in the direction of the force.
III.	When one body exerts a force on a second body, the second body exerts an equal and opposite force back on the first body.

Newtonian gravitation is sometimes called universal mutual gravitation. Newton's third law points out that forces occur in pairs. If one body attracts another, the second body must also attract the first. Thus, gravitation is mutual. Furthermore, gravity is universal. That is, all objects with mass attract all other masses in the universe. The **mass** of an object is a measure of the amount of matter in the object, usually expressed in kilograms. Mass is not the same as **weight**. An object's weight is the force that Earth's gravity exerts on the object. Thus an object in space far from Earth might have no weight, but it would contain the same amount of matter and would thus have the same mass that it has on Earth.

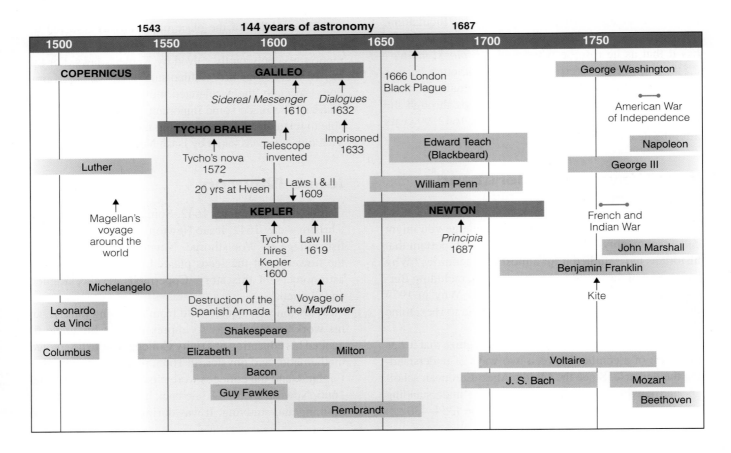

Newton also realized that the distance between the objects is important. In other words, the gravitational force between two bodies depends not only on the masses of the bodies but also on the distance between them. He recognized that the force of gravity decreases as the square of the distance between the objects increases. Specifically, if the distance from, say, Earth to the moon were doubled, the gravitational force between them would decrease by a factor of 2^2, or 4. If the distance were tripled, the force would decrease by a factor of 3^2, or

9. This relationship is known as the **inverse square relation**. Newton guessed that gravity works by an inverse square relation because he had already discovered that light behaves this way (to be discussed in more detail in Chapter 6).

To summarize, the force of gravity attracting two objects to each other equals a constant times the product of their masses divided by the square of the distance between the objects. Gravity is universal: Your mass affects the planet Neptune and the galaxy M31, and every other object in the universe, and their masses affect you—although not much, because they are so far away and your mass is relatively very small.

Orbital Motion

Newton's laws of motion and gravitation make it possible for you to understand why and how the moon orbits Earth, the planets orbit the sun, and to discover why Kepler's laws work.

To understand how an object can orbit another object, you need to see orbital motion as Newton did. Begin by noticing three important ideas:

1. An object orbiting Earth, and any orbiting object, is actually falling (being accelerated) toward Earth's center. An object in a stable orbit continuously misses Earth because of its horizontal velocity.

2. Objects orbiting each other actually revolve around their mutual center of mass.

3. Notice the difference between closed orbits and open orbits. If you want to leave Earth never to

return, you must give your spaceship a high enough velocity so it will follow an open orbit.

When the captain of a spaceship says to the pilot, "Put us into a circular orbit," the ship's computers must quickly calculate the velocity needed to achieve a circular orbit. That circular velocity depends only on the mass of the planet and the distance from the center of the planet. Once the engines fire and the ship reaches circular velocity, the engines can shut down. The ship is in orbit and will fall around the planet forever, as long as it is above the atmosphere's friction. No further effort is needed to maintain orbit, thanks to the laws Newton discovered.

Tides: Gravity in Action

Newton understood that gravity is mutual—Earth attracts the moon, and the moon attracts Earth—and that means the moon's gravity can explain the ocean tides. But Newton also realized that gravitation is universal, and that means there is much more to tides than just Earth's oceans.

Tides are caused by small differences in gravitational force. As the Earth and moon orbit around each other, they attract each other gravitationally. Because the side of Earth toward the moon is a bit closer, the moon pulls on it more strongly and that pulls up a bulge. Also, the moon pulls on Earth a bit more than it pulls on Earth's far side and

Inverse square relation: A rule that the strength of an effect (such as gravity) decreases in proportion as the distance squared increases.

Orbiting Earth

1 You can understand orbital motion by thinking of a cannonball falling around Earth in a circular path. Imagine a cannon on a high mountain aimed horizontally as shown at right. A little gunpowder gives the cannonball a low velocity, and it doesn't travel very far before falling to Earth. More gunpowder gives the cannonball a higher velocity, and it travels farther. With enough gunpowder, the cannonball travels so fast it never strikes the ground. Earth's gravity pulls it toward Earth's center, but Earth's surface curves away from it at the same rate it falls. It is in orbit. The velocity needed to stay in a circular orbit is called the **circular velocity**. Just above Earth's atmosphere, circular velocity is 7,790 m/s or about 17,400 miles per hour, and the orbital period is about 90 minutes.

A satellite above Earth's atmosphere feels no friction and will fall around Earth indefinitely.

North Pole

Earth satellites eventually fall back to Earth if they orbit too low and experience friction with the upper atmosphere.

1a A **geosynchronous satellite** orbits eastward with the rotation of Earth and remains above a fixed spot — ideal for communications and weather satellites.

A Geosynchronous Satellite

At a distance of 42,250 km (26,260 miles) from Earth's center, a satellite orbits with a period of 24 hours.

The satellite orbits eastward, and Earth rotates eastward under the moving satellite.

The satellite remains fixed above a spot on Earth's equator.

1b According to Newton's first law of motion, the moon should follow a straight line and leave Earth forever. Because it follows a curve, Newton knew that some force must continuously accelerate it toward Earth — gravity. Each second the moon moves 1,020 m (3,350 ft) eastward and falls about 1.6 mm (about 1/16 inch) toward Earth. The combination of these motions produces the moon's curved orbit. The moon is falling but nevertheless stays in its orbit.

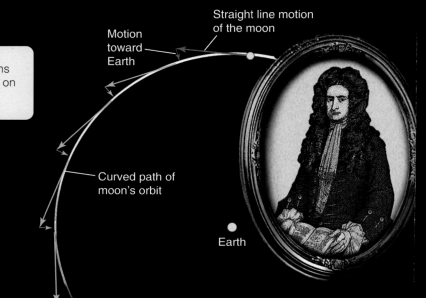

Straight line motion of the moon

Motion toward Earth

Curved path of moon's orbit

Earth

1c Astronauts in orbit around Earth feel weightless, but they are not "beyond Earth's gravity," to use a term from old science fiction movies. Like the moon, the astronauts are accelerated toward Earth by Earth's gravity, but they travel fast enough along their orbits that they continually "miss the Earth." They are literally falling around Earth. Inside or outside a spacecraft, astronauts feel weightless because they and their spacecraft are falling at the same rate. Rather than saying they are weightless, you should more accurately say they are in free fall.

2 To be precise you should not say that an object orbits Earth. Rather the two objects orbit each other. Gravitation is mutual, and if Earth pulls on the moon, the moon pulls on Earth. The two bodies revolve around their common **center of mass**, the balance point of the system.

Center of mass

2a Two bodies of different mass balance at the center of mass, which is located closer to the more massive object. As the two objects orbit each other, they revolve around their common center of mass as shown at right. The center of mass of the Earth–moon system lies only 4,700 km (2,900 miles) from the center of Earth — inside the Earth. As the moon orbits the center of mass on one side, the Earth swings around the center of mass on the opposite side.

3 **Closed orbits** return the orbiting object to its starting point. The moon and artificial satellites orbit Earth in closed orbits. Below, the cannonball could follow an elliptical or a circular closed orbit. If the cannonball travels as fast as **escape velocity**, the velocity needed to leave a body, it will enter an open orbit. An **open orbit** does not return the cannonball to Earth. It will escape.

Hyberbola

Parabola

A cannonball with a velocity greater than escape velocity will follow a hyperbola and escape from Earth.

A cannonball with escape velocity will follow a parabola and escape.

North Pole

3a As described by Kepler's Second Law, an object in an elliptical orbit has its lowest velocity when it is farthest from Earth (apogee), and its highest velocity when it is closest to Earth (perigee). Perigee must be above Earth's

Ellipse

Circle

Figure 3.7

Tides are produced by small differences in the gravitational force exerted on different parts of an object. The side of Earth nearest the moon feels a larger force than the side farthest away. Relative to Earth's center, small forces are left over, and they cause the tides. Both the moon and the sun produce tides on Earth. Tides can alter both an object's rotation and orbital motion.

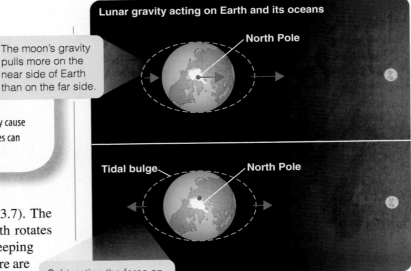

Lunar gravity acting on Earth and its oceans

North Pole

The moon's gravity pulls more on the near side of Earth than on the far side.

Tidal bulge — North Pole

Subtracting the force on Earth's center reveals the small outward forces that produce tidal bulges.

Spring tides occur when tides caused by the sun and moon add together.

Spring tides are extreme.

Full moon — New moon — To sun

Neap tides are mild. — First quarter — Neap tides occur when tides caused by the sun and moon partially cancel out.

To sun

Third quarter — Diagrams not to scale

that produces a bulge on the far side (Figure 3.7). The oceans are deeper in these bulges, and as Earth rotates and carries you into a bulge, you see the tide creeping up the beach. Because there are two bulges, there are two high tides each day, although the exact pattern of tides at any given locality depends on details such as ocean currents, the shape of the shore, etc.

The sun also produces tides on Earth, although they are smaller than lunar tides. At new and full moons, the lunar and solar tides add together to produce extra high and extra low tides that are called **spring tides**. At first and third quarter moons, the solar tides cancel out part of the lunar tides so that high and low tides are not extreme (see Figure 3.7). These are called **neap tides**.

Although the oceans flow easily into tidal bulges, the nearly rigid bulk of Earth flexes into tidal bulges and the plains and mountains rise and fall a few centimeters twice a day. Friction is gradually slowing Earth's rotation, and fossil evidence shows that Earth used to rotate faster. In the same way, Earth's gravity produces tidal bulges in the moon, and, although the moon used to rotate faster, friction has slowed it down and it now keeps the same side facing Earth.

Tides can also affect orbits. The rotation of Earth drags the tidal bulges slightly ahead of the moon, and the gravitation of the bulges of water pull the moon forward in its orbit. This makes the moon's orbit grow larger by about 4 cm a year, an effect that astronomers can measure by bouncing lasers off reflectors left on the moon by the Apollo astronauts.

Newton's Universe

Spring tide: Ocean tide of large range that occurs at full and new moon.

Neap tide: Ocean tide of small range occurring at first- and third-quarter moon.

Newton's insight gave the world a new conception of nature. His laws of motion and gravity were *general* laws that described the motions of all bodies under the action of external forces. In addition, the laws were

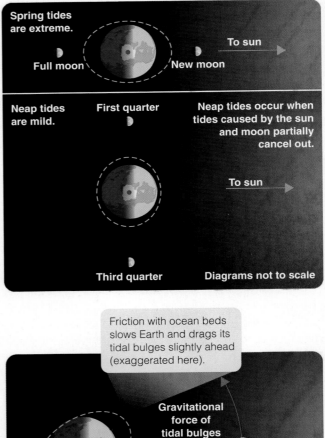

Friction with ocean beds slows Earth and drags its tidal bulges slightly ahead (exaggerated here).

Gravitational force of tidal bulges — Moon

Earth's rotation

Gravity of tidal bulges pulls the moon forward and alters its orbit.

Testing a Theory by Prediction

When you read about any science, you should notice that scientific theories face in two directions. They look back into the past and explain previously observed phenomena. For example, Newton's laws of motion and gravity explained how the planets moved. But theories also face forward in that they enable you to make predictions about what you should find as you explore further. Thus, Newton's laws allowed astronomers to calculate the orbits of comets, predict their return, and eventually understand their origin.

Scientific predictions are important in two ways. First, if a theory leads to a prediction and scientists later discover the prediction was true, the theory is confirmed, and scientists gain confidence that it is a true description of nature. But predictions are important in science in a second way. Using an existing theory to make a prediction may lead you into an unexplored avenue of knowledge. Thus, the first theories of genetics made predictions that confirmed the genetic theory of inheritance, but those predictions also created a new understanding of how living creatures evolve.

As you read about any scientific theory, think about both what it can explain and what it can predict.

recap Before Copernicus, people thought of Earth as a special place different from any of the objects in the sky. But Copernican Theory stimulated people to see Earth and humanity as part of an elegant and complex universe.

We are not in a special place ruled by mysterious planetary forces: Kepler showed that the planets move according to simple rules; Newton found laws that account for the fall of an apple, the ocean tides, and orbital motion. Earth, the sun, and all of humanity are part of a universe whose motions can be described by a few fundamental laws of motion and gravity. And those simple rules open the universe to scientific study.

Copernicus, Kepler, and Newton paved the way for astronomers to study the heavens and work out some of nature's deepest secrets. Astronomy tells us that we are special because we can study the universe and eventually understand what we are. With this new understanding, you should be able to answer these questions:

1. How did people in ancient civilizations describe Earth's place in the universe?

2. How did Copernicus revise that ancient model?

3. Why was the Copernican model gradually accepted?

4. How did Tycho Brahe and Johannes Kepler contribute to the Copernican revolution?

5. Why was Galileo condemned by the Inquisition?

6. How did Isaac Newton change humanity's view of nature?

predictive because they made possible specific calculations of predictions that could be tested by observation. For example, Newton's laws of motion can be used to derive Kepler's third law from the law of gravity. Newton's discoveries remade astronomy into an analytical science in which astronomers could measure the positions and motions of celestial bodies, calculate the gravitational forces acting on them, and predict their future motion.

One of the most often used and least often stated principles of science is *cause and effect*. You could argue that Newton's second law of motion was the first clear statement of that principle. Ancient philosophers such as Aristotle believed that objects moved because of innate tendencies; for example, objects made of earth or water had a natural tendency to move toward Earth at the center of the universe. In contrast, Newton's second law says if an object changes its motion by an acceleration, then it must have been acted on by a force. The principle of cause and effect gives scientists confidence that every effect has a cause. If the universe were not rational, then you could never expect to discover causes. Newton's second law of motion was arguably the first explicit statement that the behavior of the universe is rational and depends on causes.

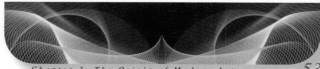

Chapter 4

Astronomical Telescopes and Instruments: Extending Humanity's Vision

Light is a treasure that links us to the sky. An astronomer's quest is to gather as much light as possible from the moon, sun, planets, stars, and galaxies in order to extract information about their natures. Telescopes, which gather and focus light for analysis, can help do that. Nearly all the interesting objects in the sky are very faint sources of light, so large telescopes like the one in Figure 4.1 are built to collect the greatest amount of light possible. This chapter's discussion of astronomical research concentrates on large telescopes and the special instruments and techniques used to analyze light. A normal telescope will gather visible light, but visible light is only one type of radiation arriving here from distant objects. Astronomers can extract information from other forms of radiation by using other types of telescopes. Radio telescopes, for example, give an entirely different view of the sky. Some of these specialized telescopes can be used from Earth's surface, but others must go high in Earth's atmosphere or even above it.

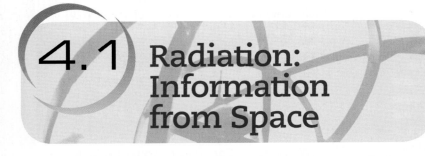

4.1 Radiation: Information from Space

Astronomers no longer study the sky by mapping constellations or charting the phases of the moon. Modern astronomers analyze light using sophisticated instruments and techniques to investigate the compositions, motions, internal processes, and evolution of celestial objects. To understand this, you need to learn about the nature of light.

Light as a Wave and as a Particle

Have you ever noticed the colors in a soap bubble? If so, then you have seen one effect of light behaving as a wave. When that same light enters the light meter on a camera, it behaves as a particle. How light behaves depends on how you treat it—light has both wavelike and particlelike properties. Sound

© Peter Arnold, Inc./Alamy

looking back

In the early chapters of this book, you looked at the sky the way ancient astronomers did, with the unaided eye. In the last chapter, you had a glimpse through Galileo's telescope that revealed wonderful things about the moon, Jupiter, and Venus. Now you can study the telescopes, instruments, and techniques of the modern astronomer.

looking ahead

Astronomy is almost entirely an observational science. Astronomers cannot visit distant galaxies and far-off worlds, so they must observe celestial objects using astronomical telescopes. Eleven chapters remain in this book, and every one will discuss information gathered by telescopes. The chapter immediately after this one will show how observations of the sun and other stars can be understood with the help of experiments in laboratories on Earth that study the interactions of light and matter.

astro

is another type of wave that you have already experienced. Sound waves are air pressure disturbances that travel through the air from source to ear. Sound requires a solid, liquid, or gas medium to carry it; so, for example, in space outside a spacecraft, there can be no sound. In contrast, light is composed of a combination of electric and magnetic waves that can travel through empty space. Unlike sound, light waves do not require a medium and thus can travel through a vacuum.

Because light is made up of both electric and magnetic fields, it is referred to as **electromagnetic radiation**. Visible light is only one form of electromagnetic radiation. Electromagnetic radiation is a wave phenomenon—that is, it is associated with a periodically repeating disturbance (a wave) that carries energy. Imagine waves in water: If you disturb a pool of water, waves spread across the surface. Now imagine placing a ruler parallel to the travel direction of the wave. The distance between peaks is the **wavelength**. The changing electric and magnetic fields of electromagnetic waves travel through space at about 300,000 kilometers per second (186,000 miles per second). That is commonly referred to as the speed of light, but it is the speed of all electromagnetic radiation.

It may seem odd to use the word *radiation* when talking about light, but radiation really refers to anything that spreads outward from a source. Light radiates from a source, so you can correctly refer to light as a form of radiation.

The Electromagnetic Spectrum

The electromagnetic spectrum is simply the types of electromagnetic radiation arranged in order of increasing wavelength. Rainbows are spectra (plural) of visible light. The colors of visible light have different wavelengths: Red has the longest wavelength and violet the shortest, as shown in the visible spectrum at the top of Figure 4.2.

Electromagnetic radiation: Changing electric and magnetic fields that travel through space and transfer energy from one place to another; examples are light or radio waves.

Wavelength: The distance between successive peaks or troughs of a wave, usually represented by a lowercase Greek lambda, λ.

Nanometer (nm): A unit of distance equaling one-billionth of a meter (10^{-9} m), commonly used to measure the wavelength of light.

Angstrom (Å): A unit of distance commonly used to measure the wavelength of light. 1 Å $= 10^{-10}$ m.

Infrared (IR): The portion of the electromagnetic spectrum with wavelengths longer than red light, ranging from 700 nm to about 1 mm, between visible light and radio waves.

Figure 4.1

The northern Gemini telescope stands over 19 m (60 ft) high when pointed straight up. Its main mirror is 8.1 m (26.5 ft) in diameter—larger than some classrooms.

Light-gathering optical surface

Telescope technician

The average wavelength of visible light is about 0.0005 mm. Fifty light waves would fit end-to-end across the thickness of a sheet of paper. It is too awkward to measure such short distances in millimeters, so physicists and astronomers describe the wavelength of light using either the unit of **nanometer (nm)**, one-billionth of a meter (10^{-9} m), or the **Ångstrom (Å)**, equal to 10^{-10} m or 0.1 nm. The wavelength of visible light ranges between about 400 nm and 700 nm, or, equivalently, 4,000 Å and 7,000 Å. Infrared astronomers often refer to wavelengths using units of microns (10^{-6} m), while radio astronomers use millimeters, centimeters, or meters. Figure 4.2 shows how the visible spectrum makes up only a small part of the electromagnetic spectrum.

Beyond the red end of the visible range lies **infrared (IR)** radiation, with wavelengths ranging from 700 nm to about 1 mm. Your eyes are not sensitive to this radiation, but your skin can sense some of it as heat. A heat lamp is nothing more than a bulb that gives off large amounts of infrared radiation. English astronomer William Herschel discovered infrared radiation—and thus, discovered that there is such a thing as invisible light.

Radio waves have even longer wavelengths than IR radiation. The radio radiation used for AM radio transmissions has wavelengths of a few hundred meters, while FM, television, and also military, governmental, and amateur radio transmissions have wavelengths from a few tens of centimeters to a few tens of meters. Microwave transmissions, used for radar and long-distance telephone communications, have wavelengths from about 1 millimeter to a few centimeters.

Look at the electromagnetic spectrum in Figure 4.2 and notice that electromagnetic waves with wavelengths shorter than violet light are called **ultraviolet (UV)**.

Electromagnetic waves with wavelengths shorter than UV light are called **X rays**, and the shortest are **gamma rays**.

The distinction between these wavelength ranges is mostly arbitrary—they are simply convenient human-invented labels. For example, the longest-wavelength infrared radiation and the shortest-wavelength microwaves are the same. Similarly, very short-wavelength ultraviolet light can be considered to be X rays. Nonetheless, it is all electromagnetic radiation, and you could say we are "making light" of it all, because all these

> **Ultraviolet (UV):** The portion of the electromagnetic spectrum with wavelengths shorter than violet light, between visible light and X rays.
>
> **X rays:** Electromagnetic waves with wavelengths shorter than ultraviolet light.
>
> **Gamma rays:** The shortest-wavelength electromagnetic waves.

Figure 4.2

The spectrum of visible light, extending from red to violet, is only part of the electromagnetic spectrum. The lower panel shows that most radiation is absorbed in Earth's atmosphere, and only radiation with certain wavelengths, such as the visual and radio windows, can easily reach Earth's surface.

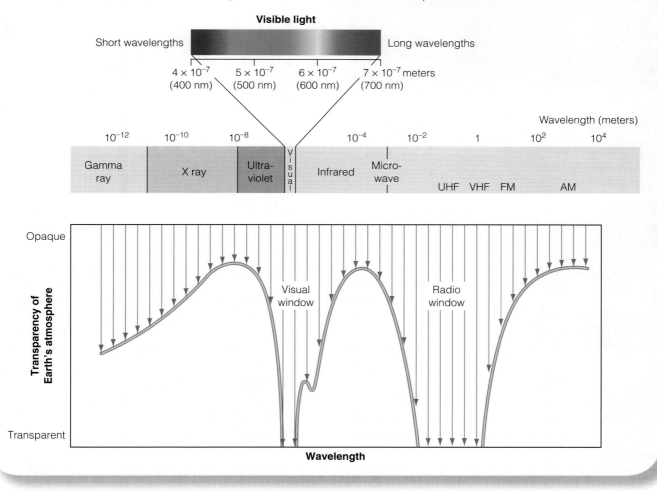

Depiction of William Herschel in 1800 discovering that sunlight contains radiation detectable by thermometers but not by human eyes, which he named infrared ("beyond red") light.

but an X-ray photon carries much more energy and a radio photon carries much less.

Astronomers are interested in electromagnetic radiation because it carries almost all available clues to the nature of planets, stars, and other celestial objects. Earth's atmosphere is opaque to most electromagnetic radiation, as shown by the graph at the bottom of Figure 4.2. Gamma rays, X rays, and some radio waves are absorbed high in Earth's atmosphere, and a layer of ozone (O_3) at an altitude of about 30 km absorbs almost all UV radiation. Water vapor in the lower atmosphere absorbs long-wavelength IR radiation. Only visible light, some short-wavelength infrared radiation, and some radio waves reach Earth's surface through what are called **atmospheric windows**. To study the sky from Earth's surface, you must look out through one of these "windows" in the electromagnetic spectrum.

4.2 Telescopes

Astronomers build optical telescopes to gather light and focus it into sharp images. This requires careful optical and mechanical designs, and it leads astronomers to build very large telescopes. To understand that, you need to learn the terminology of telescopes, starting with the different types of telescopes and why some are better than others.

Two Kinds of Telescopes

Astronomical telescopes focus light into an image in one of two ways. Either: (1) a lens bends (refracts) the light as it passes through the glass and brings it to a focus to form an image, or (2) a mirror—a curved piece of glass with a reflective surface—forms an image by bouncing light. Figure 4.3 demonstrates the difference between the two configurations. Because there are two ways to focus light, there are two types of astronomical telescopes. **Refracting telescopes** use a lens to gather and focus the light, whereas **reflecting telescopes** use a mirror.

The main lens in a refracting telescope is called the **primary lens**, and the main mirror in a reflecting telescope

Photon: A quantum of electromagnetic energy that carries an amount of energy that depends inversely on its wavelength.

Atmospheric window: Wavelength region in which our atmosphere is transparent—at visual, radio, and some infrared wavelengths.

Refracting telescope: A telescope that forms images by bending (refracting) light with a lens.

Reflecting telescope: A telescope that forms images by reflecting light with a mirror.

Primary lens: In a refracting telescope, the largest lens.

types of radiation are the same phenomenon as light: Some types your eyes can see, some types your eyes can't see.

Although light behaves as a wave, under certain conditions it also behaves as a particle. A particle of light is called a **photon**, and you can think of a photon as a minimum-sized bundle of electromagnetic waves. The amount of energy a photon carries depends on its wavelength. Shorter-wavelength photons carry more energy, and longer-wavelength photons carry less energy. A photon of visible light carries a small amount of energy,

Figure 4.3

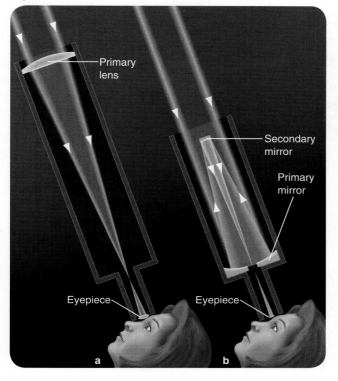

is called the **primary mirror**. Both kinds of telescopes form a small, inverted image that is difficult to observe directly, so a lens called the **eyepiece** is used to magnify the image and make it convenient to view. The **focal length** is the distance from a lens or mirror to the image it forms of a distant light source such as a star. Creating the proper optical shape to produce a good focus is an expensive process. The surfaces of lenses and mirrors must be shaped and polished to have no irregularities larger than the wavelength of light. Creating the optics for a large telescope can take months or years; involve huge, precision machinery; and employ several expert optical engineers and scientists.

Refracting telescopes have serious disadvantages. Most importantly, they suffer from an optical distortion that limits their usefulness. When light is refracted through glass, shorter wavelengths bend more than longer wavelengths. As a result, you see a color blur around every image. This color separation is called **chromatic aberration** and it can be only partially corrected. Another disadvantage is that the glass in primary lenses must

be pure and flawless because the light passes all the way through it. For that same reason, the weight of the lens can be supported only around its outer edge. In contrast, light reflects from the front surface of a reflecting telescope's primary mirror but does not pass through it, so reflecting telescopes have no chromatic aberration. Also, mirrors are less expensive to make than similarly sized lenses and the weight of telescope mirrors can be supported easily. For these reasons, every large astronomical telescope built since 1900 has been a reflecting telescope.

Optical telescopes gather visible light, but astronomers also build **radio telescopes** to gather radio radiation. Radio waves from celestial objects, like visible light waves, penetrate Earth's atmosphere and reach the ground. You can see in Figure 4.4 how the dish reflector of a typical radio

Primary mirror: In a reflecting telescope, the largest mirror.

Eyepiece: A short-focal-length lens used to enlarge the image in a telescope. The lens nearest the eye.

Focal length: The focal length of a lens or mirror is the distance from that lens or mirror to the point where it focuses parallel rays of light.

Chromatic aberration: A distortion found in refracting telescopes because lenses focus different colors at slightly different distances. Images are consequently surrounded by color fringes.

Optical telescope: Telescope that gathers visible light.

Radio telescope: Telescope that gathers radio radiation.

Figure 4.4

In most radio telescopes, a dish reflector concentrates the radio signal on the antenna. The signal is then amplified and recorded. For all but the shortest radio waves, wire mesh is an adequate reflector.

Courtesy Seth Shostak/SETI Institute

telescope focuses the radio waves so their intensity can be measured. Because radio wavelengths are so long, the disk reflector does not have to be as perfectly smooth as the mirror of a reflecting optical telescope.

> Catching light in a telescope is like catching rain in a bucket—the bigger the bucket, the more rain it catches.

The Powers of a Telescope

Astronomers struggle to build large telescopes because a telescope can help human eyes in three important ways—these are called the three powers of a telescope. The two most important of these three powers depend on the diameter of the telescope.

Most celestial objects of interest to astronomers are faint, so you need a telescope that can gather large amounts of light to produce a bright image. **Light-gathering power** refers to the ability of a telescope to collect light. Catching light in a telescope is like catching rain in a bucket—the bigger the bucket, the more rain it catches. The light-gathering power is proportional to the *area* of the primary mirror, that is, proportional to the square of the primary's diameter. A telescope with a diameter of 2 meters has four times ($4\times$) the light-gathering power of a 1-meter telescope. That is

why astronomers use large telescopes and why telescopes are ranked by their diameters. One reason radio astronomers build big radio dishes is to collect enough radio photons, which have low energies, and concentrate them for measurement.

The second power of telescopes, **resolving power**, refers to the ability of the telescope to reveal fine detail. One consequence of the wavelike nature of light is that there is an inevitable small blurring called a **diffraction fringe** around every point of light in the image, and you cannot see any detail smaller than the fringe (Figure 4.5). Astronomers can't eliminate diffraction fringes, but the fringes are smaller in larger telescopes, and that means they have better resolving power and can reveal finer detail. For example, a 2-meter telescope has diffraction fringes $\frac{1}{2}$ as large, and thus $2\times$ better resolving power, than a 1-meter telescope. The size of the diffraction fringes also depends on wavelength, and at the long wavelengths of radio waves, the fringes are large and the resolving power is poor. That's another reason radio telescopes need to be larger than optical telescopes.

One way to improve resolving power is to connect two or more telescopes in an **interferometer**, which has a resolving power equal to that of a telescope as large

Light-gathering power: The ability of a telescope to collect light; proportional to the area of the telescope's objective lens or mirror.

Resolving power: The ability of a telescope to reveal fine detail. Depends on the diameter of the telescope objective.

Diffraction fringe: Blurred fringe surrounding any image, caused by the wave properties of light. Because of this, no image detail smaller than the fringe can be seen.

Interferometer: Separated telescopes combined to produce a virtual telescope with the resolution of a much larger-diameter telescope.

Figure 4.5

(a) Stars are so far away that their images are points, but the wavelike nature of light causes each star image to be surrounded with diffraction fringes, much magnified in this computer model. (b) Two stars close to each other have overlapping diffraction fringes and become difficult to detect separately.

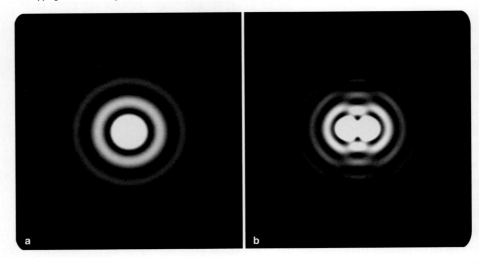

© Computer model by M. A. Seeds

as the maximum separation between the individual telescopes (Figure 4.6). The first interferometers were built by radio astronomers connecting radio dishes kilometers apart. Modern technology has allowed astronomers to connect optical telescopes to form interferometers with very high resolution.

Aside from diffraction fringes, two other factors—optical quality and atmospheric conditions—limit resolving power. A telescope must contain high-quality optics to achieve its full potential resolving power. Even a large telescope shows little detail if its optical surfaces have imperfections. In addition, when you look through a telescope, you look through miles of turbulence in Earth's atmosphere, which makes images dance and blur, a condition that astronomers call **seeing** (Figure 4.7). A related phenomenon is the twinkling of a star. The twinkles are caused by turbulence in Earth's atmosphere, and a star near the horizon, where you look through more air, will twinkle more than a star overhead. On a night when the atmosphere is unsteady, the stars twinkle, the images are blurred, and the seeing is bad. Even with good seeing, the detail visible through a large telescope is limited, not just by its diffraction fringes but by the steadiness of the air through which the observer must look. A telescope performs best on a high mountaintop where the air is thin and steady, but even at good sites atmospheric turbulence spreads star images into blobs 0.5 to 1 arc seconds in diameter. That situation can be improved by a difficult and expensive technique called **adaptive optics**, in which rapid computer calculations adjust the telescope optics and partly compensate for seeing distortions.

This limitation on the amount of information in an image is related to the limitation on the accuracy of a measurement. All measurements have some built-in uncertainty, and scientists must learn to work within those limitations.

The third and, you may be surprised to learn, least important power of a telescope is **magnifying power**, the ability to make an image large. The magnifying power of a telescope equals the focal length of the primary mirror or lens divided by the focal length of the eyepiece. For example, a telescope with a primary mirror that has a focal length of 700 mm and an eyepiece with

Figure 4.6

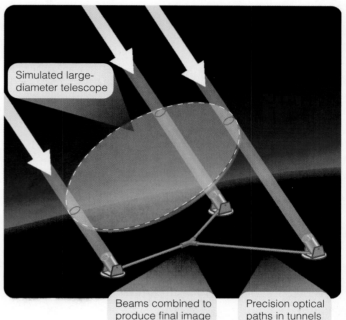

Simulated large-diameter telescope

Beams combined to produce final image

Precision optical paths in tunnels

Figure 4.7

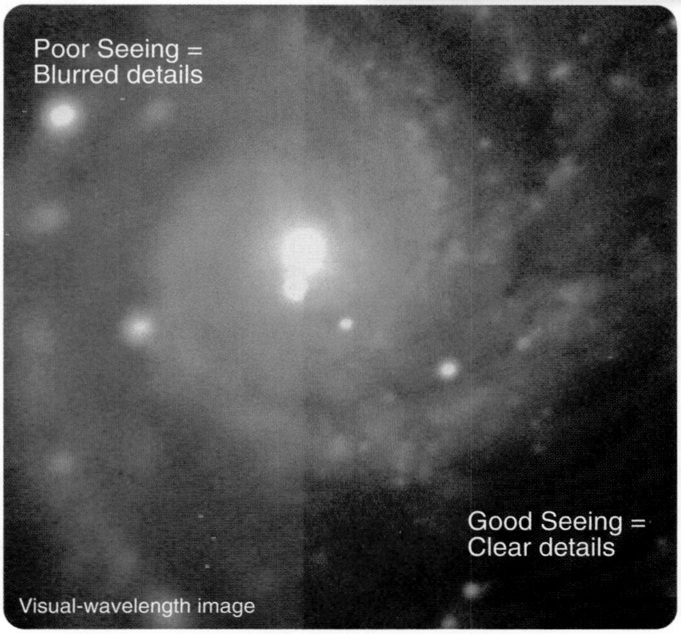

Poor Seeing = Blurred details

Good Seeing = Clear details

Visual-wavelength image

© William Keel

Seeing: Atmospheric conditions on a given night. When the atmosphere is unsteady, producing blurred images, the seeing is said to be poor.

Adaptive optics: A computer-controlled optical system in an astronomical telescope used to partially correct for seeing.

Magnifying power: The ability of a telescope to make an image larger.

Resolving Power and the Accuracy of a Measurement

Have you ever seen a movie in which the hero magnifies a newspaper photo and reads some tiny detail? It isn't really possible, because newspaper photos are made up of tiny dots of ink, and no detail smaller than a single dot will be visible, no matter how much you magnify the photo. In fact, all images are made up of elements of some sort. In an image formed by a telescope, the size of the picture element is set by seeing or diffraction. It would be foolish to try to resolve (detect) any detail smaller than this limit.

This limitation is true of all measurements in science. A zoologist might specify that a snake was 43.28932 cm long, and a sociologist might say that 98.2491 percent of people oppose drunk driving, but a critic might question the accuracy of those measurements. The resolution of the techniques may not justify the accuracy implied.

Science is based on measurement, and whenever you take a measurement, you should ask yourself how accurate that measurement can be and what its limits are.

Mars: Hi-res

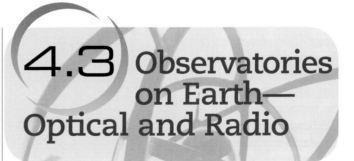

Light pollution: The illumination of the night sky by waste light from cities and outdoor lighting, which prevents the observation of faint objects.

a focal length of 14 mm has a magnifying power of 503. Higher magnifying power does not necessarily show you more detail, because the amount of detail you can see in practice is limited by a combination of the seeing conditions and the telescope's resolving power and optical quality.

A telescope's primary function is to gather light and thus make faint things appear brighter, so the light-gathering power is the most important power and the diameter of the telescope is its most important characteristic. Light-gathering power and resolving power are fundamental properties of a telescope that cannot be altered, whereas magnifying power can be changed simply by changing the eyepiece.

4.3 Observatories on Earth— Optical and Radio

Most major observatories are located far from big cities and usually on high mountains. Optical astronomers avoid cities because **light pollution**, the brightening of the night sky by light scattered from artificial outdoor lighting, can make it impossible to see faint objects. In fact, many residents of cities are unfamiliar with the beauty of the night sky because they can see only the brightest stars. Radio astronomers face a problem of

Hubble Health: check
Hubble Safety: check
Hubble Directions: sent

Figure 4.8

At Hubble's control center at the Space Telescope Science Institute in Baltimore, Md., a shift supervisor uses a series of consoles to monitor Hubble operations. All the commands transmitted to Hubble, including the instructions on recording scientific data and orders on which stars to observe, come from these consoles. The shift supervisor is responsible for overall operations and the health and safety of Hubble.

radio interference analogous to light pollution. Weak radio signals from the cosmos are easily drowned out by human radio interference—everything from automobiles with faulty ignition systems to poorly designed transmitters in communication. To avoid that, radio astronomers locate their telescopes as far from civilization as possible. Hidden deep in mountain valleys, they are able to listen to the sky protected from human-made radio noise.

As you learned previously, astronomers prefer to place optical telescopes on mountains because the air there is thin and more transparent, but, most importantly they carefully select mountains where the airflow is usually not turbulent so the seeing is good. Building an observatory on top of a high mountain far from civilization is difficult and expensive, but the dark sky and good seeing make it worth the effort.

Study page 64, and you will notice two important points:

1. Research telescopes must focus their light to positions at which cameras and other instruments can be placed.

2. Small telescopes can use other focal arrangements that would be inconvenient in larger telescopes.

Telescopes located on the surface of Earth, whether optical or radio, must move continuously to stay pointed at a celestial object as Earth turns on its axis. This is called **sidereal tracking** ("sidereal" refers to the stars).

The days when astronomers worked beside their telescopes through long, dark, cold nights are nearly gone. The complexity and sophistication of telescopes require a battery of computers, and almost all research telescopes are run from warm control rooms (Figure 4.8).

High-speed computers have allowed astronomers to build new, giant telescopes with unique designs. The European Southern Observatory has built the Very Large Telescope (VLT) high in the remote Andes Mountains of northern Chile. The VLT actually consists of four telescopes, each with a computer-controlled mirror 8.2 m in diameter and only 17.5 cm (6.9 in.) thick. The four telescopes can work singly or can combine their light to work as one large telescope. Italian and American astronomers have built the Large Binocular Telescope, which carries a pair of 8.4-m mirrors on a single mounting. The Gran Telescopio Canarias, located atop a volcanic peak in the Canary Islands, carries a segmented mirror 10.4 meters

Sidereal tracking: (pronounced "sih-dare-ee-al") The continuous movement of a telescope to keep it pointed at a star as Earth rotates.

Modern Astronomical Telescopes

1 Large astronomical telescopes must gather light and guide it to locations where it can be recorded with cameras or other instruments.

In larger telescopes, the light can be focused to a **prime focus** position high in the telescope tube, as shown at the right. Although it is a good place to image faint objects, the prime focus is inconvenient for large instruments. A **secondary mirror** can reflect the light through a hole in the primary mirror to a **Cassegrain focus**. This focal arrangement may be the most common form of astronomical telescope.

Secondary mirror

With the secondary mirror removed, the light converges at the prime focus. In large telescopes, astronomers can ride inside the prime-focus cage, although most observations are now made by instruments connected to computers in a separate control room.

The mirrors in astronomical telescopes must be ground and polished to precise shape to form the sharpest possible images. The mirrors must be carefully supported to prevent them from sagging under their own weight as the telescope moves around the sky, and many modern telescopes use computer-controlled thrusters to maintain the mirror shape.

The Cassegrain focus is convenient and has room for large instruments.

2 Smaller telescopes can use other focal arrangements. The **Newtonian focus** that Isaac Newton used in his first reflecting telescope is awkward for large telescopes, as shown at right, but is common for small telescopes.

Newtonian focus

Thin correcting lens

Schmidt-Cassegrain telescope

2a Many small telescopes such as the one on your left use a **Schmidt-Cassegrain focus**. A thin correcting plate improves the image but is too slightly curved to introduce serious chromatic aberration.

1a Shown below, the 4-meter Mayall Telescope at Kitt Peak National Observatory in Arizona can be used at either the prime focus or the Cassegrain focus. Note the human figure at lower right.

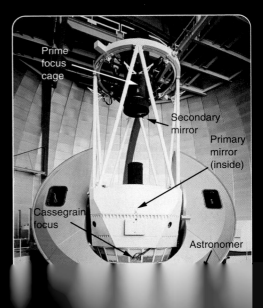

Prime focus cage

Secondary mirror

Primary mirror (inside)

Cassegrain focus

Astronomer

in diameter and holds, for the moment, the record as the largest single telescope in the world. Other giant telescopes are being planned with innovative designs.

The largest fully steerable radio telescope in the world is at the National Radio Astronomy Observatory in West Virginia. The telescope has a reflecting surface 100 meters in diameter made of 2,004 computer-controlled panels that adjust to maintain the shape of the reflecting surface. The largest radio dish in the world is 300 m (1,000 ft) in diameter, and is built into a mountain valley in Arecibo, Puerto Rico. The antenna hangs on cables above the dish, and, by moving the antenna, astronomers can point the telescope at any object that passes within 20 degrees of the zenith as Earth rotates (Figure 4.9a).

The Very Large Array (VLA) consists of 27 dishes spread in a Y-pattern across the New Mexico desert (Figure 4.9b). Operated as an interferometer, the VLA has the resolving power of a radio telescope up to 36 km

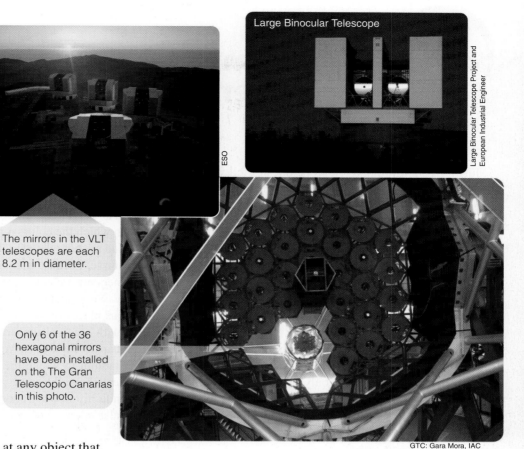

Large Binocular Telescope

The mirrors in the VLT telescopes are each 8.2 m in diameter.

Only 6 of the 36 hexagonal mirrors have been installed on the The Gran Telescopio Canarias in this photo.

ESO

Large Binocular Telescope Project and European Industrial Engineer

GTC: Gara Mora, IAC

(22 mi) in diameter. Such arrays are very powerful, and radio astronomers are now planning the Square Kilometer Array, which will consist of radio dishes spanning 6,000 km (almost 4,000 mi) and having a total collecting area of one square kilometer.

Figure 4.9

300-m radio telescope, Arecibo, Puerto Rico

a

VLA, New Mexico, USA

b

© David Parker/SPL/Photo Researchers, Inc. / © NRAO/AUI/NSF

4.4 Astronomical Instruments and Techniques

Just looking through a telescope doesn't tell you much. To learn about planets, stars, and galaxies, you must be able to analyze the light the telescope gathers. Special instruments attached to the telescope make that possible.

Imaging Systems and Photometers

The **photographic plate** was the first image-recording device used with telescopes. Brightness of objects imaged on a photographic plate can be measured with a lot of hard work, yielding only moderate precision. Astronomers also build **photometers**, sensitive light meters to measure the brightness of individual objects very precisely. Most modern astronomers use **charge-coupled devices (CCDs)** as both image-recording devices and photometers. A CCD is a specialized computer chip containing as many as a million or more microscopic light detectors arranged in an array about the size of a postage stamp. These **array detectors** can be used like a small photographic plate, but they have dramatic advantages over both photometers and photographic plates. CCDs can detect both bright and faint objects in a single exposure and are much more sensitive than a photographic plate. CCD images are **digitized**, or converted to numerical data, and can be read directly into a computer memory for later analysis. Although CCDs for astronomy are extremely sensitive and therefore expensive, less sophisticated CCDs are now used in commercial video and digital cameras.

Infrared astronomers also use array detectors similar in operation to optical CCDs. At some other wavelengths, photometers are still used for measuring brightness of celestial objects.

The digital data representing an image from a CCD or other array detector are easy to manipulate to bring out details that would not otherwise be visible. For example, astronomical images are sometimes reproduced as negatives, with the sky white and the stars dark. This makes the faint parts of the image easier to see. Astronomers also manipulate images to produce **false-color images** in which the colors represent different levels of intensity and are not related to the true colors of the object. For example, because humans can't see radio waves, astronomers must convert them into something perceptible. One way is to measure the strength of the radio signal at various places in the sky and draw a map in which contours mark areas of uniform radio intensity. Compare such a map to Figure 4.10a, a seating diagram for a baseball stadium in which the contours mark areas in which the seats have the same price. Contour maps are very common in radio astronomy and are often reproduced using false colors (Figure 4.10b).

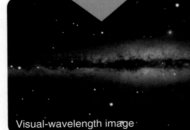
Galaxy NGC 891 as it would look to your eyes. It is edge-on and contains thick dust clouds.

Visual-wavelength image

In this image, color shows brightness. White and red are brightest, and yellow and green are dimmer.

Visual image in false color

© C. Hawk, B. Savage, N. A. Sharp NOAO/WIYN/NSF / NOAO/WIYN/NSF

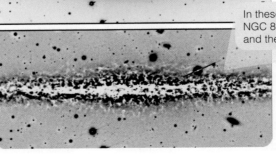
In these negative images of NGC 891, the sky is white and the stars are black.

Visual-wavelength negative images

Figure 4.10

(a) A contour map of a baseball stadium shows regions of similar admission prices. The most expensive seats are those behind home plate. (b) A false color radio map of Tycho's supernova remnant, the expanding shell of gas produced by the explosion of a star in 1572. The radio contour map has been color-coded to show intensity.

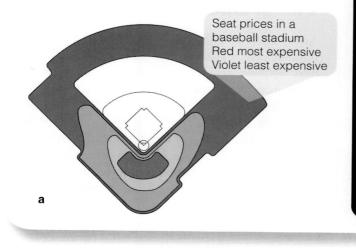

Seat prices in a baseball stadium
Red most expensive
Violet least expensive

a

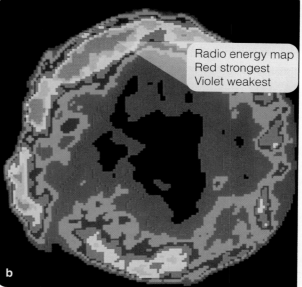

Radio energy map
Red strongest
Violet weakest

b

© NRAO

Spectrographs

To analyze light in detail, you need to spread the light out according to wavelength into a spectrum, a task performed by a **spectrograph**. You can understand how this works by reproducing an experiment performed by Isaac Newton in 1666. Boring a hole in his window shutter, Newton admitted a thin beam of sunlight into his darkened bedroom. When he placed a prism in the beam, the sunlight spread into a beautiful spectrum on the far wall. From this and similar experiments, Newton concluded that white light was made of a mixture of all the colors.

Newton didn't think in terms of wavelength, but you can use that modern concept to see that the light passing through the prism is bent at an angle that depends on the wavelength (see Figure 4.11). Violet (short-wavelength) light bends most, and red (long-wavelength) light least. Thus, the white light entering the prism is spread into what is called a **spectrum**. A typical prism spectrograph contains more than one prism to spread the spectrum wider, plus lenses to guide the light into the prism and to focus the light onto a photographic plate. Most modern spectrographs use a grating in place of a prism. A **grating** is a piece of glass with thousands of microscopic parallel lines scribed onto its surface. Different wavelengths of light reflect from the grating at slightly different angles, so white light is spread into a spectrum and can be recorded, often by a CCD camera. Recording the spectrum of a faint star or galaxy can require a long time exposure, so astronomers have developed multiobject spectrographs

© Sokolov Andrey/iStockphoto.com

DEFINITIONS

Photographic plates: The first image-recording device used with telescopes; it records the brightness of objects, but with only moderate precision.

Photometer: Sensitive astronomical instrument that measures the brightness of individual objects very precisely.

Charge-coupled device (CCD): An electronic device consisting of a large array of light-sensitive elements used to record very faint images.

Array detector: Device for collecting and recording electromagnetic radiation using multiple individual detectors arrayed on the surface of a chip; for example, a CCD electronic camera.

Digitized: Converted to numerical data that can be read directly into a computer memory for later analysis.

False-color image: A representation of graphical data with added or enhanced color to reveal detail.

Spectrograph: A device that separates light by wavelengths to produce a spectrum.

Spectrum: A range of electromagnetic radiation spread into its component wavelengths (colors); for example, a rainbow; also, representation of a spectrum as a graph showing intensity of radiation as a function of wavelength or frequency.

Grating: A piece of material in which numerous microscopic parallel lines are scribed. Light encountering a grating is dispersed to form a spectrum.

Figure 4.11

A prism bends light by an angle that depends on the wavelength of the light. Short wavelengths bend most and long wavelengths least. Thus, white light passing through a prism is spread into a spectrum.

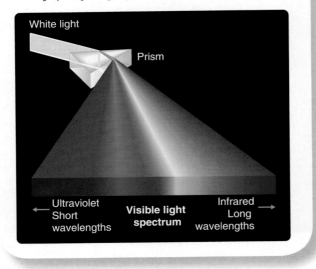

White light

Prism

Ultraviolet
Short
wavelengths

Visible light spectrum

Infrared
Long
wavelengths

that can record the spectra of as many as 100 objects simultaneously. Multiobject spectrographs automated by computers have made large surveys of many thousands of stars or galaxies possible.

Because astronomers understand how light interacts with matter, a spectrum carries a tremendous amount of information, as you will see in more detail in Chapter 5. That makes a spectrograph the astronomer's most powerful instrument. Astronomers are likely to remark, "We don't know anything about an object until we get a spectrum," and that is only a slight exaggeration.

4.5 Airborne and Space Observatories

You have learned about the observations that ground-based telescopes can make through the two atmospheric windows in the visible and radio parts of the electromagnetic spectrum. Most of the rest of the electromagnetic spectrum—infrared, ultraviolet, X-ray, and gamma-ray radiation—never reaches Earth's surface. To observe at these wavelengths, telescopes must fly above the atmosphere in high-flying aircraft, rockets, balloons, and sat-

ellites. The only exceptions are observations that can be made in what are called the near-infrared and the near-ultraviolet, almost the same wavelengths as visible light.

The Ends of the Visual Spectrum

Astronomers can observe from the ground in the near-infrared just beyond the red end of the visible spectrum because some of this radiation leaks through the atmosphere in narrow, partially open atmospheric windows ranging in wavelength from 1,200 nm to about 30,000 nm. Infrared astronomers usually describe wavelengths in micrometers or microns (10^{-6} m), so they refer to this wavelength range as 1.2 to 30 microns. In this range, most of the radiation is absorbed by water vapor and carbon dioxide molecules in Earth's atmosphere, so it is an advantage to place telescopes on the highest mountains where the air is especially thin and dry. For example, a number of important infrared telescopes are located on the summit of Mauna Kea in Hawaii at an altitude of 4,200 m (13,800 ft).

The far-infrared range, which includes wavelengths longer than 30 microns, can tell us about planets, comets, forming stars, and other cool objects, but these wavelengths are absorbed high in the atmosphere. To observe in the far-infrared, telescopes must venture to high altitudes. Remotely operated infrared telescopes suspended under balloons have reached altitudes as high as 41 km (25 mi). For many years, the NASA Kuiper Airborne Observatory (KAO) carried a 91-cm infrared telescope and crews of astronomers to altitudes of over 12 km (40,000 ft) to get above 99 percent or more of the water vapor in Earth's atmosphere. Now retired from service, the KAO will soon be replaced by the Stratospheric Observatory for Infrared Astronomy (SOFIA), a Boeing 747SP aircraft that will carry a 2.5-m (98-in.) telescope to the fringes of the atmosphere, which you can see in Figure 4.12.

If a telescope observes at far-infrared wavelengths, it must be cooled. Infrared radiation is emitted by heated objects, and if the telescope is warm it will emit many times more infrared radiation than that coming from a distant object. Imagine trying to look at a dim, moonlit scene through binoculars that are glowing brightly. In a telescope observing near-infrared radiation, only the detector, the element on which the infrared radiation is focused, must be cooled. To observe in the far-infrared, however, the entire telescope must be cooled.

At the short wavelength end of the spectrum, astronomers can observe in the near-ultraviolet. Human eyes do

NASA's Infrared Telescope Facility (IRTF), Mauna Kea, Hawaii

Figure 4.12

SOFIA, the Stratospheric Observatory for Infrared Astronomy, a 2.5-m reflecting telescope installed in a Boeing 747SP aircraft, is designed to operate for 8 hours at altitudes of 12 to 14 km (39,000 to 45,000 ft) above almost all Earth's atmospheric water vapor. It is able to observe far-infrared radiation from celestial objects that does not reach down to even the highest mountaintops. SOFIA is a joint project between NASA and the German space agency DLR.

not detect this radiation, but it can be recorded by photographic plates and CCDs. Wavelengths shorter than about 290 nm, the far-ultraviolet, X-ray, and gamma-ray ranges, are completely absorbed by the ozone layer extending from 20 km to about 40 km above Earth's surface. No mountain is that high, and no balloon or airplane can fly that high, so astronomers cannot observe far-UV, X-ray, and gamma-ray radiation without going into space.

Telescopes in Space

Earth's atmosphere is transparent in two windows in the visible and radio parts of the electromagnetic spectrum. Most of the rest of the electromagnetic radiation—infrared, ultraviolet, X ray, and gamma ray—never reaches Earth's surface; it is absorbed high in Earth's atmosphere. To observe at these wavelengths, telescopes must go above the atmosphere.

The Hubble Space Telescope, named after Edwin Hubble, the astronomer who discovered the expansion of the universe, is the most successful telescope ever to orbit Earth (Figure 4.13). It was launched in 1990 and

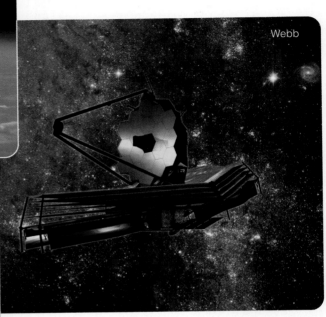

Figure 4.13

Hubble

Webb

NASA / NASA / NASA

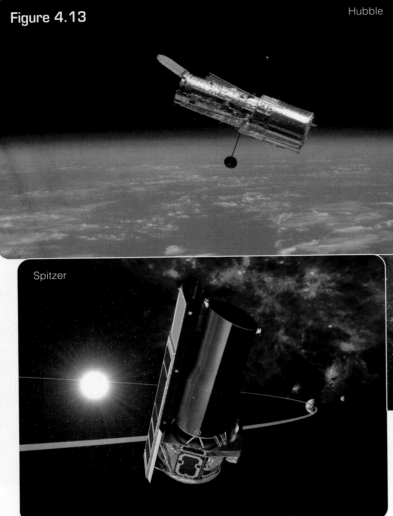

Spitzer

contains a 2.4-m (95-in.) mirror plus instruments that can observe at near-infrared, visual, and near-ultraviolet wavelengths. It is controlled from a research center on Earth and observes continuously. Nevertheless, there is time to complete only a fraction of the projects proposed by astronomers from around the world.

Most of the observations Hubble makes are at visual wavelengths, so its greatest advantage in being above Earth's atmosphere is the lack of seeing distortion. It can see faint objects and detect fine detail by concentrating light into sharp objects.

The telescope is as big as a large bus and has been visited a number of times by the space shuttle so that astronauts could maintain its equipment and install new cameras and spectrographs. Astronomers hope that it will last until it is replaced by the James Webb Space Telescope expected to launch no sooner than 2013. The Webb telescope's primary mirror will have a diameter of 6.5 m (265 in.).

Telescopes that observe in the far-infrared must be protected from heat and must get above Earth's absorbing atmosphere. They have limited lifetimes because they must carry coolant to chill their optics. The most sophisticated of the infrared telescopes put in orbit, the Spitzer Space Telescope was cooled to –269°C (–452°F). Launched in 2003, it observes from behind a sunscreen. In fact, it could not observe from Earth's orbit because Earth is such a strong source of infrared radiation, so the telescope was sent into an orbit around the sun that carried it slowly away from Earth. Named after theoretical physicist Lyman Spitzer Jr., it has made important discoveries concerning star formation, planets orbiting other stars, distant galaxies, and more. Its coolant ran out in 2009, but some of the instruments that can operate without being chilled continue to collect data.

High-energy astrophysics refers to the use of X-ray and gamma-ray observations of the sky. Making such observations is difficult but can reveal the secrets of processes such as the collapse of massive stars and eruptions of supermassive black holes.

The largest X-ray telescope to date, the Chandra X-ray Observatory, was launched in 1999 and orbits a third of the way to the moon. Chandra is named for the late Indian-American Nobel Laureate Subrahmanyan Chandrasekhar, who was a pioneer in many branches of

Figure 4.14

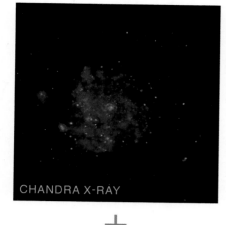

CHANDRA X-RAY

+

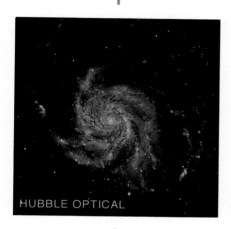

HUBBLE OPTICAL

+

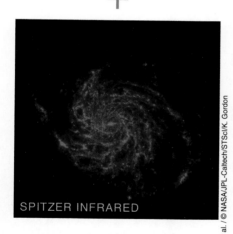

SPITZER INFRARED

=

COMPOSITE

theoretical astronomy. Focusing X rays is difficult because they penetrate into most mirrors, so astronomers devised cylindrical mirrors in which the X rays reflect from the polished inside of the cylinders and form images on special detectors. The telescope has made important discoveries about everything from star formation to monster black holes in distant galaxies (Figure 4.14).

One of the first gamma-ray observatories was the Compton Gamma Ray Observatory, launched in 1991. It mapped the entire sky at gamma-ray wavelengths. The European INTEGRAL satellite was launched in 2002 and has been very productive in the study of violent eruptions of stars and black holes. The GLAST (Gamma-Ray Large Area Space Telescope), launched in 2008, is capable of mapping large areas of the sky to high sensitivity.

Modern astronomy has come to depend on observations that cover the entire electromagnetic spectrum. More orbiting space telescopes are planned that will be more versatile and more sensitive.

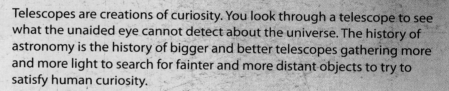

RECAP

Telescopes are creations of curiosity. You look through a telescope to see what the unaided eye cannot detect about the universe. The history of astronomy is the history of bigger and better telescopes gathering more and more light to search for fainter and more distant objects to try to satisfy human curiosity.

Curiosity is a noble trait—the mark of an active, inquiring mind. At the limits of human curiosity lies the fundamental question, "What are we?" Telescopes extend and amplify our senses, but they also extend and amplify our curiosity about the universe around us.

When people find out how something works, they say their curiosity is satisfied. Curiosity is an appetite like hunger or thirst, but it is an appetite for understanding. And, like hunger, curiosity rises again and again, pushing the boundaries of human knowledge further and further. As astronomy expands our horizons and we learn how distant stars and galaxies work, we feel satisfied because we are learning about ourselves. Apply what you have learned about the methods of satisfying human curiosity in this chapter by answering these questions:

1. What is light?

2. How do telescopes work? What are their capabilities and limitations?

3. How are observatories built, and how are good locations chosen for them?

4. What kinds of instruments and techniques do astronomers use to record and analyze light?

5. Why do astronomers sometimes use X-ray, ultraviolet, and infrared telescopes, and why must these types of telescopes operate in the upper atmosphere or in orbit?

Sun Light
and Sun Atoms

Earthbound humans knew almost nothing about the sun until the early 19th century, when the German optician Joseph von Fraunhofer studied the solar spectrum and found that it is interrupted by some 600 dark lines, representing colors that are missing from the sunlight Earth receives. As scientists realized that these **spectral lines** were related to the presence of various types of atoms in the sun's atmosphere, a window finally opened to real understanding of the sun's nature. In this chapter, you will look through that window by considering how the sun produces light and how atoms interact with light to make spectral lines. Once you understand that, you will know how astronomers have determined the chemical composition of the sun, measured motions of gas on the sun's surface and in its atmosphere, and detected magnetic fields that drive the sun's cycle of activity.

5.1 The Sun: Basic Characteristics

In its general properties, the sun is very simple. It is a great ball of hot gas held together by its own gravity. The tremendously hot gas inside the sun has such a high pressure that it would explode were it not for its own confining gravity. The same gravity would make it collapse into a small, dense body were it not so hot inside. Like a soap bubble, the sun is a simple structure balanced between opposing forces that, if unbalanced, would destroy it. These dramatic statements are also true for other stars, so you can study the sun for insight into the rest of the stars in the universe.

Another reason to study the sun is that life on Earth depends critically on the sun. Very small changes in the sun's luminosity can alter Earth's climate, and a larger change might make Earth uninhabitable. Nearly all of Earth's energy comes from the sun—the energy in oil, gasoline, coal, and even wood is merely stored sunlight. Furthermore, the sun's atmosphere of very thin gas extends past Earth's position, and changes in that atmosphere, such as eruptions or magnetic storms, can have a direct effect on Earth.

Spectral line: A line in a spectrum at a specific wavelength produced by the absorption or emission of light by certain atoms.

looking back

Up to this point, you have been thinking about what you can see with your eyes alone or aided by telescopes. In the previous chapter you read what hard work it is to gather light from distant stars using giant telescopes on mountaintops and in space. You also read how several types of instruments are used to analyze the light gathered by telescopes, including spectrographs that spread light into spectra. Now you are ready to see what all the fuss is about.

looking ahead

In the following chapters you will find that the sun is a normal star. You can use the methods of science to understand what is inside the sun and other stars, how they form and evolve, and how they finally end their existences. You can search out secrets of the stars beyond what you can see superficially.

All cannot live on the piazza, but everyone may enjoy the sun.
–Italian proverb

Distance and Size

When you watch the sun set in the west, you see a glowing disk that is 150 million kilometers (93 million miles) from Earth and has a diameter 109 times Earth's. How do you know? Recall from Chapter 3 that, thanks to the work of Johannes Kepler in the 17th century, the relative sizes of the orbits of the planets were already known in Astronomical Units—one AU being equal to the average distance of the sun from Earth. Recall from Chapter 3 (page 36) that the true distance of a nearby object can be calculated from the size of the apparent shift in its position relative to the background as seen from two viewing positions. That shift is called *parallax.*

If you and another astronomer on the opposite side of Earth agree to observe the position of a planet like Venus at the same moment relative to more distant objects, you can measure the planet's parallax shift caused by you and your colleague's different observing locations. The distance of the planet in familiar "earthly" units then can be calculated with some trigonometry. With further analysis, using a map of the solar system and Kepler's

third law, you can find the real distance to the sun. From the 17th century onward, astronomers made more and more accurate measurements of this type, especially on rare occasions, called **transits of Venus** when that planet can be seen as a tiny dot directly between Earth and the sun, as seen in Figure 5.1, so the edge of the sun's disk acts as a convenient position marker. The distance from Earth to the sun, combined with the sun's easily measured angular size, gives you the sun's diameter in familiar units.

Mass and Density

Once you know the distance to the sun, then Newton's laws (Chapter 3) tell you what the sun's mass needs to be to produce the necessary amount of gravitational force to keep Earth and the other planets in their orbits at their observed speeds. If you know the mass and diameter of the sun, you can make an easy calculation of the sun's **density** (mass per volume). Look at the Celestial Profile card again and notice the sun's mass, equivalent to 333,000 times the mass of Earth, and its average density—only a little bit denser than water. Although the sun is very large and very massive, the low density and high temperature together tell you that it must be gas from its surface to its center. When you look at the sun you see only the outer layers, the surface and atmosphere, of this vast sphere of gas.

In the following sections of this chapter you will learn how, using the results of laboratory physics experiments on Earth, you can know facts as seemingly unknowable as the temperature and composition of the sun—an object that has never been visited or touched by humans.

Figure 5.1

Comparison of Venus' path across the solar disk as observed from Australia (upper track) and Denmark (lower track) on June 8, 2004. Measurement of the small parallax shift allows the distance to Venus to be calculated based on the known distance between the two observatories. Kepler's third law can then be used to find the distance to the sun.

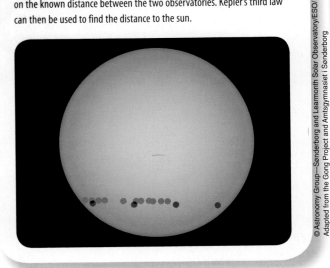

5.2 The Origin of Sunlight

The sun's surface glows for the same reason and by the same process that makes the coils in your toaster glow when they are hot. You probably have also noticed that your toaster's coils glow different colors as they heat up. If they are not too hot, the coils are deep red, but as they heat up they grow brighter and yellower. Yellow-hot is hotter than red-hot but not as hot as white-hot. Different parts of the sun's surface also glow slightly different colors depending on their temperatures.

BASIC TERMS

Atom: The smallest unit of a chemical element, consisting of a nucleus containing protons and neutrons plus a surrounding cloud of electrons.

Nucleus: The central core of an atom containing protons and neutrons that carries a net positive charge.

Proton: A positively charged atomic particle contained in the nucleus of an atom. The nucleus of a hydrogen atom.

Neutron: An atomic particle with no charge and about the same mass as a proton.

Electron: Low-mass atomic particle carrying a negative charge.

Molecule: Two or more atoms bonded together.

Heat: Energy stored in a material as agitation among its particles.

Temperature: A measure of the agitation among the atoms and molecules of a material.

Kelvin temperature scale: A temperature scale using Celsius degrees and based on zero being equal to absolute zero.

Absolute zero: The theoretical lowest possible temperature at which a material contains no extractable heat energy. Zero on the Kelvin temperature scale.

Blackbody radiation: Radiation emitted by a hypothetical perfect radiator. The spectrum is continuous, and the wavelength of maximum emission depends on the body's temperature.

Wavelength of maximum intensity: The wavelength at which a perfect radiator emits the maximum amount of energy. Depends only on the object's temperature.

Atoms and Subatomic Particles

Light from hot toaster coils and light from the sun and other stars are all produced by moving electrons. As you know, an **atom** has a massive compact **nucleus** containing positively charged **protons**, usually accompanied by electrically neutral **neutrons**. The nucleus is embedded in a large cloud of relatively low-mass, negatively charged **electrons**. These particles can also exist and move about unattached to an atom.

Charged particles, both protons and electrons, are surrounded by electric fields that they produce. Whenever you change the motion of a charged particle, the change in its electric field spreads outward at the speed of light as electromagnetic radiation. If you run a comb through your hair, you disturb electrons in both hair and comb, producing electric sparks and electromagnetic radiation, which you can sometimes hear as snaps and crackles if you are standing near a radio.

The sun is hot, and there are plenty of electrons zipping around, colliding, changing directions and speeds, and thereby making electromagnetic radiation. Protons can also make electromagnetic radiation, but because electrons are less massive, usually it is electrons that do most of the moving around.

Temperature, Heat, and Blackbody Radiation

The particles inside any object—for example, atoms linked together to form **molecules**, individual atoms, electrons inside atoms or wandering loose—are in constant motion, and in a hot object they are more agitated than in a cool object. You can refer to this agitation as *thermal energy*. When you touch a hot object, you feel **heat** as that thermal energy flows into your fingers. **Temperature** is simply a number related to the average speed of the particles, the intensity of the particle motion.

Astronomers and physicists express temperatures of the sun and other objects on the **Kelvin temperature scale**. Zero degrees Kelvin (written 0 K) is **absolute zero** (−459.7 F), the temperature at which an object contains no thermal energy that can be extracted. Water freezes at 273 K and boils at 373 K. The Kelvin temperature scale is useful in astronomy because it is based on absolute zero and consequently is related directly to the motion of the particles in an object.

Now you can understand why a hot object glows. The hotter an object is, the more motion there is among its particles. The agitated particles, including electrons, collide with each other, and when electrons are accelerated, part of the energy is carried away as electromagnetic radiation. The radiation emitted by an opaque object is called **blackbody radiation**, a name translated from a German term that refers to the way a perfectly opaque emitter and absorber of radiation would behave. At room temperature, such a perfect absorber would look black, but at higher temperatures it would visibly glow. In astronomy you will find the term *blackbody* actually refers to glowing objects.

Blackbody radiation is quite common. In fact, it is the type of light emitted by an ordinary incandescent light bulb. Electricity flowing through the filament of the bulb heats it to a high temperature, and it glows. You can also recognize the light emitted by a toaster coil as blackbody radiation. Many objects in the sky, including the sun, primarily emit blackbody radiation because they are mostly opaque.

Two simple laws describe how blackbody radiation works. A hot object radiates at all wavelengths, but there is a **wavelength of maximum intensity** at which it radiates the most energy. **Wien's Law** says that the hotter an object is, the shorter is the wavelength of its maximum intensity. This makes sense because in a hotter object, the particles travel faster, collide more violently, and emit more energetic photons, which have shorter wavelengths. This means that hot objects tend to emit radiation at shorter wavelengths and look bluer than cooler objects. Hot stars look bluer than cool stars.

The **Stefan-Boltzmann Law** says that hotter objects emit more energy than cooler objects of the same size. That makes sense, too, because the hotter an object is, the more rapidly its particles move and the more violent and more frequent are the collisions that produce photons. In later chapters, you will use these two laws to understand stars and other objects.

Figure 5.2

These graphs of blackbody radiation from three objects at different temperatures demonstrate that a hot object radiates more total energy and that the wavelength of maximum intensity is shorter for hotter objects. The hotter object here will look blue to your eyes, while the cooler object will look red.

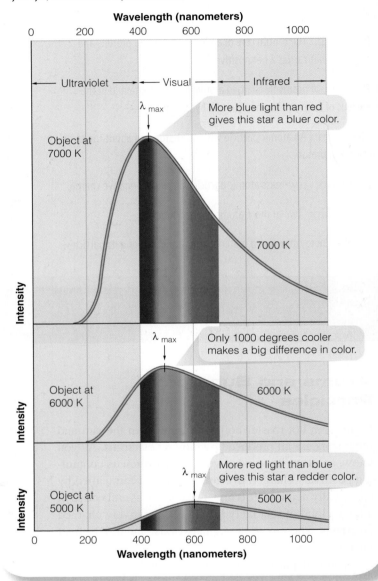

Figure 5.2 shows plots of the intensity of radiation versus wavelength for three objects with different temperatures, illustrating both Wien's Law and the Stefan–Boltzmann Law. You can see how temperature determines the color of a glowing blackbody. The hotter object has its strongest intensity at shorter wavelengths, so it emits more blue light than red and thus looks blue. The cooler object emits more red than blue light and consequently looks red. Also, the total area under each curve is proportional to the total energy emitted, so you

can see that the hotter object emits more total energy than the cooler objects.

Now you can understand why two famous stars, Betelgeuse and Rigel, have such different colors, which you saw in Chapter 2 in the section on naming stars. According to Wien's Law, Betelgeuse is cooler than the sun so it looks red, but Rigel is hotter than the sun and looks blue. A star with the same temperature as the sun would appear yellowish. According to the Stefan–Boltzmann Law, Rigel also produces more energy from each square meter of its photosphere than does the sun, which in turn produces more energy from each square meter than does Betelgeuse.

Notice that cool objects may emit little visible radiation but are still producing blackbody radiation. For example, the human body has a temperature of 310 K and emits blackbody radiation mostly in the infrared part of the spectrum. Infrared security cameras can detect burglars by the radiation they emit, and mosquitoes can track you down in total darkness by homing in on your infrared radiation. You have a wavelength of maximum intensity in the infrared part of the spectrum. At your temperature you almost never emit photons with other than infrared wavelengths.

5.3 The Sun's Surface

The sun's disk looks like a mostly smooth layer of gas. Although the sun seems to have a real surface, it is not solid. In fact, the sun is gaseous from its outer atmosphere right down to its center.

The Photosphere

The apparent surface of the sun is called the **photosphere**. The photosphere is the layer in the sun's atmosphere that is dense enough to emit plenty of light but not so dense that the light can't escape; so the photosphere is the source of most of the sunlight received by Earth. The photosphere is less than 500 km (300 mi) deep. If the sun magically shrank to the size of a bowling ball, the photosphere would be no thicker than a layer of tissue paper wrapped around the ball. When you measure the amount of light with different wavelengths coming from the pho-

tosphere and use Wien's Law, you find that the average temperature of the photosphere is about 5,800 K.

Although the photosphere appears to be substantial, it is really a very-low-density gas, 3,000 times less dense than the air you breathe. To find gases as dense as the air at Earth's surface, you would have to descend about 70,000 km below the photosphere, roughly 10 percent of the way to the sun's center. With fantastically efficient insulation, you could fly a spaceship right through the photosphere. The photosphere represents the depth where somebody outside the sun would no longer be able to see the descending spaceship; conversely, onboard the ship, you would no longer be able to see the rest of the universe.

There are regions of the photosphere that appear darker than the rest. These are called **sunspots**. They produce less light than equal-sized pieces of the normal photosphere, and their color is redder than the average, so from both the Wien and Stefan-Boltzmann laws you can conclude that sunspots are cooler than the photosphere; in fact, they are usually about 1,000 to 1,500 K cooler. Later in this chapter you will learn about the causes of sunspots and other solar dermatology problems.

Heat Flow in the Sun

At the temperature of 5,800 K, every square millimeter of the sun's photosphere must be radiating more energy than a 60-watt light bulb. Simple logic tells you that energy in the form of heat is flowing outward from the sun's interior. With all that energy radiating into space, the sun's surface would cool rapidly if energy did not flow up from inside to keep the surface hot.

As you learn more about the surface and atmosphere of the sun, you will find many phenomena that are driven by this energy flow. Like a pot of boiling soup on a hot stove, the sun is in constant activity as the heat comes up from below. In good photographs, the photosphere has a mottled appearance because it is made up of dark-edged regions called granules, and the visual pattern is called **granulation**, which you can see in Figure 5.3. Each granule is about the size of Texas and lasts for only 10 to 20 minutes before fading away. Faded granules are continuously replaced by new ones. The color and amount of light from different portions of the granules

Photosphere: The bright visible surface of the sun.

Sunspot: Relatively dark spot on the sun that contains intense magnetic fields.

Granulation: The fine structure of bright grains with dark edges covering the sun's surface.

Convection: Circulation in a fluid driven by heat. Hot material rises and cool material sinks.

Coulomb force: The electrostatic force of repulsion or attraction between charged bodies.

Ion: An atom that has lost or gained one or more electrons.

Ionization: The process in which atoms lose or gain electrons.

Binding energy: The energy needed to pull an electron away from its atom.

show, by both Wien's Law and the Stefan–Boltzmann Law, that granule centers are a few hundred degrees hotter than the edges.

Astronomers recognize granulation as the surface effects of rising and falling currents of gas in and just below the photosphere (the Doppler effect that allows astronomers to measure speeds of these gas currents is discussed later in this chapter). The centers of granules are rising columns of hot gas, and the edges of the granules are cooler, sinking gas. The presence of granulation is clear evidence that energy is flowing upward through the photosphere by a process known as **convection**, which you can see demonstrated in Figure 5.3b.

Convection occurs when hot fluid rises and cool fluid sinks, as when, for example, a convection current of hot gas rises above a candle flame. You can watch liquid convection by adding a bit of cool nondairy creamer to an unstirred cup of hot coffee. The cool creamer sinks, warms, expands, rises, cools, contracts, sinks again, and so on, creating small regions on the surface of the coffee that mark the tops of convection currents. Viewed from above, these regions look much like solar granules.

Figure 5.3

(a) This ultra-high-resolution image of the photosphere shows granulation. The largest granules here are about the size of Texas.
(b) This model explains granulation as the tops of rising convection currents just below the photosphere. Heat flows upward as rising currents of hot gas and downward as sinking currents of cool gas. The rising currents heat the solar surface in small regions seen from Earth as granules.

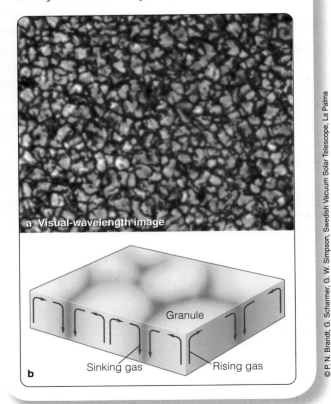

a **Visual-wavelength image**

Granule

Sinking gas   Rising gas

b

© P. N. Brandt, G. Scharmer, G. W. Simpson, Swedish Vacuum Solar Telescope, La Palma

5.4 Light, Matter, and Motion

If light did not interact with matter, you would not be able to see these words. In fact, you would not exist, because, among other problems, photosynthesis would be impossible, and there would be no grass, wheat, bread, beef, cheeseburgers, or any other kind of food. The interaction of light and matter makes your life possible, and it also makes it possible for you to understand the universe.

You should begin your study of light and matter by considering electrons that are within atoms. As you learned in the previous chapter, electrons and other charged particles produce light when they change speed or direction of their motion.

Electron Shells

The electrons are bound to the atom by the attraction between their negative charge and the positive charge on the nucleus. This attraction is known as the **Coulomb force**, after the French physicist Charles-Augustin de Coulomb (1736–1806). A positive **ion** is an atom with missing electrons, meaning, fewer electrons than protons. For an atom to go through the **ionization** process, there needs to be a certain amount of energy to pull an electron completely away from the nucleus. This energy is the electron's **binding energy**, the energy that holds it to the atom.

The size of an electron's orbit is related to the energy that binds it to the atom. If an electron orbits close to the nucleus, it is tightly bound, and a large amount of energy

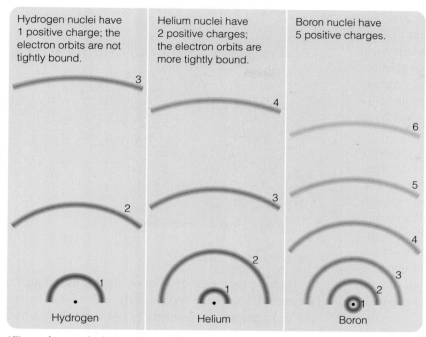

Hydrogen nuclei have 1 positive charge; the electron orbits are not tightly bound.

Helium nuclei have 2 positive charges; the electron orbits are more tightly bound.

Boron nuclei have 5 positive charges.

Hydrogen

Helium

Boron

*Figure shows only the innermost orbits

is needed to pull it away. Consequently, its binding energy is large. An electron orbiting farther from the nucleus is held more loosely, and less energy is needed to pull it away. That means it has less binding energy.

Nature permits atoms only certain amounts (quanta) of binding energy, and the laws that describe how atoms behave are called the laws of **quantum mechanics**. Much of this discussion of atoms is based on the laws of quantum mechanics.

Because atoms can have only certain amounts of binding energy, your model atom can have orbits of only certain sizes, called **permitted orbits**. These are like steps in a staircase: You can stand on the number-one step or the number-two step but not on the number-one-and-one-quarter step. The electron can occupy any permitted orbit but not orbits in between.

The arrangement of permitted orbits depends primarily on the charge of the nucleus, which in turn depends on the number of protons. The number of protons in the nucleus is unique to each element. Consequently, as shown in the above diagrams of hydrogen, helium, and boron, each kind of element has its own pattern of permitted orbits. Ionized forms of an element have orbital patterns quite different from their un-ionized forms. The arrangement of permitted orbits differs for every kind of atom and ion. **Isotopes** are versions of a given element with different numbers of neutrons. Isotopes of an element have almost—but not quite—the same pattern of permitted electron orbits as each other because they have

the same number of electrons while their nuclei have slightly different masses.

The Excitation of Atoms

Each orbit in an atom represents a specific amount of binding energy, so physicists commonly refer to the orbits as **energy levels**. Using this terminology, you can say that an electron in its smallest and most tightly bound orbit is in its lowest permitted energy level. You could move the electron from one energy level to another by supplying enough energy to make up the difference between the two energy levels. It would be like moving a flowerpot from a low shelf to a high shelf: The greater the distance between the shelves, the more energy you would need to raise the pot. The amount of energy needed to move the electron is the energy difference between the two energy levels.

If you move the electron from a low energy level to a higher energy level, you can call the atom an **excited atom**. That is, you have added energy to the atom by moving its electron. If the electron falls back to the lower energy level, that energy is released. An atom can become excited by collision. If two atoms collide, one or both may have electrons knocked into higher energy levels. This happens very commonly in hot gas, where the atoms move rapidly and collide often.

Another way an atom can get the energy that moves an electron to a higher energy level is to absorb a photon (packet) of electromagnetic radiation. Only a photon with exactly the right amount of energy corresponding to the energy difference between two levels can move the electron from one level to another. If the photon has too much or too little energy, the atom

Quantum mechanics: The study of the behavior of atoms and atomic particles.

Permitted orbit: One of the unique orbits that an electron may occupy in an atom.

Isotopes: Atoms that have the same number of protons but a different number of neutrons.

Energy level: One of a number of states an electron may occupy in an atom, depending on its binding energy.

Excited atom: An atom in which an electron has moved from a lower to a higher energy level.

cannot absorb it, and the photon passes right by. Because the energy of a photon depends on its wavelength, only photons of certain wavelengths (certain colors) can be absorbed by a given kind of atom.

The figure to the right shows the lowest four energy levels of the hydrogen atom along with three photons the atom could absorb. The longest-wavelength (reddest) photon has only enough energy to excite the electron from the first to the second energy level, but the shorter-wavelength (higher-energy, bluer) photons can excite the electron to higher levels. A photon with too much or too little energy cannot be absorbed. Because the hydrogen atom has many more energy levels than shown in the figure, it can absorb photons of many different wavelengths.

Atoms, like humans, cannot exist in an excited state forever. The excited atom is unstable and must eventually (usually within 10^{-6} to 10^{-9} seconds) give up the energy it has absorbed and return its electron to a lower energy level. The lowest energy level an electron can occupy is called the **ground state**.

When the electron drops from a higher to a lower energy level, it moves from a loosely bound level to one more tightly bound. The atom then has a surplus of energy—the energy difference between the levels—that it can emit as a photon. Study the sequence of events in Figure 5.4 to see how an atom can absorb and emit photons. Jumps of electrons from one orbit to another are sometimes called **quantum leaps**. In casual language that term has come to mean a huge change, but now you see that it represents a very small change indeed. The quantum leap represents a change of electron motion, so electromagnetic radiation is either released or absorbed in the process.

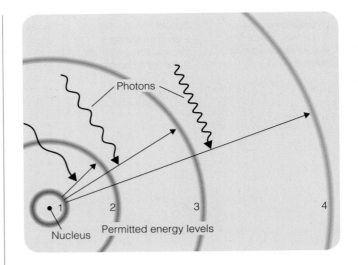

Because each type of atom or ion has its unique set of energy levels, each type absorbs and emits photons with a unique set of wavelengths. As a result, you can identify the elements in a gas by studying the characteristic wavelengths of light absorbed or emitted. It is important to note that the wavelengths (colors) emitted and absorbed by leaping electrons are determined not by the starting or ending energy level of the jump but by the difference between the levels.

The process of excitation and emission is a common sight in urban areas at night. A neon sign glows when atoms of neon gas in the glass tube are excited by electricity flowing through the tube. As the electrons in the electric current flow through the gas, they collide with the neon atoms and excite them. As you have seen, immediately after an atom is excited, its electron drops back to a lower energy level, emitting the surplus energy as a photon of a certain wavelength. The visible photons emitted by the most common electron jumps within excited neon atoms produce a reddish-orange glow. Street signs of other colors, erroneously called "neon," contain other gases or mixtures of gases instead of pure neon.

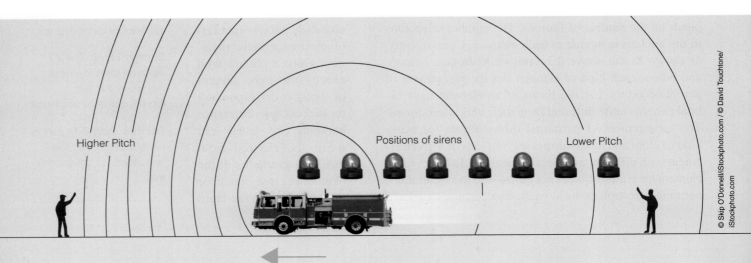

Figure 5.4

An atom can absorb a photon only if the photon has the correct amount of energy. The excited atom is unstable and within a fraction of a second returns to a lower energy level, reradiating the photon in a random direction.

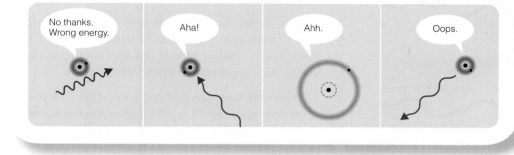

The Doppler Effect

The **Doppler effect** is an apparent change in the wavelength of radiation caused by relative motion of a source and observer. Astronomers use it to measure the speed of blobs of gas in the sun's atmosphere toward or away from Earth, as well as speeds of entire stars and galaxies.

When astronomers talk about the Doppler effect, they are talking about small shifts in the wavelength of electromagnetic radiation. The Doppler effect, however, can occur for waves of all types—sound waves, for example. You probably hear the Doppler effect in sound every day without realizing what it is. The pitch of a sound is determined by its wavelength. Sounds with long wavelengths have low pitches, and sounds with short wavelengths have higher pitches. You hear the Doppler effect every time a car or truck passes you and the pitch of its engine noise or emergency siren seems to drop. Its sound is shifted to shorter wavelengths and higher pitches while it is approaching and is shifted to longer wavelengths and lower pitches after it passes by, as demonstrated at the bottom of the previous page.

Understanding the Doppler effect for sound lets you understand the similar Doppler effect for light. Imagine a light source emitting waves continuously as it approaches you. The light will appear to have a shorter wavelength, making it slightly bluer. This is called a **blueshift**. A light source moving away from you has a longer wavelength and is slightly redder. This is a **redshift**. Redshifted and blueshifted spectra produced by a moving light source are illustrated below.

The terms *redshift* and *blueshift* are used to refer to any range of wavelengths. The light does not actually have to be red or blue, and the terms apply equally to wavelengths in the radio, X-ray, or gamma-ray parts of the spectrum. "Red" and "blue" refer to the relative direction of the shift, not to actual color. Also, note that these shifts are much too small to change the color of a star noticeably, but they are easily detected by changes in the positions of features in a star's spectrum such as spectral lines (discussed in detail in the next section). The Doppler shift, blue or red, reveals the relative motion of wave source and observer—you measure the same Doppler shift if the light source is moving and you are stationary or if the light source is stationary and you are moving, as you can see in Figure 5.5 on the next page.

The amount of change in wavelength depends on the speed of the source. A moving car has a smaller sound

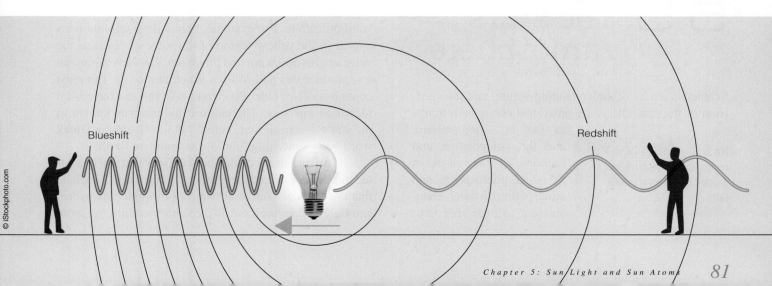

Blueshift

Redshift

Figure 5.5

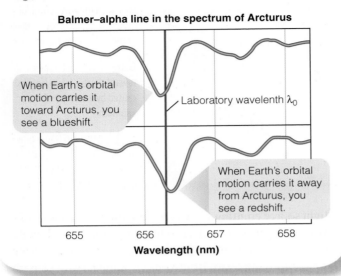

Balmer–alpha line in the spectrum of Arcturus

When Earth's orbital motion carries it toward Arcturus, you see a blueshift.

Laboratory wavelength λ_0

When Earth's orbital motion carries it away from Arcturus, you see a redshift.

655   656   657   658
Wavelength (nm)

Doppler shift than a jet plane, and a slow-moving star has a smaller light Doppler shift than one that is moving at high velocity. You can measure the speed of a star toward or away from you by measuring the size of the Doppler shift of its spectral lines.

Note that the Doppler shift is sensitive only to the part of the velocity directed away from you or toward you, called the **radial velocity** (V_r). A star moving perpendicular to your line of sight, to the left for instance, would have no blueshift or redshift because its distance from Earth would not be decreasing or increasing. Police radar guns use the Doppler effect to measure the speeds of cars. The police park next to the highway and aim their "hair dryers" directly along the road because they can measure only radial velocities, and they want to measure your full velocity along the highway.

5.5 The Sun's Atmosphere

Science is a way of understanding nature, and the spectrum of the sun tells you a great deal about such things as the sun's temperature and the composition and motions of the gases in the solar photosphere and atmosphere. In later chapters you will use spectra to

Radial velocity (V_r): That component of an object's velocity directed away from or toward Earth.

study stars, galaxies, and planets, but you can begin with spectra of the sun.

Formation of Spectra

Spectra of the sun and other stars are formed as light passes from their photospheres outward through their atmospheres. Study "Atomic Spectra" on pages 84–85. Notice three important properties of spectra:

1. There are three kinds of spectra described by three simple rules. When you see one of these types of spectra, you can recognize the arrangement of matter that emitted the light. Dark (absorption) lines in the sun's spectrum are caused by atoms in the sun's (or Earth's) atmosphere between you and the sun's photosphere. The photosphere itself produces a blackbody (continuous) spectrum.

2. The wavelengths of the photons that are absorbed by a given type of atom are the same as the wavelengths of the photons emitted by that type of atom; both are determined by the electron energy levels in the atom. The emitted photons coming from a hot cloud of hydrogen gas have the same wavelengths as the photons absorbed by hydrogen atoms in the sun's atmosphere. Although the hydrogen atom produces many spectral lines from the ultraviolet to the infrared, only three hydrogen lines are visible to human eyes.

3. Most modern astronomy books display spectra as graphs of intensity versus wavelength. Be sure you see the connection between dark absorption lines and dips in the graphed spectrum.

The Sun's Chemical Composition

Identifying the elements in the sun's atmosphere by identifying the lines in its spectrum is a relatively straightforward procedure. For example, two dark absorption lines appear in the yellow region of the solar spectrum at the wavelengths 589.0 nm and 589.6 nm. The only atom that can produce this pair of lines is sodium, so the sun must contain sodium. Over 90 elements in the sun have been identified this way. The element helium was known in the sun's spectrum first, before helium (from the Greek word *helios,* meaning "sun") was found on Earth.

However, just because spectral lines that are characteristic of an element are missing, you cannot conclude that the element itself is absent. For example, the hydrogen Balmer lines are weak in the sun's spectrum, yet more

Table 5.1 The Most Abundant Elements in the Sun

Element	Percentage by Number of Atoms	Percentage by Mass
Hydrogen	91.0	70.9
Helium	8.9	27.4
Carbon	0.03	0.3
Nitrogen	0.008	0.1
Oxygen	0.07	0.8
Neon	0.01	0.2
Magnesium	0.003	0.06
Silicon	0.003	0.07
Sulfur	0.002	0.04
Iron	0.003	0.1

than 90 percent of the atoms in the sun are hydrogen. The reason for this apparent paradox is that the sun is too cool to produce strong Balmer lines. At the sun's photosphere temperature, atoms do not usually collide violently enough to knock electrons in hydrogen atoms into the second energy level, which is the necessary starting place for Balmer line absorptions. Spectral lines of other varieties of atoms (for example, ionized calcium) are especially easy to observe in the sun's spectrum because the sun is the right temperature to excite those atoms to the energy levels that can produce visible spectral lines, even though those atoms are not very common in the sun.

The effect of temperature on the visibility of spectral lines was first understood by Cecila Payne (later, Payne-Gaposchkin), who was an astronomer doing Ph.D. research work at Harvard Observatory in the 1920s. She used the new techniques of quantum mechanics to derive accurate chemical abundances for the sun and other stars and so was the first person to know that the sun is mostly composed of hydrogen, even though its visible-wavelength hydrogen spectral lines are only moderately strong.

Astronomers must use the physics that describes the interaction of light and matter to analyze a star's spectrum, taking into account the star's temperature, to calculate correctly the amounts of each element present in the star. Such results show that nearly all stars have compositions similar to the sun—about 90 percent of the atoms are hydrogen, and about 9 percent are helium, with small traces of heavier elements (Table 5.1). It is fair to say that Cecilia Payne, whose thesis has been called the most important doctoral work in the history of astronomy, figured out the true chemical composition of the universe. You will use her results in later chapters when you study the life stories of the stars, the history of our galaxy, and the origin of the universe.

The Chromosphere

Above the photosphere lies the **chromosphere**. Solar astronomers define the lower edge of the chromosphere as lying just above the visible surface of the sun with its upper regions blending gradually with the atmosphere's outermost layer, the **corona**, which is described later in this chapter.

You can think of the chromosphere as being an irregular layer with a depth on average less than Earth's diameter, as seen in Figure 5.6 on page 86. Because the chromosphere is roughly 1,000 times fainter than the photosphere, you can see it with your unaided eyes only during a total solar eclipse when the moon covers the brilliant photosphere. Then, the chromosphere flashes into view as a thin line of pink just above the photosphere. The word *chromosphere* comes from the Greek word *chroma*, meaning "color." The pink color is produced by combined light of three bright emission lines—the red, blue, and violet Balmer lines of hydrogen.

Astronomers know a great deal about the chromosphere from its spectrum. The chromosphere produces an emission spectrum, and Kirchhoff's second law (page 84) tells you the chromosphere must be an excited, transparent low-density gas viewed with a dark, cold background. The density is about 10^8 times less than that of the air you breathe.

Atoms in the lower chromosphere are ionized, and atoms in the higher layers of the chromosphere are even more highly ionized. That is, they have lost more electrons. From this, astronomers can find the temperature in different parts of the chromosphere. Just above the photosphere, the temperature falls to a minimum of about 4,500 K and then rises to the extremely high temperatures of the corona.

Solar astronomers can take advantage of some of the physics of spectral line formation to study the chromosphere. A photon with a wavelength corresponding to one of the solar atmosphere's strong absorption lines is very unlikely to escape from deeper layers and reach Earth. A **filtergram** is a photograph made

Chromosphere: Bright gases just above the photosphere of the sun.

Corona: The faint outer atmosphere of the sun, composed of low-density, high-temperature gas.

Atomic Spectra

1 To understand how to analyze a spectrum, begin with a simple incandescent lightbulb. The hot filament emits blackbody radiation, which forms a **continuous spectrum**.

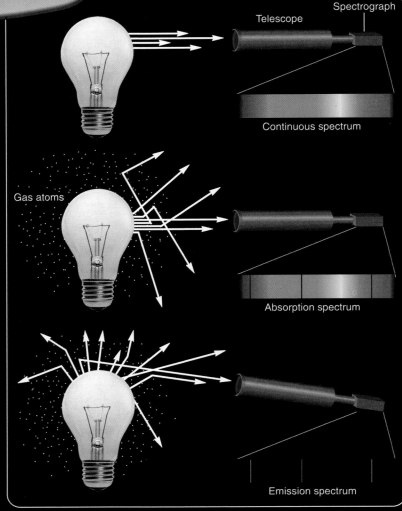

Telescope
Spectrograph

Continuous spectrum

An **absorption spectrum** results when radiation passes through a cool gas. In this case you can imagine that the lightbulb is surrounded by a cool cloud of gas. Atoms in the gas absorb photons of certain wavelengths, which are missing from the spectrum, and you see their positions as dark **absorption lines**. Such spectra are sometimes called **dark-line spectra**.

Gas atoms

Absorption spectrum

An **emission spectrum** is produced by photons emitted by an excited gas. You could see **emission lines** by turning your telescope aside so that photons from the bright bulb did not enter the telescope. The photons you would see would be those emitted by the excited atoms near the bulb. Such spectra are also called **bright-line spectra**.

Emission spectrum

1a The spectrum of a star is an absorption spectrum. The denser layers of the photosphere emit blackbody radiation. Gases in the atmosphere of the star absorb their specific wavelengths and form dark absorption lines in the spectrum.

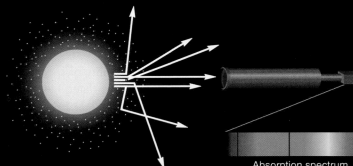

Absorption spectrum

GUSTAV ROBERT KIRCHHOFF
1824
1887
$\Sigma I_v = 0$
$\Sigma U_v = 0$
30
DEUTSCHE BUNDESPOST BERLIN

KIRCHHOFF'S LAWS

Law I: The Continuous Spectrum

A solid, liquid, or dense gas excited to emit light will radiate at all wavelengths and thus produce a continuous spectrum.

Law II: The Emission Spectrum

A low-density gas excited to emit light will do so at specific wavelengths and thus produce an emission spectrum.

Law III: The Absorption Spectrum

If light comprising a continuous spectrum passes through a cool, low-density gas, the result will be an absorption spectrum.

1b In 1859, long before scientists understood atoms and energy levels, the German scientist Gustav Kirchhoff formulated three rules, now known as **Kirchhoff's Laws**, that describe the three types of spectra.

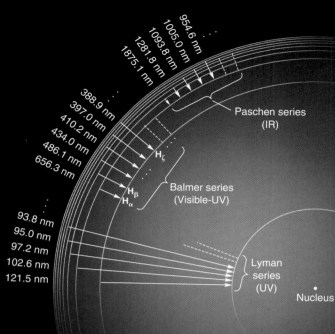

954.6 nm
1005.0 nm
1093.8 nm
1281.8 nm
1875.1 nm

388.9 nm
397.0 nm
410.2 nm
434.0 nm
486.1 nm
656.3 nm

93.8 nm
95.0 nm
97.2 nm
102.6 nm
121.5 nm

Paschen series
(IR)

H_ζ

H_β
H_α

Balmer series
(Visible-UV)

Lyman
series
(UV)

Nucleus

2 The electron orbits in the hydrogen atom are shown here as energy levels. When an electron makes a **transition** from one orbit to another, it changes the energy stored in the atom. In this diagram, arrows pointed inward represent transitions that result in the emission of a photon. If the arrows pointed outward, they would represent transitions that result from the absorption of a photon. Long arrows represent large amounts of energy and correspondingly short-wavelength photons.

2a Transitions in the hydrogen atom can be grouped into series—the **Lyman series**, **Balmer series**, **Paschen series**, and the like. Transitions and the resulting spectral lines are identified by Greek letters. Only the first few transitions in the first three series are shown at left.

2b In this drawing (right) of the hydrogen spectrum, emission lines in the infrared and ultraviolet are shown as gray. Only the first three lines of the Balmer series are visible to human eyes.

2c Excited clouds of gas in space emit light at all of the Balmer wavelengths, but you see only the red, blue, and violet photons blending to create the pink color typical of ionized hydrogen.

The shorter-wavelength lines in each series blend together.

Visual-wavelength image

AURA/NDAO/NSF

3 Modern astronomers rarely work with spectra as bands of light. Spectra are usually recorded digitally, so it is easy to represent them as graphs of intensity versus wavelength. Here the artwork above the graph suggests the appearance of a stellar spectrum. The graph below reveals details not otherwise visible and allows comparison of relative intensities. Notice that dark absorption lines in the spectrum appear as dips in the curve of intensity.

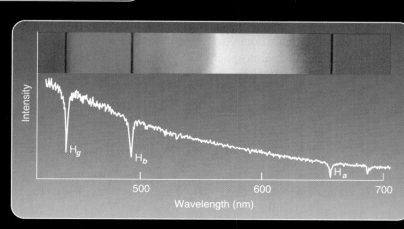

Intensity

H_g

H_b

H_a

500   600   700
Wavelength (nm)

2000 nm

Paschen lines

1500 nm

Infrared

1000 nm

H_α

500 nm

Visible

Balmer lines

H_β

H_γ

Ultraviolet

Lyman lines

100 nm

Figure 5.6

(a) A cross section at the edge of the sun shows the relative thickness of the photosphere and chromosphere. Earth is shown for scale. On this scale, the disk of the sun would be more than 1.5 m (5 ft) in diameter. (b) The corona extends from the top of the chromosphere to a great distance above the photosphere. This photograph, made during a total solar eclipse, shows only the inner part of the corona.

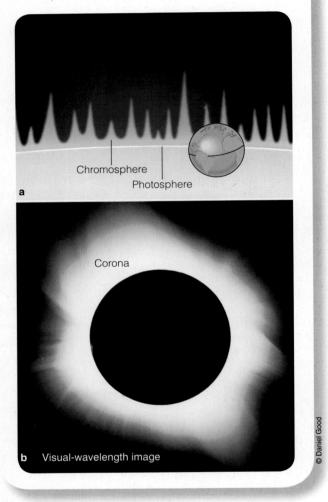

Chromosphere
Photosphere

a

Corona

b Visual-wavelength image

© Daniel Good

Figure 5.7

H_α filtergrams reveal complex structures in the chromosphere, including long, dark filaments and spicules springing from the edges of supergranules larger than twice the diameter of Earth.

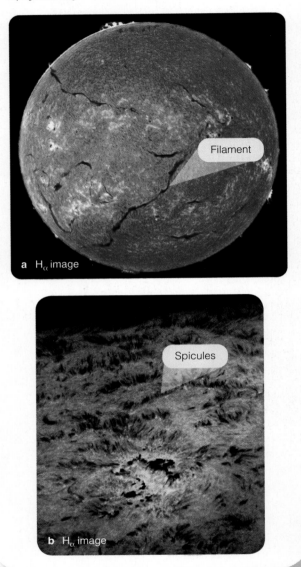

Filament

a H_α image

Spicules

b H_α image

© NOAA/SEL/USAF / © 1971 NOAO/NSO

Filtergram: A photograph (usually of the sun) taken in the light of a specific region of the spectrum—for example, an H_α filtergram.

Filament: A solar eruption, seen from above, silhouetted against the bright photosphere.

Spicule: A small, flamelike projection in the chromosphere of the sun.

using light in one of those dark absorption lines. Those photons can only come from high in the solar atmosphere. In this way, filtergrams reveal detail in the upper layers of the chromosphere. In a similar way, an image recorded in the far-ultraviolet or in the X-ray part of the spectrum reveals other structures in the hottest parts of the solar atmosphere.

Figure 5.7a shows a filtergram made at the wavelength of the H_α Balmer line. This image shows complex structure in the chromosphere, including long, dark **filaments** silhouetted against the brighter surface. **Spicules** are flamelike jets of gas extending upward into the chromosphere and lasting 5 to 15 minutes. Figure 5.7b shows how, at the limb of the sun's disk, the spicules blend together and look like flames covering a burning prairie, but they are not flames at all. Spectra show that spicules are cooler gas from the lower chromosphere extending upward into hotter regions.

Spectroscopic analysis of the chromosphere shows that it is a low-density gas in constant motion where the temperature increases rapidly with height. Just above the chromosphere lies even hotter gas.

The Corona

The outermost part of the sun's atmosphere is called the *corona*, after the Greek word for "crown." The corona is so dim that it is not visible in Earth's daytime sky because of the glare of scattered light from the sun's brilliant photosphere. During a total solar eclipse, however, when the moon covers the photosphere, you can see the innermost parts of the corona, as shown in Figure 5.6b. Observations made with specialized telescopes called **coronagraphs** on Earth or in space can block the light of the photosphere and record the corona out beyond 20 solar radii, almost 10 percent of the way to Earth. Such images reveal that sunspots are linked with features in the chromosphere and corona.

The spectrum of the corona can tell you a great deal about the coronal gases and simultaneously illustrate how astronomers analyze a spectrum. Some of the light from the corona produces a continuous spectrum that lacks absorption lines. Superimposed on the corona's continuous spectrum are emission lines of highly ionized gases. In the lower corona, the atoms are not as highly ionized as they are at higher altitudes, and this tells you that the temperature of the corona rises with altitude. Just above the chromosphere, the coronal temperature is about 500,000 K, but in the outer corona the temperature can be 2,000,000 K or more. Despite that very high temperature, the corona does not produce much light because its density is very low, only 10^6 atoms/cm^3 or less. That is about one-trillionth the density of the air you breathe.

Astronomers have wondered for years how the corona and chromosphere can be so hot. Heat flows from hot regions to cool regions, never from cool to hot. So how can the heat from the pho-

tosphere, with a temperature of only 5,800 K, flow out into the much hotter chromosphere and corona? Observations made by the SOHO satellite show a **magnetic carpet** of looped magnetic fields extending up through the photosphere. Remember that the gas of the chromosphere and corona has a very low density, so it can't resist movement in the magnetic fields. Turbulence below the photosphere seems to flick the magnetic loops back and forth and whip the gas about. That heats the gas. In this instance, energy appears to flow outward in the form of agitation of the magnetic fields. Solar magnetic phenomena will be discussed more thoroughly later in this chapter.

Not all of the sun's magnetic field loops back toward the sun; some of the field lines lead outward into space. Gas from the solar atmosphere follows along the magnetic fields that point outward and flows away from the sun in a breeze called the **solar wind** that can be considered an extension of the corona. The low-density gases of the solar wind blow past Earth at 300 to 800 km/s with gusts as high as 1,000 km/s.

Because of the solar wind, the sun is slowly losing mass, but this is a minor loss for an object as massive as the sun. The sun loses about 10^7 tons per second, but that is only 10^{-14} of a solar mass per year. Later in life, the sun, like many other stars, will lose mass rapidly. You will see in future chapters how this affects the evolution of stars.

Do other stars have chromospheres, coronae, and stellar winds like the sun? Ultraviolet and X-ray observations suggest that the answer is yes. The spectra of many stars contain emission lines in the far-ultraviolet that could only have formed in the low-density, high-temperature gases of

> **Coronagraph:** A telescope designed to capture images of faint objects such as the corona of the sun that are near relatively bright objects.
>
> **Magnetic carpet:** The network of small magnetic loops that covers the solar surface.
>
> **Solar wind:** Rapidly moving atoms and ions that escape from the solar corona and blow outward through the solar system.

© SOHO/ESA/NASA / © Nicholas Belton/iStockphoto.com

Visual-wavelength image

Ultraviolet

Two nearly simultaneous images show sunspots in the photosphere and excited regions in the chromosphere above the sunspots.

Visual image

Sun hidden behind mask

Twisted streamers in the corona suggest magnetic fields.

Visual image

Sun hidden behind mask

The corona extends far from the disk.

a chromosphere or corona. Also, many stars are sources of X rays that appear to have been produced by the high temperature gas in coronae. This observational evidence gives astronomers good reason to believe that the sun, for all its complexity, is a typical star.

5.6 Solar Activity

The sun is not quiet. It is home to slowly changing spots larger than Earth and rapid, vast eruptions that dwarf human imagination. All of these seemingly different forms of solar activity have one thing in common—magnetic fields.

Observing the Sun

Solar activity is often visible with even a small telescope, but you should be very careful about observing the sun. It is not safe to look directly at the sun, and it is even more dangerous to look at the sun through any optical instrument such as a telescope, binoculars, or even the viewfinder of a camera. These concentrate sunlight and can cause severe injury. You can safely project an image of the sun onto a screen, or you can use specially designed solar blocking filters.

In the early 17th century, Galileo observed the sun and saw spots on its surface; day by day, he saw the spots moving across the sun's disk. He correctly concluded that the sun is a sphere and is rotating (Figure 5.8).

Sunspots

The dark sunspots that you see at visible wavelengths only hint at the complex processes that go on in the sun's atmosphere. To explore those processes, you must turn to the analysis of images and spectra at a wide range of wavelengths.

Study "Sunspots and the Sunspot Cycle" on pages 90–91 and notice five important points:

1. Sunspots are cool spots on the sun's surface caused by strong magnetic fields.

2. Sunspots follow an 11-year cycle not only in the number of spots visible but also in their location on the sun.

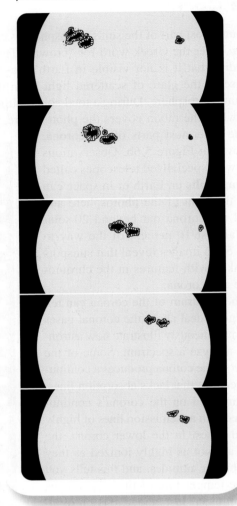

Figure 5.8
If you sketch the location and structure of sunspots on successive days, you will see the rotation of the sun and gradual changes in the size and structure of sunspots just as Galileo did in 1610.

3. The Zeeman effect gives astronomers a way to measure the strength of magnetic fields on the sun.

4. Characteristics of the sunspot cycle vary over centuries and appear to affect Earth's climate.

5. Finally, there is clear evidence that sunspots are part of a larger magnetic process that involves other layers of the sun's atmosphere and parts of its interior.

The sunspot groups are merely the visible traces of magnetically active regions. But what causes this magnetic activity? The answer appears to be linked to the cyclical strengthening and weakening of the sun's overall magnetic field.

Figure 5.9

Helioseismology: The sun can vibrate in millions of different patterns or modes, and each mode corresponds to a different vibration wavelength that penetrates to a different level in the solar interior. By measuring Doppler shifts produced as the photosphere moves gently up and down, astronomers can map the inside of the sun.

© AURA/NOAO/NSF

Insight into the Sun's Interior

Almost no light emerges from below the photosphere, so you can't see into the solar interior. However, solar astronomers can use the vibrations in the sun to explore its depths in a type of analysis called **helioseismology**. Random motions in the sun constantly produce vibrations. Astronomers can detect these vibrations by observing Doppler shifts in the solar surface. These waves make the photosphere move up and down by small amounts—roughly plus or minus 15 km. This covers the surface of the sun with a pattern of rising and falling regions that can be mapped, as in Figure 5.9. In the sun, a vibration with a period of 5 minutes is strongest, but other periods are observed ranging from 3 to 20 minutes. Just as geologists can study Earth's interior by analyzing vibrations from earthquakes, so solar astronomers can use helioseismology to map the temperature, density, and rate of rotation in the sun's invisible interior.

The Sun's Magnetic Cycle

Sunspots are magnetic phenomena, so the 11-year cycle of sunspots must be caused by cyclical changes in the sun's magnetic field. To explore that idea, you can begin with the sun's rotation.

*Figure 5.10

(a) In general, the photosphere of the sun rotates faster at the equator than at higher latitudes. If you started five sunspots in a row, they would not stay lined up as the sun rotates. (b) Detailed analyses of the sun's interior rotation from helioseismology reveal regions of slow rotation (blue) and rapid rotation (red). Such studies show that differential rotation occurs inside the sun as well as on its surface.

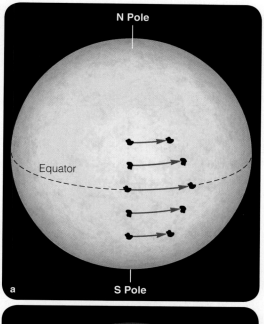

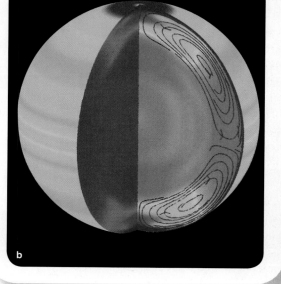

© NASA/SOI

The sun does not rotate as a rigid body. It is a gas from its outermost layers down to its center, and some parts of the sun rotate faster than other parts. Figure 5.10

Helioseismology:
The study of the interior of the sun by the analysis of its modes of vibration.

Sunspots and the Sunspot Cycle

1 The dark spots that appear on the sun are only the visible traces of complex regions of activity. Observations over many years and at a range of wavelengths tell you that sunspots are clearly linked to the sun's magnetic field.

Spectra show that sunspots are cooler than the photosphere with a temperature of about 4,200 K. The photosphere has a temperature of about 5,800 K. Because the total amount of energy radiated by a surface depends on its temperature raised to the fourth power, sunspots look dark in comparison. Actually, a sunspot emits quite a bit of radiation. If the sun were removed and only an average-size sunspot were left behind, it would be brighter than the full moon.

Visual wavelength image

Umbra

Penumbra

A typical sunspot is about twice the size of Earth, but there is a wide range of sizes. They appear, last a few weeks to as long as 2 months, and then shrink away. Usually, sunspots occur in pairs or complex groups.

Earth to scale

Sunspots are not shadows, but astronomers refer to the dark core of a sunspot as its umbra and the outer, lighter region as the penumbra.

Streamers above a sunspot suggest a magnetic field.

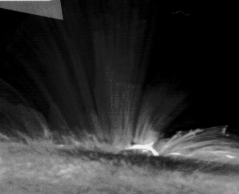

2 The number of spots visible on the sun varies in a cycle with a period of 11 years. At maximum, there are often over 100 spots visible. At minimum, there are very

(Graph: Number of sunspots vs. Year, 1950–2010, labeled "Sunspot minimum" and "Sunspot maximum")

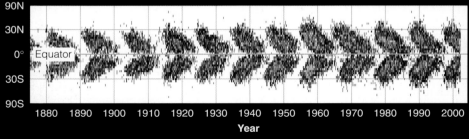

(Maunder butterfly diagram: latitude 90N–90S vs. Year 1880–2000)

2a Early in the cycle, spots appear at high latitudes north and south of the sun's equator. Later in the cycle, the spots appear closer to the sun's equator. If you plot the latitude of sunspots versus time, the graph looks like butterfly wings, as shown in this **Maunder butterfly diagram,** named after E. Walter Maunder of Greenwich Observatory.

3 ▸ Astronomers can measure magnetic fields on the sun using the **Zeeman effect** as shown below. When an atom is in a magnetic field, the electron orbits are altered, and the atom is able to absorb a number of different wavelength photons even though it was originally limited to a single wavelength. In the spectrum, you see single lines split into multiple components, with the separation between the components proportional to the strength of the magnetic field.

Slit allows light from sunspot to enter spectrograph.

Visual

Spectral line split by Zeeman effect

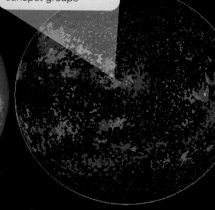

Sunspot groups

Magnetic fields around sunspot groups

Ultraviolet filtergram

Magnetic image

Simultaneous images

3a Images of the sun above show that sunspots contain magnetic fields a few thousand times stronger than Earth's. The strong fields are believed to inhibit motion below the photosphere; consequently, convection is reduced below the sun and the surface there is cooler. Heat prevented from emerging through the sunspo deflected and emerges around the sunspot, which can be detected in ultraviolet an infrared images.

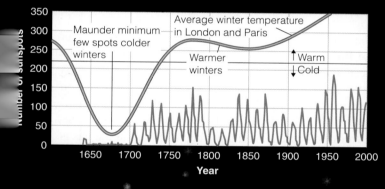

Maunder minimum few spots colder winters

Average winter temperature in London and Paris

Warmer winters

↑ Warm
↓ Cold

4 Historical records show that there were very few sunspot from about 1645 to 1715, a phenomenon known as the **Maunder minimum.** This coincides with a period called the "littl ice age," a period of unusually cool weather in Europe and North America from about 1500 to about 1850, as shown in the graph left. Other such periods of cooler climate are known. The eviden suggests that there is a link between solar activity and the amou of solar energy Earth receives. This link has been confirmed by measurements made by spacecraft above Earth's atmosphere.

Magnetic fields can reveal themselves by their shape. For example, iron filings sprinkled over a bar magnet reveal an arched shape.

M. Seeds

-UV image

5 Observations at nonvisible wavelengths reveal that the chromosphere and corona above sunspots are violently disturbed in what astronomers call **active regions.** Spectrographic observations show that active regions contain powerful magnetic fields. Arched structures above an active region are evidence of gas trapped in magnetic fields.

The complexity of an active region becomes visible at short wavelengths.

Visual-wavelength image

Simultaneous images

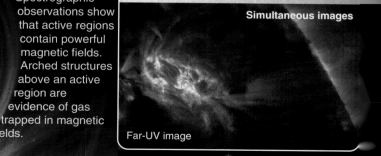

Far-UV image

demonstrates how the equatorial region of the photosphere rotates faster than do regions at higher latitudes. At the equator, the photosphere rotates once every 25 days, but at latitude 45° one rotation takes 28 days. Helioseismology can map the rotation throughout the interior (Figure 5.10b). This phenomenon is called **differential rotation**, and it is clearly linked with the magnetic cycle.

The sun's magnetic field appears to be powered by the energy flowing outward through the moving currents of gas. The gas is highly ionized, so it is a very good conductor of electricity. When an electrical conductor rotates rapidly and is stirred by convection, it can convert some of the energy flowing outward as convection into a magnetic field. This process is called the **dynamo effect**, and it is believed to operate also in the liquid metal of Earth's core to produce Earth's magnetic field. Helio-

seismology observations have found evidence that the sun's magnetic field is generated by the dynamo effect at the bottom of the **convective zone** deep under the photosphere. The sun's magnetic cycle is clearly related to how its magnetic field is created.

The magnetic behavior of sunspots provides an insight into how the magnetic cycle works. Sunspots tend to occur in groups or pairs, and the magnetic field around the pair resembles that around a bar magnet with one end magnetic north and the other end magnetic south. At any one time, sunspot pairs south of the sun's equator have reversed polarity compared to those north of the sun's equator. Figure 5.11 illustrates this by showing sunspot pairs south of the sun's equator with magnetic south poles leading and sunspots north of the sun's equator with magnetic north poles leading. At the end of an 11-year sunspot cycle, the new spots appear with reversed magnetic polarity.

The Babcock Model for Solar Activity

The sun's magnetic cycle is not fully understood, but the **Babcock model** (named for its inventor) explains the magnetic cycle as a progressive tangling and then untangling of the solar magnetic field, demonstrated in Figure 5.12 on page 94. Because the electrons in an ionized gas are free to move, the gas is a very good conductor of electricity, and any magnetic field in the gas is "frozen" (firmly embedded) in the gas. If the gas moves, the magnetic field must move with it. The sun's magnetic field is frozen into its gases, and the differential rotation wraps this field around the sun like a long string caught in a rotating wheel. Rising and sinking gas currents twist the field into ropelike tubes, which tend to float upward. Where these magnetic tubes burst through the sun's surface, sunspot pairs occur.

The Babcock model explains the reversal of the sun's magnetic field from cycle to cycle. As the magnetic field becomes tangled, adjacent regions of the sun's surface are dominated by magnetic fields that point in different directions. After about 11 years of tangling, the field becomes so complex that adjacent regions of the surface are forced to change their magnetic field directions to align with neighboring regions. The entire field quickly rearranges itself into a simpler pattern, and differential rotation begins winding it up to start a new cycle. The newly organized field is reversed, and the next sunspot cycle begins with magnetic north replaced by magnetic south. Consequently, the complete magnetic cycle is 22 years long, whereas the sunspot cycle is 11 years long.

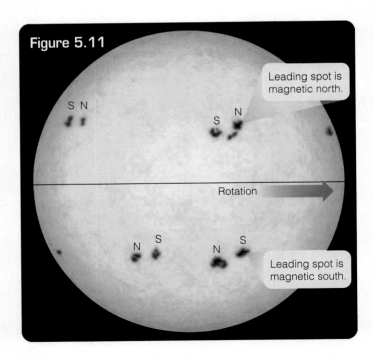

Figure 5.11

Leading spot is magnetic north.

Rotation

Leading spot is magnetic south.

CONFIRMATION, CONFIDENCE, AND CONSOLIDATION

Science is not always about upending paradigms and dramatically changing the way people view nature. Many experiments and observations are simply carried out to confirm hypotheses that have already been tested. The biologist knows that all worker bees in a hive are sisters, but a careful study of the DNA from different workers further confirms that hypothesis. By repeatedly confirming a hypothesis, scientists build confidence and extend its application. Much of the daily grind of science is confirmation.

Repeated confirmation of hypotheses increases confidence in those hypotheses. Confidence in well-tested scientific principles helps scientists avoid rushing to faulty judgments. For example, claims for perpetual motion machines occasionally crop up in the news, but the world's scientists don't instantly abandon the laws of energy and motion pending an analysis of the latest claim. Because the known laws of energy and motion have been well tested and no perpetual motion machine has ever been successful, scientists know which way to bet.

Confirmation and confidence enable scientists to broaden the application of hypotheses and link them to other phenomena through consolidation. Chemists may understand certain kinds of carbon molecules shaped like rings, but by repeated study they find a carbon molecule shaped like a hollow sphere. To consolidate their findings, they must show that the chemical bonding in the two molecules follows the same rules and that the molecules have certain properties in common.

The Babcock model of the solar magnetic cycle is an astronomical example of the scientific process. Solar astronomers know that the model has shortcomings, but they work through confirmation and consolidation to better understand how the solar magnetic cycle works and how it is related to cycles in other stars.

This magnetic cycle seems to explain the Maunder butterfly diagram (see page 90). As a sunspot cycle begins, the twisted tubes of magnetic force first begin to float upward and produce sunspot pairs at higher latitude. Later in the cycle, when the field is more tightly wound, the tubes of magnetic force arch up through the surface closer to the equator. As a result, the later sunspot pairs in a cycle appear closer to the equator.

Notice the power of a scientific model. The Babcock model may in fact be incorrect in some details, but it provides a framework on which to organize your thinking about all the complex solar activity. Even though the models of the sky in Chapter 2 and of atomic energy levels in this chapter are only partially correct, they serve as organizing themes to guide your explorations. Similarly, although the precise details of the solar magnetic cycle are not yet understood, the Babcock model gives you a general picture of the behavior of the sun's magnetic field.

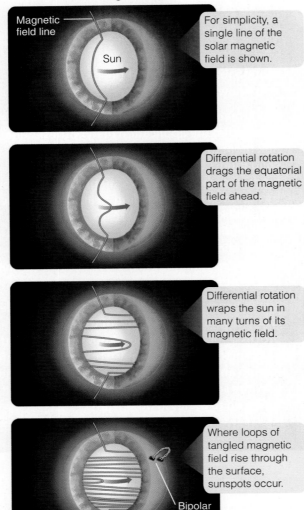

*Figure 5.12
The Babcock model of the solar magnetic cycle explains the sunspot cycle as a consequence of the sun's differential rotation gradually winding up the magnetic field near the base of the sun's outer, convective layer.

The Solar Magnetic Cycle

Magnetic field line

Sun

For simplicity, a single line of the solar magnetic field is shown.

Differential rotation drags the equatorial part of the magnetic field ahead.

Differential rotation wraps the sun in many turns of its magnetic field.

Where loops of tangled magnetic field rise through the surface, sunspots occur.

Bipolar sunspot pair

If the sun is truly a representative star, you might expect to find similar magnetic cycles on other stars, but they are too distant for spots to be directly visible. Some stars, however, vary in brightness over a period of days, in a way revealing that they are marked with dark spots believed to resemble sunspots. Other stars have features in their spectra that vary over periods of years, suggesting that they are subject to magnetic cycles much like the sun's cycle. Once again, the evidence tells you that the sun is a normal star.

Chromospheric and Coronal Activity

The solar magnetic fields extend high into the chromosphere and corona, where they produce beautiful and powerful phenomena. Study "Magnetic Solar Phenomena" on pages 96–97 and notice three important points:

1. All solar activity is magnetic. You do not experience such events on Earth because Earth's magnetic field is weak and Earth's atmosphere is not ionized, so it is free to move independent of the magnetic field. On the sun, however, the "weather" is a magnetic phenomenon.

2. Tremendous energy can be stored in arches of magnetic fields. These are visible near the edge of the solar disk as prominences, and, seen from above, as filaments. When that stored energy is released, it can trigger powerful eruptions; and, although these eruptions occur far from Earth, they can affect Earth in dramatic ways, including auroral displays. Auroras, the eerie and pretty northern and southern lights, are produced when gases in Earth's upper atmosphere glow from energy delivered by the solar wind.

3. In some regions of the solar surface, the magnetic field does not loop back. High-energy gas from these regions flows outward and produces much of the solar wind.

You needed the entire range of physical principles presented in this chapter—parallax, Wien's Law and the Stefan–Bolzmann Law for blackbody radiation, atomic structure, Kirchhoff's Laws for the formation of spectra, the Doppler effect, the Zeeman effect—to realize how the sun's surface temperature and composition are known, and to understand solar activity cycles and their effects on Earth. In the next chapter you will learn how the sun fits into the context of other stars.

recap

We live very close to a star and depend on it for survival. All of our food comes from sunlight that was captured by plants on land or by plankton in the oceans. We either eat those plants directly or eat the animals that feed on those plants. Whether you had salad, seafood, or a cheeseburger for supper last night, you dined on sunlight, thanks to photosynthesis. Almost all of the energy that powers human civilization comes from the sun through photosynthesis in ancient plants that were buried and converted to coal, oil, and natural gas. New technology is making energy from plant products like corn, soy beans, and sugar. It is all stored sunlight. Windmills generate electrical power, and the wind blows because of heat from the sun. Photocells make electricity directly from sunlight. Even our bodies have adapted to use sunlight to manufacture vitamin D.

Our planet is warmed by the sun, and without that warmth the oceans would be ice and much of the atmosphere would be a coating of frost. Books often refer to the sun as "our sun" or "our star." It is ours in the sense that we live beside it and by its light and warmth, but we can hardly say it belongs to us. It is more correct to say that we belong to the sun.

This chapter helped you realize how little astronomers would know about the sun were it not for analyses of its spectrum. Just a bit of spectroscopic ingenuity reveals that the brilliance of the sun hides a complex atmosphere of hot gases, churned by powerful storms which affect our lives on Earth. From your reading, you should be able to answer these questions:

1. How do you know the distance, size, mass, and density of the sun?

2. How does matter produce light?

3. What do astronomers see when they observe the sun?

4. How does matter interact with light to produce spectral lines?

5. What can you learn from the sun's spectrum?

6. Why does the sun have a cycle of activity, and how does that affect Earth?

Magnetic Solar Phenomena

1 Magnetic phenomena in the chromosphere and corona, like magnetic weather, result as constantly changing magnetic fields on the sun trap ionized gas to produce beautiful arches and powerful outbursts. Some of this solar activity can affect Earth's magnetic field and atmosphere.

This ultraviolet image of the solar surface was made by the NASA TRACE spacecraft. It shows hot gas trapped in magnetic arches extending above active regions. At visual wavelengths, you would see sunspot groups in these active regions.

Sacramento Peak Observatory

H_α filtergram

1a A **prominence** is composed of ionized gas trapped in a magnetic arch rising up through the photosphere and chromosphere into the lower corona. Seen during total solar eclipses at the edge of the solar disk, prominences look pink because of the three Balmer emission lines. The image above shows the arch shape suggestive of magnetic fields. Seen from above against the sun's bright surface, prominences form dark **filaments**.

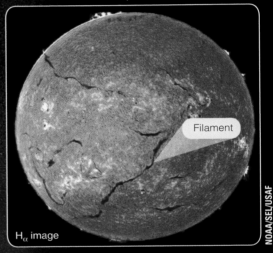

Filament

H_α image

NOAA/SEL/USAF

1b Quiescent prominences may hang in the lower corona for many days, whereas eruptive prominences burst upward in hours. The eruptive prominence below is many Earth diameters long.

The gas in prominences may be 60,000 to 80,000 K, quite cold compared with the low-density gas in the corona, which may be as hot as a million Kelvin.

Trace/NASA

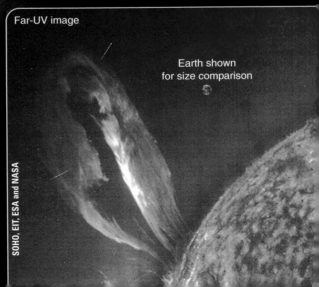

Far-UV image

Earth shown for size comparison

SOHO, EIT, ESA and NASA

2 Solar **flares** rise to maximum in minutes and decay in an hour. They occur in active regions where oppositely directed magnetic fields meet and cancel each other out in what astronomers call **reconnections**. Energy stored in the magnetic fields is released as short-wavelength photons and as high-energy protons and electrons. X-ray and ultraviolet photons reach Earth in 8 minutes and increase ionization in our atmosphere, which can interfere with radio communications. Particles from flares reach Earth hours or days later as gusts in the solar wind, which can distort Earth's magnetic field and disrupt navigation systems. Solar flares can also cause surges in electrical power lines and damage to Earth satellites.

This multiwavelength image shows a sunspot interacting with a neighboring magnetic field to produce a solar flare.

Hinode JAXA/NASA

2a At right, waves rush outward at 50 km/sec from the site of a solar flare 40,000 times stronger than the 1906 San Francisco earthquake. The biggest solar flares can be a billion times more powerful than a hydrogen bomb.

Helioseismology image

SOHO/MDI, ESA, and NASA

2b The solar wind, enhanced by eruptions on the sun, interacts with Earth's magnetic field and can create electrical currents up to a million megawatts. Those currents flowing down into a ring around Earth's magnetic poles excite atoms in Earth's upper atmosphere to emit photons as shown below. Seen from Earth's surface, the gas produces glowing clouds and curtains of **aurora**.

Auroras occur about 130 km (80 mi) above the Earth's surface.

© Jan Curtis

Coronal mass ejection

Ring of aurora around the north magnetic pole

NSSDC, Holzworth and Meng

2c Magnetic reconnections can release enough energy to blow large amounts of ionized gas outward from the corona in **coronal mass ejections (CMEs)**. If a CME strikes Earth, it can produce especially violent disturbances in Earth's magnetic field.

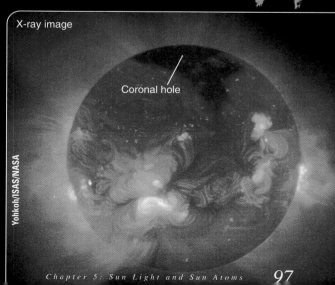

X-ray image

Coronal hole

Yohkoh/ISAS/NASA

3 Much of the solar wind comes from **coronal holes**, where the magnetic field does not loop back into the sun. These open magnetic fields allow ionized gas in the corona to flow away as the solar wind. The dark area in this X-ray image at right is a coronal hole.

Chapter 6

The Family
of Stars

Shakespeare compared love to a star that can be seen easily and even used for guidance, but whose real nature is utterly unknown. He lived at about the same time as Galileo and had no idea what stars actually are. To understand the history of the universe, the origin of Earth, and the nature of our human existence, you will discover what people in Shakespeare's time did not know—the real nature of the stars. Unfortunately, it is quite difficult to find out what a star is like. When you look at a star even through a telescope, you see only a point of light. Real understanding of stars requires careful analysis of starlight. This chapter concentrates on five goals: knowing how far away stars are, how much energy they emit, what their surface temperatures are, how big they are, and how much mass they contain. By the time you finish this chapter, you will know the family of stars well.

6.1 Star Distances

Distance is the most important and the most difficult measurement in astronomy, and astronomers have many different ways to find the distances to stars. Each of those ways depends on a simple and direct geometrical method that is much like the method surveyors would use to measure the distance across a river they cannot cross. You can begin by reviewing that method and then apply it to stars.

The Surveyor's Triangulation Method

To measure the distance across a river, a team of surveyors begins by driving two stakes into the ground. The distance between the stakes is called the *baseline*. The surveyors then choose a landmark on the opposite side of the river, a tree perhaps, thus establishing a large triangle marked by the two stakes and the tree, as in Figure 6.1 on page 100. Using their instruments, the surveyors sight the tree from the two ends of the baseline and measure the two angles on their side of the river. Knowing two angles of this large triangle and the length of the baseline side between them, the surveyors can find the distance across the river by simple trigonometry. For example, if the baseline is 50 meters and the angles are 66° and 71°, they would calculate that the distance from the baseline to the tree is 64 meters.

looking back

Science is based on measurement, but measurement in astronomy is very difficult. In Chapter 4 you studied the complex telescopes and instruments that astronomers use. In Chapter 5 you learned about the information that can be found in spectra and how astronomers use their understanding of the interaction of light and matter to study a typical star, the sun.

looking ahead

Once you know how to find the basic properties of stars, you will be ready to trace the history of the stars from birth to death, a story that begins in the next chapter.

astro

I wonder how far away the Eskimo Nebula is?

Figure 6.1

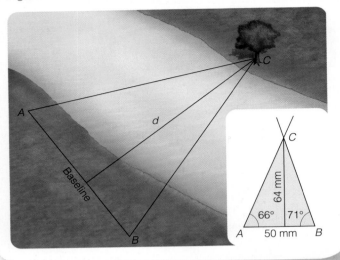

The Astronomer's Triangulation Method

To find the distance to a star, astronomers use a very long baseline, the size of Earth's orbit, as you can see in Figure 6.2. If you take a photograph of a nearby star and then wait 6 months, Earth will have moved halfway around its orbit. You can then take another photograph of the star from that slightly different location in space. When you examine the photographs, you will discover that the star is not in exactly the same place in the two photographs.

In Chapter 3 you learned that *parallax* is the term that refers to the common experience of an apparent shift in the position of a foreground object due to a change in the location of the observer's viewpoint. Your thumb, held at arm's length, appears to shift position against a distant background when you look first with one eye and then with the other (see page 36). In that case, the baseline is the distance between your eyes, and the parallax is the angle through which your thumb appears to move when you switch eyes. The farther away you hold your thumb,

the smaller the parallax. In chapter 5 you learned that Venus shows a parallax when observed from different locations on Earth.

Because the stars are so distant, their parallaxes are very small angles, usually expressed in arc seconds. The quantity that astronomers call **stellar parallax (*p*)** is conventionally defined as half the total shift of the star shown in Figure 6.2, in other words, the shift seen across a baseline of 1 AU rather than 2 AU. Astronomers measure the parallax, and surveyors measure the angles at the ends of the baseline, but both measurements tell the same thing—the shape and size of the triangle and thus the distance to the object in question.

Measuring the parallax *p* is very difficult because it is such a small angle. The nearest star, Alpha Centauri, has a parallax of only 0.76 arc seconds, and more distant stars have even smaller parallaxes. To see how small these angles are, imagine a dime two miles away from you. That dime covers an angle of about 1 arc second. Stellar parallaxes are so small that the first successful measurement of one did not happen until 1838, more than 200 years after the invention of the telescope.

The distances to the stars are so large that it is not convenient to use kilometers or Astronomical Units. When you measure distance via parallax, it is better to use the unit of distance called a **parsec (pc)**. The word *parsec* was created by combining *par*allax and arc *sec*ond. One parsec equals the distance to an imaginary star that has a parallax of 1 arc second.

A parsec is 206,265 AU, which equals roughly 3.26 ly (light-years). Parsec units are used more often than light-years by astronomers because parsecs are more directly related to the process of measurement of star distances. However, there are instances in which the light-year is also useful. You will notice that following chapters use either parsecs or light-years as convenience and custom dictate.

The blurring caused by Earth's atmosphere makes star images appear to be about 1 arc second in diameter, and that makes it difficult to measure parallax. Even if you average together many observations made from Earth's surface, you cannot measure parallax with an uncertainty smaller than about 0.002 arc seconds. Thus, if you measure a parallax of 0.006 arc seconds from an observatory on the ground, your uncertainty will be about 30 percent. If you consider, as astronomers generally do, that 30 percent is the maximum acceptable level of uncertainty, then ground-based

> [Love] is the star
> to every wandering bark,
> Whose worth's unknown,
> although his height be taken.
> –*William Shakespeare*
> *Sonnet 116*

Stellar parllax (p): The small apparent shift in position of a nearby star relative to distant background objects due to Earth's orbital motion.

Parsec (pc): The distance to a hypothetical star whose parallax is 1 second of arc.

astronomers can't accurately measure parallaxes that are smaller than about 0.006 arc seconds. That parallax corresponds to a distance of about 170 pc (550 ly).

In 1989, the European Space Agency launched the satellite Hipparcos to measure stellar parallaxes from above the blurring effects of Earth's atmosphere. That small space telescope observed for four years, and the data were used to produce two parallax catalogs in 1997. One catalog contains 120,000 stars with parallaxes 20 times more accurate than ground-based measurements. The other catalog contains over a million stars with parallaxes as accurate as ground-based parallaxes. Knowing accurate distances from the Hipparcos observations has given astronomers new insights into the nature of stars.

6.2 Apparent Brightness, Intrinsic Brightness, and Luminosity

Your eyes tell you that some stars look brighter than others, and in Chapter 2 you learned about the apparent magnitude scale that refers to stellar apparent brightness. The scale of apparent magnitudes tells you only how bright stars appear to you on Earth, however. To know the true nature of a star you must know its **intrinsic brightness**, a measure of the amount of light the star produces. An intrinsically very bright star might appear faint if it is far away. Thus, to know the intrinsic brightness of a star, you must take into account its distance.

Brightness and Distance

When you look at a bright light, your eyes respond to the visual-wavelength energy falling on your retinas, which tells you how bright

the object appears. Thus, brightness is related to the flux of energy entering your eye. Astronomers and physicists define **flux** as the energy in Joules (J) per second falling on 1 square meter. Recall that a Joule is about the amount of energy released when an apple falls from a table onto the floor. A flux of 1 Joule per second is also known as 1 Watt. The wattage of a light bulb tells you its *intrinsic* brightness. Compare that with the *apparent* brightness of a light bulb, which depends on its distance from you.

If you placed a screen 1 meter square near a light bulb, a certain amount of flux would fall on the screen. If you moved the screen twice as far from the bulb, the light that previously fell on the screen would be spread to cover an area four times larger, and the screen would receive only one-fourth as much light. If you tripled the distance to the screen, it would receive only one-ninth as much light. Thus, the flux you receive from a light source is inversely proportional to the square of the distance to the source. This is known as the inverse square relation, which you can see in Figure 6.3. You first encountered

> **Intrinsic brightness:** A measure of the amount of light a star produces.
>
> **Flux:** A measure of the flow of energy out of a surface. Usually applied to light.

Figure 6.2

You can measure the parallax of a nearby star by photographing it from two points along Earth's orbit. For example, you might photograph it now and again in six months. Half of the star's total change in position from one photograph to the other is defined as its stellar parallax, *p*.

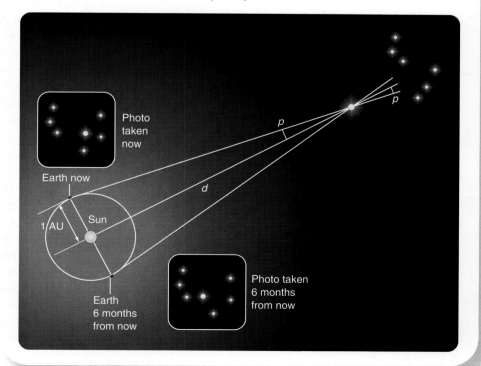

Figure 6.3

The inverse square relation: A light source is surrounded by imaginary spheres with radii of 1 unit and 2 units. The light falling on an area of 1 m² on the inner sphere spreads to illuminate an area of 4 m² on the outer sphere. Thus, the brightness of the light source is inversely proportional to the square of the distance.

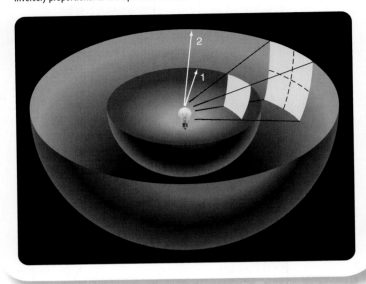

the inverse square relation in Chapter 3, where it was applied to the strength of gravity.

Now you can understand how the apparent brightness of a star depends on its distance. If astronomers know the apparent brightness of a star and its distance from Earth, they can use the inverse square law to correct for distance and find the intrinsic brightness of the star.

Absolute Visual Magnitude

If all the stars were the same distance away, you could compare one with another and decide which was intrinsically brighter or fainter. Of course, the stars are scattered at different distances, and you can't move them around to line them up for comparison. If, however, you know the distance to a star, you can use the inverse square relation to calculate the brightness the star would have at some standard distance. Astronomers use 10 pc as the standard distance and refer to the intrinsic brightness of the star as its **absolute visual magnitude** (M_v), which is the apparent visual magnitude that star would have if it were 10 pc away. The subscript

> **Absolute visual magnitude (M_v):** Intrinsic brightness of a star. The apparent visual magnitude the star would have if it were 10 pc away.
>
> **Luminosity (L):** The total amount of energy a star radiates per second at all wavelengths.

V tells you it is a visual magnitude, referring only to the wavelengths of light your eye can see. Other magnitude systems are based on other parts of the electromagnetic spectrum, such as infrared and ultraviolet radiation.

The sun's absolute magnitude is easy to calculate because its distance and apparent magnitude are well known. The absolute visual magnitude of the sun is about 4.8. In other words, if the sun were only 10 pc (33 ly) from Earth (not a great distance in astronomy), it would have an apparent magnitude of 4.8 and look no brighter to your eye than the faintest star in the handle of the Little Dipper.

This path to find the distance to stars has led you to absolute magnitude, a measure of the intrinsic brightness of the stars. You are now ready to reach the next of your five goals in this chapter.

Luminosity

The second goal for this chapter is to find out how much energy the stars emit. With the absolute magnitudes of the stars in hand, you can now compare stars using our sun as a standard. The intrinsically brightest stars have absolute magnitudes of about –8. If such a star were 10 pc away from Earth, it would seem nearly as bright as the moon. Such stars emit over 100,000 times more visible light than the sun.

Absolute visual magnitude refers to visible light, but you want to know the total output including all types of radiation. Hot stars emit a great deal of ultraviolet radiation that you can't see, and cool stars emit plenty of infrared radiation. To add in the energy you can't see, astronomers make a mathematical correction that depends on the temperature of the star. With that correction, astronomers can find the total electromagnetic energy output of a star, which they refer to as its **luminosity (L)**.

Astronomers know the luminosity of the sun because they can send satellites above Earth's atmosphere and measure the amount of energy arriving from the sun, adding up radiation of every wavelength, including the types blocked by the atmosphere. Of course, they also know the distance from Earth to the sun very accurately, which is necessary to calculate luminosity. The luminosity of the sun is about 4×10^{26} Watts (Joules per second).

You can express a star's luminosity in two ways. For example, you can say that the star Capella (Alpha Aurigae) is 100 times more luminous than the sun. You can also express this in real energy units by multiplying by the luminosity of the sun. The luminosity of Capella is 4×10^{28} Watts.

When you look at the night sky, the stars look much the same, yet your study of distances and luminosities reveals an astonishing fact. Some stars are almost a million times more luminous than the sun, and some are almost a million times less luminous. Clearly, the family of stars is filled with interesting characters.

6.3 Star Temperatures

Your third goal in this chapter is to learn about the temperatures of stars. The surprising fact is that stellar spectral lines can be used as a sensitive star thermometer.

From the discussion of blackbody radiation in Chapter 5, you know that temperatures of stars can be estimated from their color—red stars are cool, and blue stars are hot. The relative strengths of various spectral lines, however, give much greater accuracy in measuring star temperatures. Recall from Chapter 5 that, for stars, the term *surface* refers to the photosphere, which is the limit of our vision into the star from outside but is not an actual solid surface. Stars typically have surface temperatures of a few thousand or tens of thousands of degrees Kelvin. As you will discover in Chapter 7, the centers of stars are much hotter than their surfaces—many millions of degrees hotter—but the spectra tell only about the outer layers from which the light you see departed.

As you learned in Chapter 5, hydrogen Balmer absorption lines are produced by hydrogen atoms with electrons initially in the second energy level. The strength of these spectral lines can be used to gauge the temperature of a star because scientists know from lab experiments with gasses and radiation, and also from theoretical calculations, that:

- If the surface of a star is as cool as the sun or cooler, there are few violent collisions between atoms to excite the electrons, and most atoms will have their electrons in the ground (lowest) state. These atoms can't absorb photons in the Balmer series. As a result, you should expect to find weak hydrogen Balmer absorption lines in the spectra of very cool stars.

- In the surface layers of stars hotter than about 20,000 K, there are many violent collisions between atoms, exciting electrons to high energy levels or knocking the electrons completely out of most atoms so they become ionized. In this situation, few hydrogen atoms will have electrons in the second energy level to form Balmer absorption lines. As a result, you should also find weak hydrogen Balmer absorption lines in the spectra of very hot stars.

- At an intermediate temperature, roughly 10,000 K, the collisions have the correct amount of energy to excite large numbers of electrons into the second energy level. With many atoms excited to the second level, the gas absorbs Balmer-wavelength photons well and thus produces strong hydrogen Balmer lines.

The strength of the hydrogen Balmer lines thus depends on the temperature of the star's surface layers. Both hot and cool stars have weak Balmer lines, but medium-temperature stars have strong Balmer lines.

Figure 6.4 shows the relationship between gas temperature and strength of spectral lines for hydrogen and other substances. Each type of atom or molecule produces spectral lines that are weak at high and low temperatures and strong at some intermediate temperature, although the temperature at which the lines reach maximum strength is different for each type of atom or molecule. Theoretical calculations of the type first made by Cecilia Payne (discussed in Chapter 5) can predict just how strong various spectral lines should be for stars of different temperatures. Astronomers can determine a star's temperature by comparing the strengths of its spectral lines with the predicted strengths.

From stellar spectra, astronomers have found that the hottest stars have surface temperatures above 40,000 K and the coolest about 2,500 K. Compare these with the surface temperature of the sun, which is about 5,800 K.

Figure 6.4

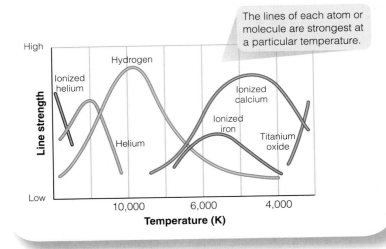

The strength of spectral lines can tell you the temperature of a star. Using the lines of several atoms or molecules gives you the most accurate results.

The lines of each atom or molecule are strongest at a particular temperature.

Temperature Spectral Classification

You have seen that the strengths of spectral lines depend on the surface (photosphere) temperature of the star. From this you can predict that all stars of a given temperature should have similar spectra. Learning to recognize the pattern of spectral lines produced in the atmospheres of stars of different temperatures means there is no need to do a full analysis every time each type of spectrum is encountered. Time can be saved by classifying stellar spectra rather than analyzing each spectrum individually. Astronomers classify stars by the lines and bands in their spectra, as shown in Table 6.1. For example, if it has weak Balmer lines and lines of ionized helium, it must be an O star.

The star classification system now used by astronomers was devised at Harvard during the 1890s and 1900s. One of the astronomers there, Annie J.

Cannon, personally inspected and classified the spectra of over 250,000 stars. The spectra were first classified in groups labeled A through Q, but some groups were later dropped, merged with others, or reordered. The final classification includes seven main **spectral classes** or types that are still used today: O, B, A, F, G, K, M. This set of star types, called the **spectral sequence**, is important because it is a temperature sequence. The O stars are the hottest, and the temperature continues to decrease down to the M stars, the coolest. For further precision, astronomers divide each spectral class into 10 subclasses. For example, spectral class A consists of the subclasses A0, A1, A2, . . . A8, A9. Next come F0, F1, F2, and so on. These finer divisions define a star's temperature to a precision of about 5 percent. The sun, for example, is not just a G star, but a G2 star, with a temperature of 5,800 K. Generations of astronomy students have remembered the spectral sequence by using mnemonics such as "Oh Boy, An F Grade Kills Me," or "Only Bad Astronomers Forget Generally Known Mnemonics." Recently, astronomers have added two more spectral classes, L and T, representing objects cooler than stars, called brown dwarfs, which you will encounter in the next chapter.

Figure 6.5 shows color images of 13 stellar spectra ranging from the hottest at the top to the coolest at the bottom. Notice how spectral features change gradually from hot to cool stars. Although these spectra are attractive, astronomers rarely work with spectra as color images. Rather, they display spectra as graphs of intensity versus wavelength with dark absorption lines as dips in the graph in Figure 6.6 on page 106. Such graphs show

Spectral class: A star's position in the temperature classification system O, B, A, F, G, K, M, based on the appearance of the star's spectrum.

Spectral sequence: The arrangement of spectral classes (O, B, A, F, G, K, M) ranging from hot to cool.

Table 6.1 **Spectral Classes**

Spectral Class	Approximate Temperature (K)	Hydrogen Balmer Lines	Other Spectral Features	Naked-Eye Example
O	40,000	Weak	Ionized helium	Meissa (O8)
B	20,000	Medium	Neutral helium	Achernar (B3)
A	10,000	Strong	Ionized calcium weak	Sirius (A1)
F	7,500	Medium	Ionized calcium weak	Canopus (F0)
G	5,500	Weak	Ionized calcium medium	Sun (G2)
K	4,500	Very weak	Ionized calcium strong	Arcturus (K2)
M	3,000	Very weak	TiO strong	Betelgeuse (M2)

Figure 6.5

These spectra show stars from hot O stars at top to cool M stars at bottom. The hydrogen Balmer lines are strongest for about type A0, but the two closely spaced lines of sodium in the yellow are strongest for very cool stars. Helium lines appear only in the spectra of the hottest stars. Notice that the helium line visible in the top spectrum has nearly—but not exactly—the same wavelength as the sodium lines visible in cooler stars. Bands produced by the molecule titanium oxide are strong in the spectra of the coolest stars.

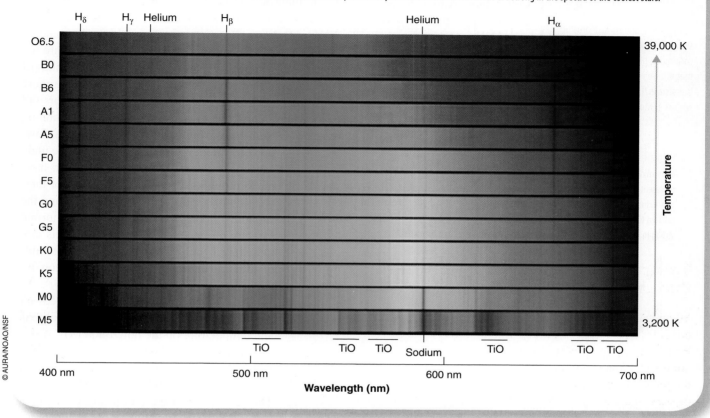

more detail than photographs. Notice also that the overall curves are similar to blackbody curves. The wavelength of maximum is in the infrared for the coolest stars and in the ultraviolet for the hottest stars. Compare Figures 6.5 and 6.6 and notice how the strength of spectral lines depends on temperature, as indicated in the previous discussion regarding Figure 6.4. It is straightforward to determine a star's temperature from the details of its spectrum.

6.4 Star Sizes

Your fourth goal in this chapter is to learn about the sizes of stars. Do they all have the same diameter as the sun, or are some larger and some smaller? You certainly can't see their sizes through a telescope; the images of the stars are much too small for you to resolve their disks and measure their diameters. There is a way, however, to find out how big stars really are.

Luminosity, Temperature, and Diameter

The luminosity of a glowing object like a star is determined by its temperature (Stefan-Boltzmann Law, Chapter 5) and its surface area. For example, you can eat dinner by candlelight because the candle flame has a small surface area, and, although it is very hot, it cannot radiate much heat because it is small; it has a low luminosity. However, if the candle flame were 12 ft tall, it would have a very large surface area from which to radiate, and, although it might be the same temperatures as a normal candle flame, its luminosity would drive you from the table.

In a similar way, a star's luminosity is proportional to its surface area. A hot star may not be very luminous if it has a small surface area, but it could be highly luminous if it were larger. Even a cool star could be luminous if it had a large surface area. You can use stellar luminosities to determine the diameters of stars if you can separate the effects of temperature and surface area.

The **Hertzsprung-Russell (H-R) diagram,** named after its originators,

Ejnar Hertzsprung in the Netherlands and Henry Norris Russell in the United States, is a graph that separates the effects of temperature and surface area on stellar luminosities and enables astronomers to sort the stars according to their diameters.

Before discussing the details of the H-R diagram, let's look at a similar diagram you might use to compare automobiles. You can plot a diagram such as Figure 6.7 to show horsepower versus weight for various makes of cars. In so doing, you will find that in general the more a car weighs, the more horsepower it has. Most cars fall somewhere along the normal sequence of cars running from heavy, high-powered cars to light, low-powered models. You might call this the main sequence of cars. Some cars, however, have much more horsepower than normal for their weight—the sport or racing models—and the economy models have less power than normal for cars of the same weight. Just as this diagram helps you understand the different kinds of autos, so the H-R diagram can help you understand different kinds of stars.

An H-R diagram, seen in Figure 6.8 on page 108, has luminosity on the vertical axis and temperature on the horizontal axis. A star is represented by a point on the graph that tells you its luminosity and temperature. Note that in astronomy the symbol $\odot$ refers to the sun. Thus $L_\odot$ refers to the luminosity of the sun, $T_\odot$ refers to the temperature of the sun, and so on. The H-R diagram in Figure 6.8 also contains a scale of spectral types across the top. The spectral type of a star is determined by its temperature, so you can use either spectral type or temperature as the horizontal axis of an H-R diagram.

Astronomers use H-R diagrams so often that they usually skip the words "the point that represents the star." Rather, they will say that a star is located in a certain place in the diagram. Of course, they mean the point that represents the luminosity and temperature of the star and not the star itself. The location of a star in the H-R diagram has nothing to do with the location of the star in space. Furthermore, a star may move in the H-R diagram as it ages and its luminosity and temperature change, but such motion in the diagram has nothing to do with the star's motion in space.

In an H-R diagram, the location of a point representing a star tells you a great deal about the star. Points near the top of the diagram

Figure 6.6

Modern digital spectra show how stellar spectra depend on spectral class. Here spectra are represented by graphs of intensity versus wavelength, and dark absorption lines appear as sharp dips in intensity. Hydrogen Balmer lines are strongest about A0, while lines caused by ionized calcium (Ca II) are strong in K stars. Bands produced by molecules such as TiO (titanium oxide) are strong in the coolest stars. Compare these spectra with Figures 6.4 and 6.5.

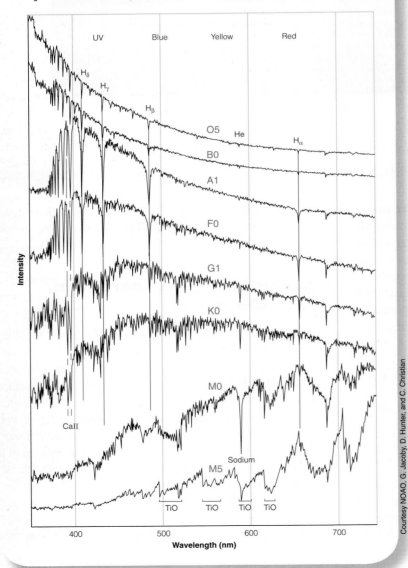

Courtesy NOAO, G. Jacoby, D. Hunter, and C. Christian

represent very luminous stars, and points near the bottom represent very low luminosity stars. Also, points near the right edge of the diagram represent very cool stars, and points near the left edge of the diagram represent very hot stars. Notice in Figure 6.8 how the artist has used color to represent temperature. Cool stars are red, and hot stars are blue.

The **main sequence** is the region of the H-R diagram running from upper left to lower right. It includes roughly 90 percent of all normal stars, represented by a curved line with dots for stars plotted along it in Figure 6.8. As you might expect, the hot main-sequence stars are more luminous than the cool main-sequence stars. Look again at the classification diagram for cars in Figure 6.7. Vehicles not on the car main sequence have different kinds of engines than main-sequence cars. In Chapter 7 you will find that stars not on the main sequence have different nuclear reactions as their power sources than do main-sequence stars. In addition to temperature, size is important in determining the luminosity of a star. Notice in the H-R diagram that some cool stars lie above the main sequence. Although they are cool, they are luminous, and that must mean they are larger—have more surface area—than main-sequence stars of the same temperature. These are called **giant** stars, and they are roughly 10 to 100 times larger than the sun. The **supergiant** stars at the top of the H-R diagram are as large as 1,000 times the sun's diameter. In contrast, **red dwarfs** at the lower end of the main sequence are not only cool but also small, giving them low luminosities.

At the bottom of the H-R diagram are the "economy models," stars that are very low in luminosity because they are very small. The **white dwarfs** lie in the lower left of the H-R diagram, and although some white dwarfs are among the hottest stars known, they are so small they have very little surface area from which to radiate, and that limits them to low luminosities.

A simple calculation can be made to draw precise lines of constant radius across the H-R diagram based on the luminosities and temperatures at each point as shown in Figure 6.8. For example, locate the line labeled $1R_\odot$ (1 solar radius) and notice that it passes through the point representing the sun. Any star whose point is located along this line has a radius equal to the sun's radius. Notice also that the lines of constant radius slope downward to the right, because cooler stars are always fainter than hotter stars of the same size, following the Stefan-Boltzman Law. These lines of constant radius show dramatically that the supergiants and giants are much larger than the sun. In contrast, white dwarf stars fall near the

Figure 6.7

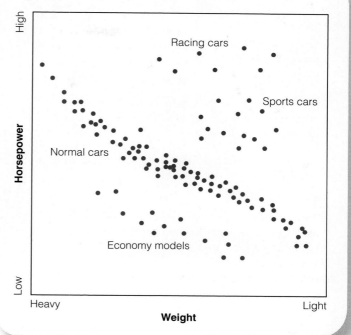

line labeled $0.01\ R_\odot$. They all have about the same radius—approximately the size of Earth!

Notice the great range of sizes among stars. The largest stars are 100,000 times larger than the tiny white dwarfs. If the sun were a tennis ball, the white dwarfs would be grains of sand, and the largest supergiant stars would be as big as football fields.

Luminosity Spectral Classification

A star's spectrum also contains clues as to whether it is a main-sequence star, a giant, or a supergiant. The larger a star is, the less dense its atmosphere is. The widths of spectral lines are partially determined by the density of the gas: If the atoms collide often in a dense gas, their energy levels become distorted, and the spectral lines are broadened. For example, in the spectrum of a main-sequence star, the hydrogen Balmer lines are broad because the star's atmosphere is dense and the hydrogen atoms collide often. In the spectrum of a giant star, the lines are narrower (Figure 6.9), because the giant star's atmosphere is less dense and the hydrogen atoms collide less often. In the spectrum of a supergiant star, the Balmer lines are very narrow.

Thus, an astronomer can look closely at a star's spectrum and tell roughly how big it is. Size categories

TYPES OF STARS

Main sequence: The region of the H-R diagram running from upper left to lower right, which includes roughly 90 percent of all stars generating energy by nuclear fusion.

Giant: Large, cool, highly luminous star in the upper right of the H-R diagram, typically 10 to 100 times the diameter of the sun.

Supergiant: Exceptionally luminous star whose diameter is 100 to 1,000 times that of the sun.

Red dwarf: A faint, cool, low-mass, main-sequence star.

White dwarf: Dying star at the lower left of the H-R diagram that has collapsed to the size of Earth and is slowly cooling off.

 Stars are categorized by luminosity classes based on the widths of lines in their spectra.

Figure 6.8

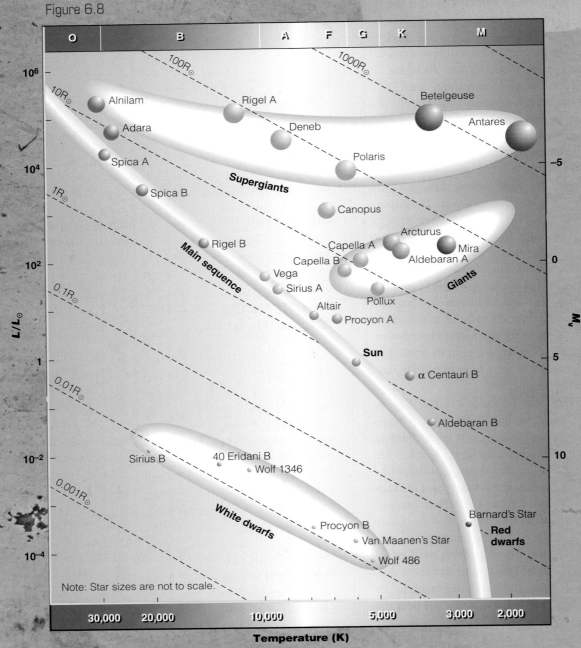

Figure 6.9

These schematic spectra show how the widths of spectral lines reveal a star's luminosity classification. Supergiants have very narrow spectral lines, and main-sequence stars have broad lines. Some spectral lines are more sensitive to this effect than others. Examining the details of a star's spectrum can determine its luminosity classification.

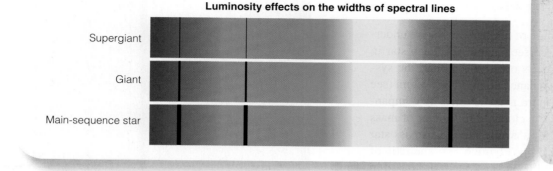

Luminosity effects on the widths of spectral lines

Supergiant

Giant

Main-sequence star

Luminosity class: A category of stars of similar luminosity, determined by the widths of lines in their spectra.

Spectroscopic parallax: The method of determining a star's distance by comparing its apparent magnitude with its absolute magnitude as estimated from its spectrum.

Binary star: Pairs of stars that orbit around their common center of mass.

are called **luminosity classes**, because the size of the star is the dominating factor in determining luminosity. Supergiants, for example, are very luminous because they are very large.

The luminosity classes are represented by the Roman numerals I through V, with supergiants further subdivided into types Ia and Ib. For example, you can distinguish between a luminous supergiant (Ia) such as Rigel (Beta Orionis) and a regular supergiant (Ib) such as Polaris, the North Star (Alpha Ursa Minoris). The star Adhara (Epsilon Canis Majoris) is a luminous giant (II), Capella (Alpha Aurigae) is a giant (III), and Altair (Alpha Aquilae) is a subgiant (IV). The sun is a main-sequence star (V). The luminosity class notation appears after the spectral type, as in G2V for the sun. White dwarfs don't enter into this classification because their spectra are very different from the other types of stars.

The approximate positions of the main sequence, giant, and supergiant luminosity classes are shown on the H-R diagram in Figure 6.8. Luminosity classification is subtle and not very accurate, but it is an important technique in modern astronomy because it provides clues to distance

Most stars are too distant to have measurable parallaxes, but astronomers can find the distances to these stars if they can record the stars' spectra and determine their luminosity classes. From spectral type and luminosity class, astronomers can estimate the star's absolute magnitude, compare with its apparent magnitude, and compute its distance. Although this process finds distance and not true parallax, it is called **spectroscopic parallax**.

For example, Betelgeuse (Alpha Orionis) is classified M2 Ia, and its apparent magnitude averages about 0.4 (Betelgeuse is somewhat variable). You can plot this star in an H-R diagram such as Figure 6.8, where you would find that a temperature class of M2 and luminosity class of Ib (supergiant) corresponds to a luminosity of about 30,000 $L_\odot$. That information, combined with the star's apparent brightness, allows astronomers to estimate that Betelgeuse is about 190 pc from Earth. The Hipparcos satellite finds the actual distance to be 131 pc, so the distance from the spectroscopic parallax technique is only approximate. Spectroscopic parallax does give a good first estimate of the distances of stars so far away that their parallax can't easily be measured.

6.5 Star Masses— Binary Stars

Your fifth goal is to find out how much matter stars contain, that is, to know their masses. Do they all contain about the same mass as the sun, or are some more massive and others less?

Gravity is the key to determining mass. Matter produces a gravitational field. Astronomers can figure out how much matter a star contains if they watch an object such as another star move through the star's gravitational field.

Finding the masses of stars involves studying **binary stars**, pairs of stars that orbit each other. Many of the familiar stars in the sky are actually pairs of stars orbiting each other, as in Figure 6.10. Binary systems are common; more than half of all stars are members of binary star systems. Few, however, can be analyzed completely. Many are so far apart that their periods are much too long for practical mapping of their orbits. Others are so close together they are not visible as separate stars.

Binary Stars in General

The key to finding the mass of a binary star is an understanding of orbital motion (see Chapter 3). Each star in a binary system moves in its own orbit around the system's center of mass, the balance point of the system (see page 53). If one star is more massive than its companion, then the massive star is closer to the center of mass and travels in a smaller orbit, while the lower-mass star whips around in a larger orbit (Figure 6.11). The ratio of the masses of the stars in the binary system portrayed in Figure 6.11 is M_A/M_B, which equals r_B/r_A, the inverse of the ratio of the radii of the orbits. For example, if one star in a binary system has an orbit twice as large as the other star's orbit, then it must be half as massive. Getting the

Figure 6.10

(a) At the bend of the handle of the Big Dipper lies a pair of stars, Mizar and Alcor. A telescope reveals that Mizar is a member of a visual binary system. (b) Spectra of these stars recorded at different times show that Mizar, its faint companion, and the nearby star Alcor are all spectroscopic binaries.

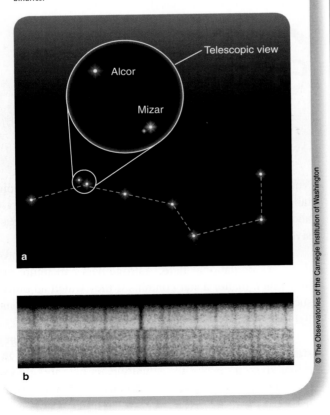

As stars in a binary star system revolve around each other, the line connecting them always passes through the center of mass, and the more massive star is always closer to the center of mass.

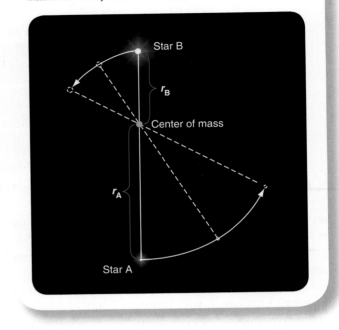

ratio of the masses is easy, but that doesn't tell you the individual masses of the stars, which is what you really want to know. That takes one more step.

To find the total mass of a binary star system, you must know the size of the orbits and the orbital period—the length of time the stars take to complete one orbit. The smaller the orbits are and the shorter the orbital period is, the stronger the stars' gravity must be to hold each other in orbit. From the sizes of the orbits and the orbital period, astronomers can use Kepler's third law (see Chapter 3) to figure out how much mass the stars contain in total. Combining that information with the ratio of the masses found from the relative sizes of the orbits reveals the individual masses of the stars.

Finding the mass of a binary star system is easier said than done. One difficulty is that the true sizes of the star orbits must be measured in units like meters or Astronomical Units in order to find the masses of the stars in units like kilograms or solar masses. Measuring the true sizes of the orbits in turn requires knowing the distance to the binary system. Therefore, the only stars whose masses astronomers know for certain are in binary systems with orbits that have been determined and distances from Earth that have been measured. Other complications are that the orbits of the two stars may be elliptical; also, the plane of their orbits can be tipped at an angle to your line of sight, distorting the apparent shapes of the

LEARNING ABOUT NATURE

Learning about Nature through Chains of Inference

Scientists can rarely observe the things they really want to know, so they must construct chains of inference. You can't observe the mass of stars directly, so you must find a way to use what you can observe— orbital period and angular separation—and figure out step-by-step the parameters you need to reach your goal.

Chains of inference can be mathematical, as when geologists use earthquakes to calculate the temperature and density of Earth's interior. There is no way to physically record Earth's internal temperature, but the speed of vibrations from distant earthquakes depends on the temperature and density of the rock they pass through. Geologists measure the delay in the arrival time of the vibrations at different locations on the surface, and that allows them to work their

way back to the seismic wave speed and, finally, to the temperature and density deep inside Earth.

Chains of inference can also be nonmathematical. Biologists studying the migration of whales can't follow individual whales for years at a time, but they observe feeding and mating in different locations and take into consideration food sources, ocean currents, and water temperatures in order to construct a chain of inference that leads back to the seasonal migration pattern for whales.

Almost all sciences use chains of inference, especially astronomers. It is primarily through chains of inference that they can describe what stars are like.

A Visual Binary Star System

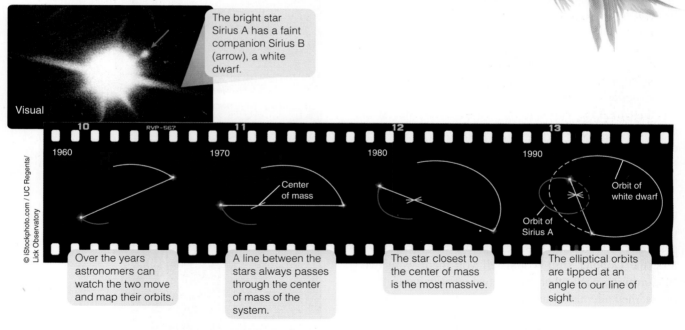

The bright star Sirius A has a faint companion Sirius B (arrow), a white dwarf.

Visual

RVP-567

1960 · 1970 · 1980 · 1990

Center of mass

Orbit of white dwarf

Orbit of Sirius A

Over the years astronomers can watch the two move and map their orbits.

A line between the stars always passes through the center of mass of the system.

The star closest to the center of mass is the most massive.

The elliptical orbits are tipped at an angle to our line of sight.

orbits. Astronomers must find ways to correct for these complications. Notice that finding the masses of binary stars requires a number of steps to get from what can be observed to what astronomers really want to know—the masses. Constructing such sequences of steps is an important part of science.

Three Kinds of Binary Systems

Although there are many different kinds of binary stars, three types are especially important for determining stellar masses. Studying binary stars is also preparation for

Visual binary system: A binary star system in which the two stars are separately visible in the telescope.

Spectroscopic binary system: A star system in which the stars are too close together to be visible separately. We see a single point of light, and only by taking a spectrum can we determine that there are two stars.

knowing how to find planets around stars other than our sun, because a star with a planet orbiting around it is like a binary star system with one very small component. Each type of binary star system corresponds to a different technique for finding planets, as you will learn in Chapter 12.

In a **visual binary system**, the two stars are separately visible in the telescope and astronomers can watch the stars orbit each other over periods of years or decades, as the series of illustrations of Sirius A and B on the previous page demonstrates. From that, astronomers can find the orbital period and, if the distance of the system from Earth can be found, the size of the orbits. That is enough to find the masses of the stars.

Many visual binaries have such large orbits their orbital periods are hundreds or thousands of years, and astronomers have not yet seen them complete an entire orbit. Also, many binary stars orbit so close to each other they are not visible as separate stars. Such systems can't be analyzed as a visual binary.

If the stars in a binary system are close together, a telescopic view, limited by diffraction and atmospheric seeing, shows a single point of light. Only by looking at a spectrum, which is formed by light from both stars and contains spectral lines from both, can astronomers tell that there are two stars present and not one. Such a system is a **spectroscopic binary system**. Familiar examples of spectroscopic binary systems are the stars Mizar and Alcor in the handle of the Big Dipper (Figure 6.10b).

Figure 6.12 shows a pair of stars orbiting each other; the circular orbit appears elliptical because you see it nearly edge-on. If this were a true spectroscopic binary system, you would not see the separate stars. Nevertheless, the Doppler shift would tell you there were two stars orbiting each other. As the two stars move in their orbits, they alternately approach toward and recede from Earth and their spectral lines are Doppler shifted alternately toward blue and then red wavelengths. Noticing pairs of spectral lines moving back and forth across each other would alert you that you were observing a spectroscopic binary. Although spectroscopic binaries are very common, they are not as useful as visual binaries.

Figure 6.12

From Earth, a spectroscopic binary looks like a single point of light, but the Doppler shifts in its spectrum reveal the orbital motion of the two stars. If the plane of their orbits is oriented very close to edge-on, the stars can pass in front of each other as they orbit and the system will be an eclipsing binary, as illustrated in Figure 6.13.

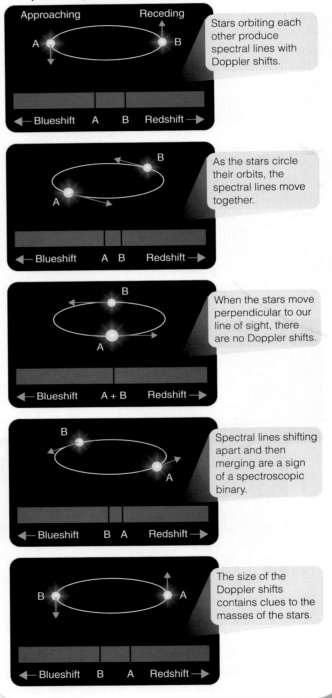

A Spectroscopic Binary Star System

Stars orbiting each other produce spectral lines with Doppler shifts.

As the stars circle their orbits, the spectral lines move together.

When the stars move perpendicular to our line of sight, there are no Doppler shifts.

Spectral lines shifting apart and then merging are a sign of a spectroscopic binary.

The size of the Doppler shifts contains clues to the masses of the stars.

Figure 6.13

From Earth, an eclipsing binary looks like a single point of light, but changes in brightness reveal that two stars are eclipsing each other. The light curve, shown here as magnitude versus time, combined with Doppler shift information from spectra, can reveal the size and mass of the individual stars.

An Eclipsing Binary Star System

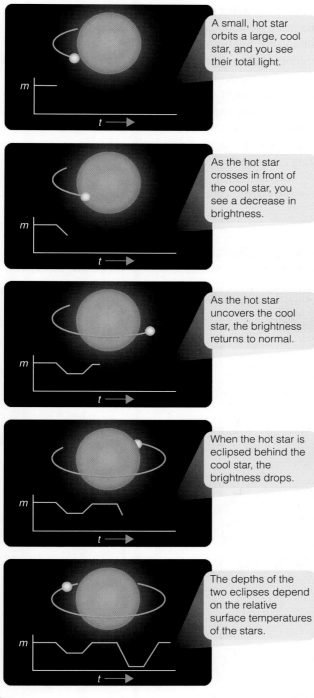

A small, hot star orbits a large, cool star, and you see their total light.

As the hot star crosses in front of the cool star, you see a decrease in brightness.

As the hot star uncovers the cool star, the brightness returns to normal.

When the hot star is eclipsed behind the cool star, the brightness drops.

The depths of the two eclipses depend on the relative surface temperatures of the stars.

Astronomers can find the orbital period easily, but they can't find the true size of the orbits because there is no way to find the angle at which the orbits are tipped. That means they can't find the true masses of a spectroscopic binary. All they can find is a lower limit to the masses.

Eclipsing binary system: A binary star system in which the stars cross in front of each other as seen from Earth.

Light curve: A graph of brightness versus time commonly used in analyzing variable stars and eclipsing binaries.

If the plane of the orbits is nearly edge-on to Earth, then the stars can cross in front of each other as seen from Earth. When one star moves in front of the other, it blocks some of the light, and the star is eclipsed. Such a system is called an **eclipsing binary system**.

Seen from Earth, the two stars are not visible separately. The system looks like a single point of light. But, when one star moves in front of the other star, part of the light is blocked, and the total brightness of the point of light decreases. Figure 6.13 shows a smaller star moving in an orbit around a larger star, first eclipsing the larger star and then being eclipsed as it moves behind. The resulting variation in the brightness of the system is shown as a graph of brightness versus time, a **light curve**.

The light curves of eclipsing binary systems contain plenty of information about the stars, but the curves can be difficult to analyze. Figure 6.13 shows an idealized example of such a system. Once the light curve of an eclipsing binary system has been accurately observed, astronomers can construct a chain of inference that leads to the masses of the two stars. They can find the orbital period easily and can get spectra showing the Doppler shifts of the two stars. They can find the orbital speed because they don't have to correct for the inclination of the orbits; you know the orbits are nearly edge-on, or there would not be eclipses. From that, astronomers can find the size of the orbits and the masses of the stars.

Earlier in this chapter you learned that luminosity and temperature can be used to determine the radii of stars, but eclipsing binary systems give a way to check those calculations by measuring the sizes of a few stars directly. The light curve shows how long it takes for the stars to cross in front of each other, and multiplying these time intervals by the orbital speeds gives the diameters of the stars. There are complications due to the inclination and eccentricity of orbits, but often these effects can be taken into account, so observations of an eclipsing binary system can directly tell you not only the masses of its stars but also their diameters.

From the study of binary stars, astronomers have found that the masses of stars range from roughly 0.1 solar masses to nearly 100 solar masses. The most massive stars ever found in a binary system have masses of 83 and 82 solar masses. A few other stars are believed to be more massive, 100 solar masses to 150 solar masses, but they do not lie in binary systems, so astronomers can only estimate their masses.

6.6 Typical Stars

You have achieved the five goals set at the start of this chapter. You know how to find the distances, luminosities, temperatures, diameters, and masses of stars. Now you can put those data together to paint a family portrait of the stars. As in human family portraits, both similarities and differences are important clues to the history of the family.

Luminosity, Mass, and Density

The H-R diagram is filled with patterns that give you clues as to how stars are born, how they age, and how they die. When you add your data, you see traces of those patterns.

If you label an H-R diagram with the masses of the stars determined by observations of binary star systems, you will discover that the main-sequence stars are ordered by mass, which you can see in Figure 6.14. The most massive main-sequence stars are the hot stars. As you run your eye down the main sequence, you will find lower-mass stars, and the lowest-mass stars are the coolest, faintest main-sequence stars.

Stars that do not lie on the main sequence are not in order according to mass. Some giants and supergiants are massive, while others are no more massive than the sun. All white dwarfs have about the same mass, usually in the narrow range of 0.5 to about 1.0 solar masses.

Because of the systematic ordering of mass along the main sequence, these main-sequence stars obey a **mass-luminosity relation**—the more massive a star is, the more luminous it is (Figure 6.15). In fact, the mass-luminosity relation can be expressed as: Luminosity is proportional to mass to the 3.5 power. For example, a star with a mass of 4.0 $M_\odot$ can be expected to have a luminosity of about $4^{3.5}$ or 128 $L_\odot$. Giants, supergiants, and white dwarfs do not follow the mass-luminosity relation. In the next chapters, the main sequence mass-luminosity relation will help you understand how main sequence stars generate their energy.

Though mass alone does not reveal any pattern among giants, supergiants, and white dwarfs, density

Figure 6.14

The masses of the plotted stars are labeled on this H-R diagram. Notice that the masses of main-sequence stars decrease from top to bottom but that masses of giants and supergiants are not arranged in any ordered pattern.

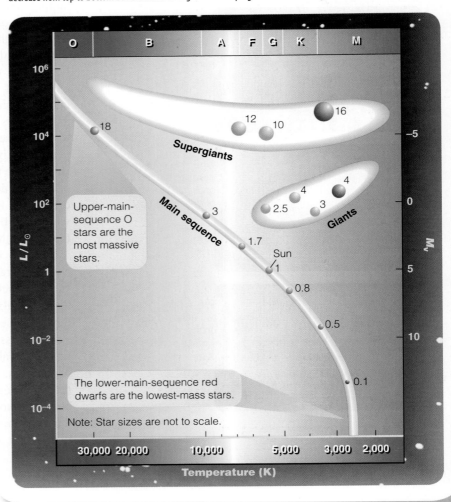

does. Once you know a star's mass and diameter, you can calculate its average density by dividing its mass by its volume. Stars are not uniform in density but are most dense at their centers and least dense near their surface. The center of the sun, for instance, is about 100 times as dense as water; its density near the photosphere is about 3,400 times less dense than Earth's atmosphere at sea level. A star's average density is intermediate between its central and surface densities. The sun's average density is approximately 1 gram per cubic centimeter—about the density of water.

Main-sequence stars have average densities similar to the sun's density. As you learned earlier in the discussion about luminosity classification, giant stars are much larger in diameter than the main-sequence stars but not much larger in mass, so giants have low average densities, ranging from 0.1 to 0.01 g/cm³. The enormous supergiants have still lower densities, ranging from 0.001 to 0.000001 g/cm³. These densities are thinner than the air you breathe, and if you could insulate yourself from the heat, you could fly an airplane through these stars. Only near the center would you be in any danger; astronomers calculate that the material there is very dense.

The white dwarfs have masses about equal to the sun's but are very small—only about the size of Earth. Thus, the matter is compressed to densities of 3,000,000 g/cm³ or more. On Earth, a teaspoonful of this material would weigh about 15 tons.

Density divides stars into three groups. Most stars are main-sequence stars with densities similar to the sun's. Giants and supergiants are very-low-density stars, and white dwarfs are high-density stars. You will see in later chapters that these densities reflect different stages in the evolution of stars.

Surveying the Stars

If you want to know what the average person thinks about a certain subject, you take a survey. If you want to know what the average star is like, you can survey the stars. Such surveys reveal important relationships among the family of stars. Over the years, many astronomers have added their results to the growing collection of data on star distances, luminosities, temperatures, sizes, and masses. They can now analyze those data to search for relationships between these and other parameters.

As the 21st century begins, astronomers are deeply involved in extensive surveys. Remember the earlier mention of the Hipparcos satellite that surveyed the entire sky measuring the parallax of over a million stars? Powerful computers to control instruments and analyze

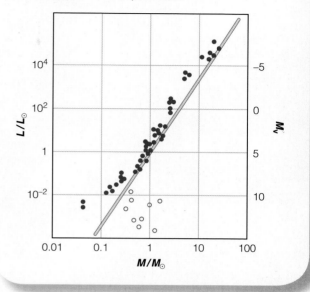

Figure 6.15

The mass–luminosity relation shows that the more massive a main-sequence star is, the more luminous it is. The open circles represent white dwarfs, which do not obey the relation.

data make such immense surveys possible. For example, the Sloan Digital Sky Survey mapped a quarter of the sky, measuring the position and brightness of 100 million stars and galaxies. Also, the Two Micron All Sky Survey (2MASS) has mapped the entire sky at three near-infrared wavelengths. A number of other sky surveys are underway. Astronomers will "mine" these mountains of data for decades to come.

What could you learn about stars from a survey of the stars near the sun? The evidence astronomers have is that the sun is in a typical place in the universe. Therefore, such a survey could reveal general characteristics of the stars. Study "The Family of Stars" on pages 116–117 and notice three important points:

1. Taking a survey is difficult because you must be sure to get an honest sample. If you don't survey enough stars, or if you miss some types of stars, your results can be biased.

2. M dwarfs and white dwarfs are so faint they are difficult to find even near Earth and may be undercounted in surveys.

3. Luminous stars, although they are rare, are easily visible even at great distances. Typical nearby stars have lower luminosity than our sun.

The night sky is a beautiful carpet of stars. Some are giants and supergiants, and some are dwarfs. The family

The Family of Stars

1 What is the most common kind of star? Are some rare? Are some common? To answer those questions you must survey the stars. To do so you must know their spectral class, their luminosity class, and their distance. Your census of the family of stars produces some surprising demographic results.

1a You could survey the stars by observing every star within 62 pc of Earth. A sphere 62 pc in radius encloses a million cubic parsecs. Such a survey would tell you how many stars of each type are found within a volume of a million cubic parsecs.

62 pc

Earth

2 Your survey faces two problems.

1. The most luminous stars are so rare you find few in your survey region. There are no O stars at all within 62 pc of Earth.

2. Lower-main-sequence M stars, called red dwarfs, and white dwarfs are so faint they are hard to locate even when they are only a few parsecs from Earth. Finding every one of these stars in your survey sphere is a difficult task.

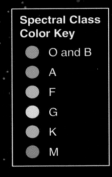

Spectral Class Color Key

- O and B
- A
- F
- G
- K
- M

The star chart in the background of these two pages shows most of the constellation Canis Major; stars are represented as dots with colors assigned according to spectral class. The brightest stars in the sky tend to be the rare, highly luminous stars, which look bright even though they are far away. Most stars are of very low luminosity, so nearby stars tend to be very faint red dwarfs.

Red dwarf
15 pc

o² Canis Majoris
B3Ia   790 pc

Red dwarf
17 pc

δ Canis Majoris
F8Ia   550 pc

σ Canis Majoris
M0Iab   370 pc

η Canis Majoris
B5Ia   980 pc

ε Canis Majoris
B2II   130 pc

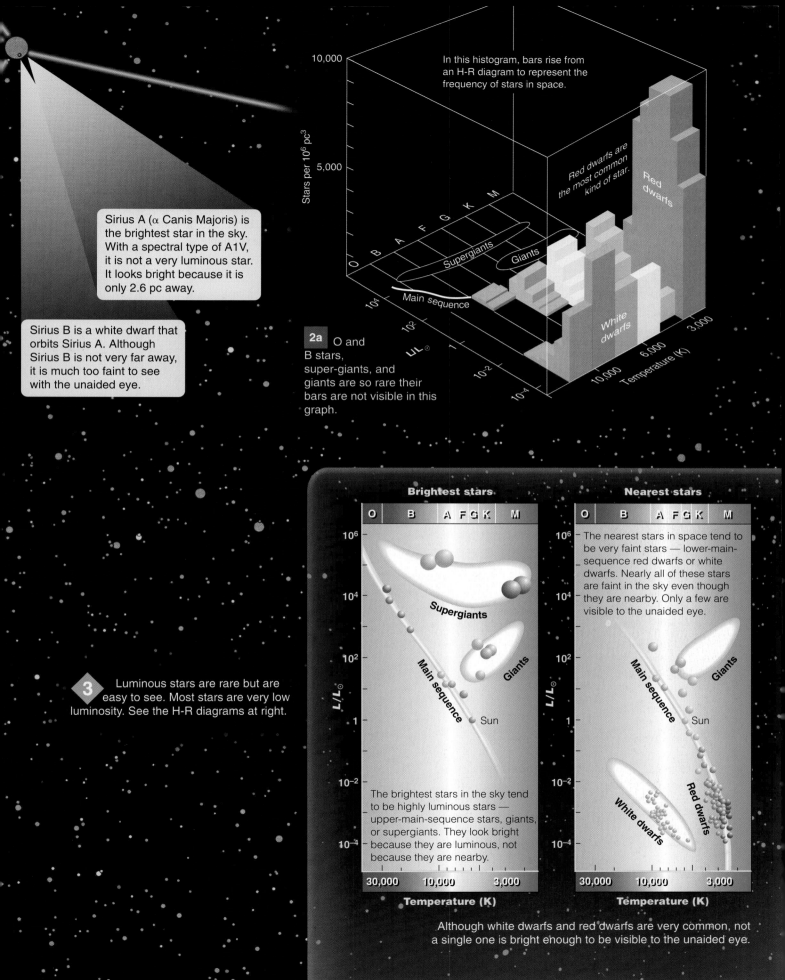

In this histogram, bars rise from an H-R diagram to represent the frequency of stars in space.

Stars per 10^6 pc^3

Red dwarfs are the most common kind of star.

Red dwarfs

Supergiants

Giants

Main sequence

White dwarfs

$L/L_\odot$

Temperature (K)

2a O and B stars, super-giants, and giants are so rare their bars are not visible in this graph.

Sirius A (α Canis Majoris) is the brightest star in the sky. With a spectral type of A1V, it is not a very luminous star. It looks bright because it is only 2.6 pc away.

Sirius B is a white dwarf that orbits Sirius A. Although Sirius B is not very far away, it is much too faint to see with the unaided eye.

3 Luminous stars are rare but are easy to see. Most stars are very low luminosity. See the H-R diagrams at right.

Brightest stars

O B A F G K M

Supergiants

Main sequence

Giants

Sun

$L/L_\odot$

The brightest stars in the sky tend to be highly luminous stars — upper-main-sequence stars, giants, or supergiants. They look bright because they are luminous, not because they are nearby.

30,000 10,000 3,000

Temperature (K)

Nearest stars

O B A F G K M

The nearest stars in space tend to be very faint stars — lower-main-sequence red dwarfs or white dwarfs. Nearly all of these stars are faint in the sky even though they are nearby. Only a few are visible to the unaided eye.

Main sequence

Giants

Sun

$L/L_\odot$

White dwarfs

Red dwarfs

30,000 10,000 3,000

Temperature (K)

Although white dwarfs and red dwarfs are very common, not a single one is bright enough to be visible to the unaided eye.

of stars is rich in its diversity. Pretend space is like a very clear ocean in which you are swimming. When you look at the night sky you are seeing mostly "whales," the rare very large and luminous stars, mostly far away. If instead you cast a net near your location, you would mostly catch "sardines," very low luminosity M dwarfs that make up most of the stellar population of the universe.

In this chapter you explored the basic properties of stars. Once you found the distance to the stars, you were able to find their luminosities. Knowing their luminosi-ties and temperatures gives you their diameters. Studying binary stars gives you their masses. These are all rather mundane data. But you have now discovered a puzzling situation. The largest and most luminous stars are so rare you might joke that they hardly exist, and the average stars are such small low-mass things they are hard to see even if they are near Earth in space. Why does nature make stars in this peculiar way? To answer that question, you can explore the birth, life, and death of stars. That quest begins in the next chapter.

RECAP

We humans experience medium-sized things, such as trees, flowers, and small insects. But we cannot see the beauty of the microscopic world without instruments and special methods. Similarly, we can sense the grandeur of a mountain range, but larger objects, such as stars, are too big for our senses.

We live between the microscopic world and the astronomical world, and science enriches our lives by revealing the parts of the universe beyond our daily experience. We may enjoy experiencing a flower by admiring its color and shape and by smelling its fragrance. But to truly appreciate the flower, we need to understand it, to understand how it serves its plant and how the plant came to create such a beautiful blossom.

Humans have a natural drive to understand as well as experience. You have experienced the stars in the night sky, and now you are beginning to understand them as objects ranging from hot O stars to cool red dwarfs. It is natural for you to wonder why these stars are so different, but although you have medium senses, you can understand the stars, and answer the following questions:

1. How far away are the stars?

2. How much energy do stars make?

3. How can you tell a star's temperature using its spectrum?

4. How big are stars?

5. How much matter do stars contain?

6. What is the typical star like?

Chapter 7

The Structure
and Formation
of Stars

The stars are not eternal. When you look at the sky, you see hundreds of points of light. Each is an object like the sun held together by its gravity and generating tremendous energy in its core through nuclear reactions. The stars you see tonight are the same stars your parents, grandparents, and great-grandparents saw. Stars change hardly at all in a human lifetime, but they are not eternal. Stars are born, and stars die. This chapter begins that story. How can you know what the internal processes and life cycles of stars are when you can't see inside them and humans don't live long enough to see them evolve? The answer lies in the methods of science. By constructing theories that describe how nature works and then testing those theories against evidence from observations, you can unravel some of nature's greatest secrets. In this chapter, you will see how the flow of energy from inside to the surfaces of the stars balances gravity, making the stars stable, and how nuclear reactions inside stars generate that energy. You will also see how gravity creates new stars from the thin dusty gas between the stars.

7.1 Stellar Structure

If there is a single idea in stellar astronomy that can be called crucial, it is the concept of balance. In this section you will discover that stars are simple, elegant power sources held together by their own gravity, balanced by the support of their internal heat and pressure.

Using the basic concept of balance, you can consider the structure of a star. What is meant here by structure is the variation in temperature, density, pressure, and so on between the surface of the star and its center. It will be easier to think about stellar structure if you imagine that the star is divided into concentric shells like those in an onion. Knowing the four basic laws of stellar structure (Table 7.1) will also help you understand stars. With these laws you can consider the temperature, pressure, and density in each shell. These helpful shells exist only in the imagination; stars do not actually have such separable layers.

looking back

Throughout the preceding chapters you have been building a general understanding of the sky and familiarity with the methods of astronomy. The last chapter provided a detailed family portrait of the stars. You can now describe the basic *external* properties of different kinds of stars.

looking ahead

The most important of the questions discussed in this chapter is "What is the evidence that the theory of star formation is correct?" Testing theories against evidence is the basic skill required of all scientists, and you will use it over and over in the eight chapters that follow this one. In the next chapter you will learn about what happens at the ends of stars' lives, when they begin to run out of fuel.

The Laws of Conservation of Mass and Energy

The first two laws of stellar structure have something in common—they are both what astronomers and physicists call *conservation* laws. Conservation laws say that certain things cannot be created out of nothing or vanish into nothing. Such conservation laws are powerful aids to help you understand how nature works.

The law of **conservation of mass** is a basic law of nature that can be applied to the structure of stars. It says that the total mass of a star must equal the sum of the masses of its shells. This is like saying the weight of a cake must equal the sum of the weight of its layers.

The law of **conservation of energy** is another basic law of nature. It says that the amount of energy flowing out of the top of a layer in the star must be equal to the amount of energy coming in at the bottom, plus whatever energy is generated within the layer. That means that the energy leaving the surface of the star, its luminosity, must equal the sum of the energies generated in all the layers inside the star. This is like saying that all the new cars driving out of a factory must equal the sum of all the cars made on each of the production lines.

These two laws may seem so familiar or so obvious that they hardly need to be stated, but they are important clues to the structure of stars. The third law of stellar structure is a law about balance.

Hydrostatic Equilibrium

The weight of each layer of a star must be supported by the layer below. (The words *down* or *below* are conven-

tionally used to refer to regions closer to the center of a star.) Picture a pyramid of people in a circus stunt—the people in the top row do not have to hold up anybody else; the people in the next row down are holding up the people in the top row; and so on. In a star that is stable, the deeper layers must support the weight of all of the layers above. The inside of a star is made up of gas, so the weight pressing down on a layer must be balanced by gas pressure in that layer. If the pressure is too low, the weight from above will compress and push down the layer; and if the pressure is too high, the layer will expand and lift the layers above.

This balance between weight and pressure is called **hydrostatic equilibrium**. *Hydro* (from the Greek word for water) tells you the material is a fluid, which by definition includes the gases of a star, and *static* tells you the fluid is stable, neither expanding nor contracting.

The first figure on the next page shows this hydrostatic balance in the imaginary layers of a star. The weight pressing down on each layer is shown by lighter red arrows, which grow larger with increasing depth because the weight grows larger. The pressure in each layer is shown by darker red arrows, which must grow larger with increasing depth to support the weight.

The pressure in a gas depends on the temperature and density of the gas. Near the surface, there is little weight pressing down, so the pressure does not need to be high for stability. Deeper in the star, the pressure must be higher, which means that the temperature and density of the gas must also be higher. Hydrostatic equilibrium tells you that stars must have high temperature, pressure, and density inside to support their own weight and be stable.

definitions

Although the law of hydrostatic equilibrium can tell you some things about the inner structure of stars, you need one more law to completely describe a star. You need to know how energy flows from the inside to the outside.

Energy Transport

The surface of a star radiates light and heat into space and would quickly cool if that energy were not replaced. Because the inside of the star is hotter than the surface, energy must flow outward to the surface, where it radiates away. This flow of energy through

---

Table 7.1 The Four Laws of Stellar Structure

I. Conservation of mass	Total mass equals the sum of shell masses.
II. Conservation of energy	Total luminosity equals the sum of energy generated in all of the layers.
III. Hydrostatic equilibrium	The weight on each layer is balanced by the pressure in that layer.
IV. Energy transport	Energy moves from hot to cool regions by conduction, radiation, or convection.

each shell determines its temperature, which, as you saw previously, determines how much weight that shell can balance. To understand the structure of a star, you need to understand how energy moves from the center to the surface.

The law of **energy transport** says that energy must flow from hot regions to cooler regions either by conduction, convection, or radiation. *Conduction* is the most familiar form of heat flow. If you hold the bowl of a spoon in a candle flame, the handle of the spoon grows warmer as heat, in the form of motion among the atoms of the spoon, is conducted from atom to atom up the handle. Conduction requires close contact between the atoms, so it is not an important cause of heat flow in normal stars because radiation or convection are much more efficient, even in their centers; it is only important in rare types of stars with extremely high densities.

The transport of energy by *radiation* is another familiar experience. Put your hand near a candle flame, and you can feel the heat. What you actually feel are infrared photons—packets of energy—radiated by the flame and absorbed by your hand. Radiation is the principal means of energy transport in the interiors of most stars. Photons are absorbed and reemitted in random directions over and over as energy works its way from the hot interior toward the cooler surface.

The flow of energy by radiation depends on how difficult it is for photons to move through the gas. The **opacity** of the gas, its resistance to the flow of radiation, depends strongly on its temperature: A hot gas is more transparent than a cool gas. If the opacity is high, radiation cannot flow through the gas easily, and it backs up like water behind a dam. When enough heat builds up, the gas begins to churn as hot gas rises upward and cool gas sinks downward. This heat-driven circulation of a fluid is *convection*, the third way energy can move in a star. You are familiar with convection; the rising wisp of smoke above a candle flame is carried by convection. Energy is carried upward in these convection currents as rising hot gas and also as sinking cool gas. The cross section of the sun to the right indicates in which zones radiative and convective energy are located.

The four laws of stellar structure, when properly understood, not only tell you about the insides of stars, but also how stars are born, how they live, and how they die.

Stellar Models

The laws of stellar structure, described in general terms in the previous sections, can be written as mathematical equations. By solving those equations, astronomers can

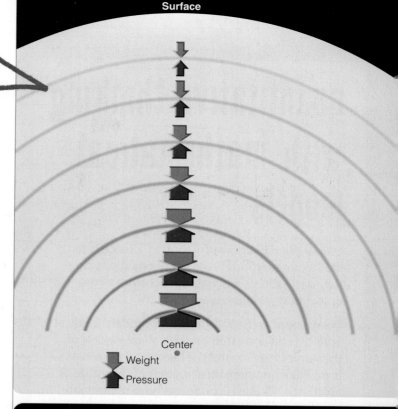

Surface

Center

Weight

Pressure

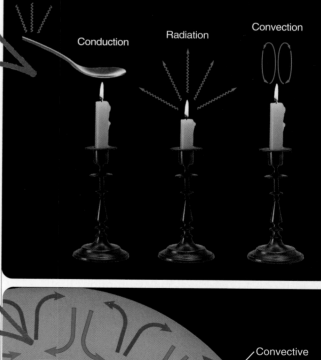

Conduction

Radiation

Convection

Convective zone

Radiative zone

Core energy generation

© Valery Seleznev/iStockphoto.com / © iStockphoto.com

quantitative thinking with mathematical models

One of the most powerful tools in science is the mathematical model, a group of equations carefully designed to describe the behavior of the object that scientists want to study.

Many sciences use mathematical models. Medical scientists have created mathematical models of the nerves that control the heart, and physicists have made mathematical models of the inside of an atomic nucleus. Economists have mathematical models of certain aspects of systems such as the municipal bond market, and Earth scientists have mathematical models of Earth's atmosphere. In each case, the mathematical model allows the scientists to study something that is difficult to study directly. The model can reveal regions scientists cannot observe, speed up a slow process, slow down a fast process, or allow scientists to perform experiments that would be impossible in reality. Astronomers, for example, can change the abundance of different chemical elements in a model star to see how its structure depends on its composition. Mathematical modeling is one of the most important research methods in astronomy, even though it often requires exceptionally large and fast computers.

As is true for any scientific model, a mathematical model is only as reliable as the assumptions that go into its creation. The celestial sphere (Chapter 2) is an adequate model of the sky for some purposes, but it doesn't work if you take it too literally. The same is true of mathematical models. You can think of a mathematical model as a numerical expression of one or more theories. Although such models can be very helpful, they are always based on theories and assumptions and so must be compared with the real world at every opportunity. As always in science, there is no substitute for a careful comparison with reality.

make a mathematical model of the inside of a star. To start one of these models requires knowing the type of information you learned in Chapter 6 about the external properties of a star: its diameter, luminosity, mass, surface temperature, atmospheric density, and composition.

If you want to build a model of a star, you could divide the star into, say, 100 concentric shells and then write down the four equations of stellar structure for each shell. You would then have 400 equations that would have 400 unknowns, namely, the temperature, density, mass, and energy flow in each shell. Solving 400 equations simultaneously is not easy, and the first such solutions, done by hand before the invention of electronic computers, took months of work. Now a properly programmed computer can solve the equations in a few seconds and print a table of numbers that represent the conditions in each shell of the star. Such a table is called a **stellar model**.

The table shown in Figure 7.1 is a model of the sun. The bottom line, for radius equal to 0.00, represents the center of the sun, and the top line, for radius equal to 1.00, represents the surface. The other lines tell you the temperature and density in each shell, the mass inside each shell, and the fraction of the sun's luminosity flowing outward through the shell. The bottom line tells you that the temperature at the center of the sun must be over 15 million Kelvin in order for the sun to be stable. At such a high temperature, the gas is very transparent, and energy flows as radiation. Nearer the surface, the temperature is lower, the gas is more opaque, and energy moves by convection. This is confirmed by evidence of convection visible on the sun's surface (see Chapter 5).

Stellar models also let you look into a star's past and future. In fact, you could use models as time machines to follow the evolution of stars over billions of years. To look into a star's future, for instance, you could use a stellar model to determine how fast the star uses its fuel in each shell. As the fuel is consumed, the chemical composition of the gas changes, the opacity changes, and the amount of energy generated declines. By calculating the rate of these changes, you could predict what the star will look like at some time in the future. You could repeat that process over and over and follow the evolution of the star step-by-step as it ages.

Stellar astronomy has made great advances since the 1960s, when high-speed computers that could calculate models of the structure and evolution of stars began to be available. The summary of star formation later in this chapter is based on thousands of stellar models. You will continue to rely on theoretical models as you study main-sequence stars later in this chapter and the deaths of stars in the next chapter.

7.2 Nuclear Fusion in the Sun and Stars

Astronomers often use the wrong words to describe energy generation in the sun and stars. Astronomers will say, "The star ignites hydrogen burning." You normally use the word *ignite* to mean "catch on fire" and *burn* to mean "consume by fire." What goes on inside stars isn't really burning in the usual sense.

The sun is a normal star, and it is powered by nuclear reactions that occur near its center. Later in this section you will learn some of the evidence for this. The energy produced keeps a star's interior hot, and the gas is totally ionized. That is, the electrons are not attached to the atomic nuclei, and the gas is a soup of rapidly moving particles colliding with each other at high velocity.

When astronomers discuss nuclear reactions inside stars, they refer to atomic nuclei and not to atoms. How exactly can the nucleus of an atom yield energy? The answer lies in the forces that hold the nuclei together.

Nuclear Binding Energy

The sun generates its energy by breaking and reconnecting the bonds between the particles *inside* atomic nuclei. That is quite different from the way we generate energy by burning wood in a fireplace. The process of burning extracts energy by breaking and reconnecting chemical bonds between atoms in the fuel. Chemical bonds are formed by electrons on the outsides of atoms, and you saw in Chapter 5 that the electrons are bound to the atoms by the electromagnetic force. Thus chemical energy originates in the electromagnetic force.

There are only four known forces in nature: the force of gravity, the electromagnetic force, and the strong and weak **nuclear forces**. The weak nuclear force is involved in the radioactive decay of certain kinds of nuclear particles, and the strong force binds together atomic nuclei. Thus nuclear energy originates in the strong nuclear force.

Nuclear power plants on Earth generate energy through **nuclear fission** reactions that split uranium nuclei into less massive fragments. A uranium nucleus contains a total of 235 protons and neutrons, and it splits into a range of fragments containing roughly half as many particles. Because the fragments produced are more tightly bound than the uranium nuclei, energy is released during uranium fission, as demonstrated in Figure 7.2.

> **Stellar model:** A table of numbers representing the conditions in various layers within a star.
>
> **Nuclear forces:** The two forces of nature that only affect the particles in the nuclei of atoms.
>
> **Nuclear fission:** Reactions that break the nuclei of atoms into fragments.

Figure 7.1

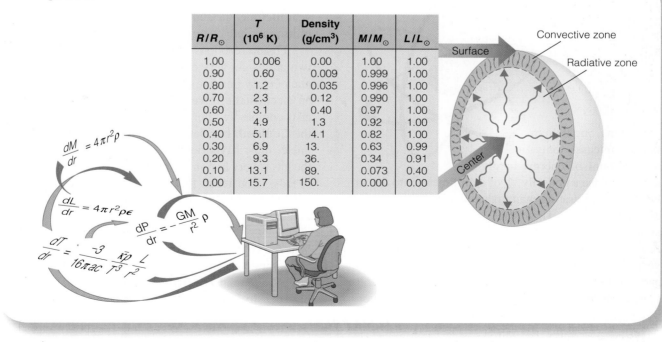

$R/R_\odot$	T (10^6 K)	Density (g/cm^3)	$M/M_\odot$	$L/L_\odot$
1.00	0.006	0.00	1.00	1.00
0.90	0.60	0.009	0.999	1.00
0.80	1.2	0.035	0.996	1.00
0.70	2.3	0.12	0.990	1.00
0.60	3.1	0.40	0.97	1.00
0.50	4.9	1.3	0.92	1.00
0.40	5.1	4.1	0.82	1.00
0.30	6.9	13.	0.63	0.99
0.20	9.3	36.	0.34	0.91
0.10	13.1	89.	0.073	0.40
0.00	15.7	150.	0.000	0.00

$$\frac{dM}{dr} = 4\pi r^2 \rho$$

$$\frac{dL}{dr} = 4\pi r^2 \rho \epsilon$$

$$\frac{dP}{dr} = -\frac{GM}{r^2}\rho$$

$$\frac{dT}{dr} = \frac{-3}{16\pi ac}\frac{\bar{\kappa}\rho}{T^3}\frac{L}{r^2}$$

Surface

Center

Convective zone

Radiative zone

Stars make energy in **nuclear fusion** reactions that combine light nuclei into heavier nuclei. The most common reaction, including that in the sun, fuses hydrogen nuclei (single protons) into helium nuclei (two protons and two neutrons). Because the nuclei produced are more tightly bound than the original nuclei, energy is released. Notice in Figure 7.2 that both fusion and fission reactions move downward in the diagram, meaning they release the binding energy of nuclei and create more tightly bound nuclei.

Hydrogen Fusion

The sun fuses four hydrogen nuclei to make one helium nucleus. Because one helium nucleus contains 0.7 percent less mass than four hydrogen nuclei, some mass vanishes in the process. In fact, that mass is converted to energy, and you could figure out how much by using Einstein's famous equation, $E = mc^2$. For example, converting 1 kg (2.2 lb) of matter completely into energy would produce an enormous amount of energy, 9×10^{16} Joules, comparable to the amount of energy a big city like Philadelphia uses in a year.

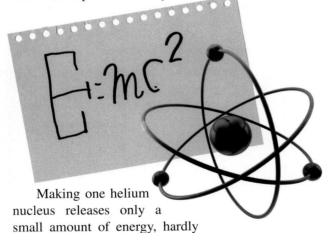

Making one helium nucleus releases only a small amount of energy, hardly enough to raise a housefly $1/40$ of a millimeter into the air. Only by concentrating many reactions in a small region can nature produce significant amounts of energy. The sun has a voracious appetite and needs 10^{38} reactions per second, transforming 5 million tons of mass into energy every second, just to replace the energy pouring into space from its surface. That might sound as if the sun is losing mass at a furious rate, but in its entire 10-billion-

Figure 7.2

The red line in this graph shows the binding energy (the energy that holds an atomic nucleus together) for different types of atoms as it depends on atomic mass (the number of protons and neutrons in an atom's nucleus). Both fission and fusion nuclear reactions convert fuel nuclei to product nuclei that are lower in the diagram (arrows), with more tightly bound nuclei. Iron has the most tightly bound nucleus, so no nuclear reactions can begin with iron and release energy.

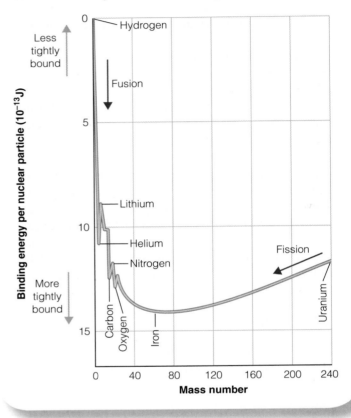

year lifetime, the sun will convert less than 0.07 percent of its mass into energy.

Fusion reactions can occur only when the nuclei of two atoms get very close to each other. Because atomic nuclei carry positive charges, they repel each other with an electrostatic force called the Coulomb force. Physicists commonly refer to this repulsion between nuclei as the **Coulomb barrier**. To overcome this barrier, atomic nuclei must have violent collisions that are rare unless the gas is very hot, in which case the nuclei move at high speeds. (You recall that an object's temperature is just a measure of the average speed with which its particles move.)

Thus nuclear reactions in the sun take place only near the center, where the gas is hot and dense. A high temperature ensures that collisions between nuclei are violent enough to overcome the Coulomb barrier, and a high density ensures that there are enough collisions, and thus enough reactions, to produce enough energy to keep the sun stable.

You can symbolize this process with a simple nuclear reaction:

$$4\,^1\text{H} \rightarrow {}^4\text{He} + \text{energy}$$

In this equation, ^1H represents a proton, the nucleus of a hydrogen atom, and ^4He represents the nucleus of a helium atom. The superscripts indicate the total numbers of protons and neutrons in each nucleus.

The actual steps in the process are more complicated than this simple summary suggests. Instead of waiting for four hydrogen nuclei to collide simultaneously, which would be a highly unlikely event, the process can proceed by steps in a series of reactions called the proton–proton chain.

The **proton–proton chain** is a series of three nuclear reactions that builds a helium nucleus by adding protons one at a time. This process is efficient at temperatures above 10,000,000 K. The sun, for example, manufactures its energy in this way. Recall from the previous section that models of the interior of the sun based on its overall stability indicate the central temperature is about 15,000,000 K.

The three reactions in the proton–proton chain are:

$$^1\text{H} + {}^1\text{H} \rightarrow {}^2\text{H} + e^+ + \nu$$

$$^2\text{H} + {}^1\text{H} \rightarrow {}^3\text{He} + \gamma$$

$$^3\text{He} + {}^3\text{He} \rightarrow {}^4\text{He} + {}^1\text{H} + {}^1\text{H}$$

In the first reaction, two hydrogen nuclei (two protons) combine, and one changes into a neutron, to result in a heavy hydrogen nucleus called **deuterium**, while emitting a particle called a *positron* (a positively charged electron, symbolized by e^+) and another called a **neutrino** (ν). In the second reaction, the heavy hydrogen nucleus absorbs another proton and, with the emission of a gamma ray (γ), becomes a lightweight helium nucleus (^3He). Finally, two light helium nuclei combine to form a normal helium nucleus and two hydrogen nuclei. Because the last reaction needs two ^3He nuclei, the first and second reactions must occur twice for each ^4He produced, as

in Figure 7.3. The net result of this chain reaction is the transformation of four hydrogen nuclei into one helium nucleus plus energy.

All main-sequence stars fuse hydrogen into helium to generate energy. The sun and smaller stars fuse hydrogen by the proton–proton chain. Upper-main-sequence stars, more massive than the sun, fuse hydrogen by a more efficient process called the **CNO(carbon-nitrogen-oxygen) cycle**. The CNO cycle begins with a carbon nucleus and transforms it first into a nitrogen nucleus, then into an oxygen nucleus, and then back to a carbon nucleus. The carbon is unchanged in the end, but along the way four hydrogen nuclei are fused to make a helium nucleus plus energy, just as in the proton–proton chain. A carbon nucleus has six times more positive electric charge than hydrogen, so the Coulomb barrier is higher than for combining two protons. Temperatures higher than 16,000,000 K are required to make the CNO cycle work: The center of the sun is not quite hot enough for this reaction. In

Proton-proton chain: A series of three nuclear reactions that builds a helium atom by adding together protons. The main energy source in the sun.

Deuterium: An isotope of hydrogen in which the nucleus contains a proton and a neutron.

Neutrino: A neutral, nearly massless atomic particle that travels at or nearly at the speed of light.

CNO (carbon-nitrogen-oxygen) cycle: A series of nuclear reactions that use carbon as a catalyst to combine four hydrogen nuclei to make one helium nucleus plus energy, effective in stars more massive than the sun.

Figure 7.3

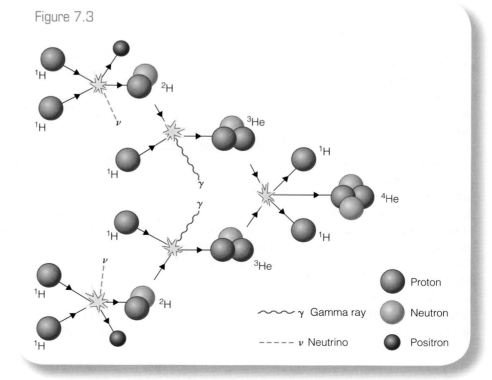

stars more massive than about 1.1 solar masses, the cores are hot enough and the CNO cycle dominates over the slower proton–proton chain.

In both the proton–proton chain and the CNO cycle, energy appears in the form of gamma rays, positrons, and neutrinos. The gamma rays are photons that are absorbed by the surrounding gas before they can travel more than a few centimeters. This heats the gas. The positrons produced in the first reaction combine with free electrons, and both particles vanish, converting their mass into more gamma rays. Thus, the positrons also help keep the center of the star hot. The neutrinos, however, are particles that travel at nearly the speed of light and almost never interact with other particles. The average neutrino could pass unhindered through a lead wall more than 1 light-year thick. Consequently, the neutrinos do not help heat the gas but race out of the star, carrying away roughly 2 percent of the energy produced.

It is time to ask the critical question that lies at the heart of science: What is the evidence to support this theoretical explanation of how the sun and other stars make energy? The search for that evidence will introduce you to one of the great triumphs of modern astronomy.

Neutrinos from the Sun's Core

Nuclear reactions in the sun's core produce floods of neutrinos that rush out of the sun and off into space. If you could detect these neutrinos, you could probe the sun's interior.

Because neutrinos almost never interact with atoms, neutrinos are extremely hard to detect. You never feel the flood of over 10^{12} solar neutrinos that flows through your body every second. Even at night, neutrinos from the sun rush through Earth as if it weren't there, up through your bed, through you, and onward into space. Certain particle reactions, however, can be triggered by a neutrino of the right energy.

Beginning in the 1960s astronomers found several ways to detect solar neutrinos. Initial results, however, detected too few neutrinos—about one-third as many as predicted by models of the sun's interior. Later experiments were finally able to solve the puzzle and confirm that the original models were correct. Physicists know of three kinds of neutrinos, which they call "flavors." The first neutrino experiments could detect (or taste) only one flavor, electron-neutrinos. Theory hinted that the electron-neutrinos produced in the core of the sun might change back and forth among the three flavors as they rushed out through the sun and across space to Earth. Observations begun in

Figure 7.4
The Sudbury Neutrino Observatory is a globe 12 meters in diameter containing water rich in deuterium in place of ordinary hydrogen. Buried 2,100 m (6,800 ft) deep in an Ontario mine, it can detect all three flavors of neutrinos and confirms that neutrinos change back and forth between the flavors.

Photo courtesy of SNO

2000, using the facility in Figure 7.4, confirm this explanation. Two-thirds of the electron neutrinos produced in the sun transform into muon- and tau-neutrinos, which most detectors cannot count, before they reach Earth. You can be confident that models of the sun's interior are basically correct because they predict the true rate of proton–proton fusion in the sun's core.

The Pressure-Temperature Thermostat

Nuclear reactions in stars manufacture energy and heavy atoms under the supervision of a natural thermostat that keeps the reactions from erupting out of control. That thermostat is the relation between gas pressure and temperature.

In a star, the nuclear reactions generate just enough energy to balance the inward pull of gravity. Consider what would happen if the reactions began to produce too much energy. The star balances gravity by generating energy, so the extra energy flowing out of the star would force it to expand. The expansion would lower the central temperature and density and slow the nuclear reactions until the star regained stability. Thus the star has a built-in regulator that keeps the nuclear reactions from occurring too rapidly.

The same thermostat keeps the reactions from dying down. Suppose the nuclear reactions began making too little energy. Then the star would contract slightly, increasing the central temperature and density and increasing the nuclear energy generation.

The overall stability of a star depends on the relation between gas pressure and temperature. If the material of the star has the property of normal gases—for which an increase or decrease in temperature produces a corresponding change in pressure—then the nuclear reaction pressure–temperature thermostat can function properly and contribute to the stability of the star. In the next section of this chapter you will see how this thermostat accounts for the relation between mass and luminosity for main-sequence stars.

7.3 Main-Sequence Stars

You can understand the deepest secrets of the stars by looking at the most obvious feature of the H-R diagram—the main sequence. Roughly 90 percent of all the stars that make energy by nuclear fusion are main-sequence stars, and they obey a simple relationship that is the key to understanding the life and death of stars.

The Mass–Luminosity Relation

You learned in Chapter 6 that observations of the temperature and luminosity of stars show that main-sequence stars obey a simple rule—the more massive a star is, the more luminous it is. That rule is the key to understanding the stability of main-sequence stars. In fact, the mass–luminosity relation is predicted by the theories of stellar structure, giving astronomers direct observational confirmation of those theories.

To understand the mass–luminosity relation, you can consider both the law of hydrostatic equilibrium, which says that pressure balances weight, and the pressure–temperature thermostat, which regulates energy production. A star that is more massive than the sun has more weight pressing down on its interior, so the interior must have a high pressure to balance that weight. That means the massive star's automatic pressure–temperature thermostat must keep the gas in its interior hot and the pressure high. A star less massive than the sun has less weight on its interior and thus needs less internal pressure; therefore, its pressure–temperature thermostat is set lower. In other words, massive stars are more luminous because they must make more energy to support their larger weight. If they were not so luminous, they would not be stable.

> **Brown dwarf:** A stellar object with such low mass that it cannot raise its central temperature high enough to sustain hydrogen fusion.

Brown Dwarfs

The mass–luminosity relation also tells you why the main sequence has a lower end, a minimum mass. Objects with masses less than about 0.08 of the sun's mass cannot raise their central temperature high enough to sustain hydrogen fusion. Called **brown dwarfs**, such objects are only about ten times the diameter of Earth—about the same size as Jupiter—and may still be warm from the processes of formation, but they do not generate energy by hydrogen fusion. They have contracted as much as they can and are slowly cooling off.

Brown dwarfs fall in the gap between low-mass M stars and massive planets like Jupiter. They would look dull orange or red to your eyes—which is why they are labeled "brown"—because they emit most of their energy in the infrared. The warmer brown dwarfs fall in spectral class L and the cooler in spectral class T (see Chapter 6). Because they are so small and cool, brown dwarfs are very low-luminosity objects and thus are difficult to find. Nevertheless, hundreds are known, and they may be as common as M stars.

The Life of a Main-Sequence Star

While a star is on the main sequence, it is stable, so you might think its life would be totally uneventful. But a main-sequence star balances its gravity by fusing hydrogen, and as the star gradually uses up its fuel, that balance

Figure 7.5

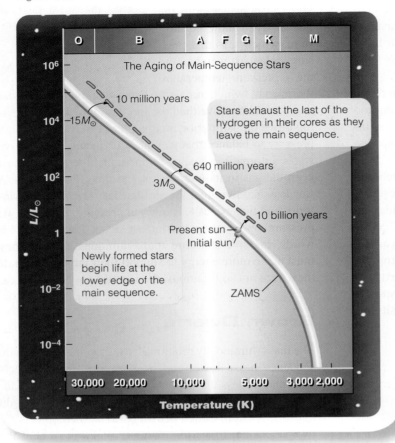

The Aging of Main-Sequence Stars

10 million years

−15M☉

Stars exhaust the last of the hydrogen in their cores as they leave the main sequence.

640 million years

3M☉

10 billion years

Present sun

Initial sun

Newly formed stars begin life at the lower edge of the main sequence.

ZAMS

Temperature (K)

These gradual changes during the lifetimes of main-sequence stars mean that the main sequence is not a sharp line across the H-R diagram, but rather a band. Stars begin their stable lives fusing hydrogen on the lower edge of this band, which is known as the **zero-age main sequence (ZAMS)**, but gradual changes in luminosity and surface temperature move the stars upward and slightly to the right, as shown in Figure 7.5. By the time they reach the upper edge of the main sequence, they have exhausted nearly all the hydrogen in their centers. If you precisely measure and plot the luminosity and temperature of main-sequence stars on the H-R diagram, you will find them at various positions within the main-sequence band, indicating how much hydrogen has been converted to helium in their cores. Therefore, you can use the position of a star in the band, combined with stellar evolution models, as one way to estimate the star's age.

The sun is a typical main-sequence star; and, as it undergoes these gradual changes, Earth will suffer. When the sun began its main-sequence life about 5 billion years ago, it was only about 70 percent as luminous as it is now, and by the time it leaves the main sequence in another 5 billion years, the sun will have twice its present luminosity. Long before that, the rising luminosity of the sun will raise Earth's average temperature, melt the polar caps, modify Earth's climate, and ultimately boil away and destroy Earth's oceans. Life on Earth will probably not survive these changes in the sun, but humans have a billion years or more to prepare.

The average star spends 90 percent of its life on the main sequence. This explains why 90 percent of all true stars are main-sequence stars—you are most likely to see a star during that long, stable period while it is on the main sequence. The number of years a star spends on the main sequence depends on its mass (Table 7.2). Massive stars consume fuel rapidly and live short lives, but low-mass stars conserve their fuel and shine for billions of years. For example, a 25-solar-mass star will exhaust its hydrogen and die in only about 7 million years. The sun has enough fuel to last about 10 billion years. The red dwarfs, although they have little fuel, use it up very slowly and may be able to survive for 100 billion years or more.

must change. Thus, even the stable main-sequence stars are changing as they consume their hydrogen fuel.

Recall that hydrogen fusion combines four nuclei into one. As a main-sequence star fuses its hydrogen, the total number of particles in its interior decreases. Each newly made helium nucleus can exert only the same pressure as a hydrogen nucleus. Because the gas has fewer nuclei, its total pressure is less. This unbalances the gravity–pressure stability, and gravity squeezes the core of the star more tightly. As the core contracts, its temperature increases and the nuclear reactions run faster, releasing more energy. This additional energy flowing outward forces the outer layers to expand. Therefore, as a main-sequence star slowly turns hydrogen into helium in its core, the core contracts and heats up while the outer parts of the star become larger, the star becomes more luminous, and its surface cools down.

Zero-age main sequence (ZAMS): The location in the H-R diagram where stars first reach stability as hydrogen-burning stars.

Table 7.2 Main-Sequence Stars

Spectral Type	Mass (Sun = 1)	Luminosity (Sun = 1)	Approximate Years on Main Sequence
O5	40	400,000	1×10^6
B0	15	13,000	11×10^6
A0	3.5	80	440×10^6
F0	1.7	6.4	3×10^9
G0	1.1	1.4	8×10^9
K0	0.8	0.46	17×10^9
M0	0.5	0.08	56×10^9

7.4 The Birth of Stars

The key to understanding star formation is the correlation between young stars, short-lived stars, and interstellar clouds of gas and dust. Where you find the youngest groups of stars, as well as stars with very short lifetimes, you also find large clouds of gas and dust illuminated by the hottest and brightest of the new stars, as you can see on the next page in Figure 7.6. This should lead you to suspect that stars form from such clouds.

The Interstellar Medium

A common misperception is to imagine that space is empty—a vacuum. In fact, if you glance at Orion's sword on a winter evening, you can see the Great Nebula in Orion, a glowing cloud of gas and dust. That cloud and others are prominent examples of the low-density gas and dust between the stars called the **interstellar medium (ISM)**.

About 75 percent of the mass of interstellar gas is hydrogen, and 25 percent is helium; there are also traces of carbon, nitrogen, oxygen, calcium, sodium, and heavier atoms. Notice that this is very similar to the composition of the sun (see Chapter 5) and other stars. Roughly 1 percent of the ISM's mass is made up of microscopic particles called **interstellar dust**. The dust particles are

about the size of the particles in cigarette smoke. The dust seems to be made mostly of carbon and silicates (rocklike minerals) mixed with or coated with frozen water plus some organic compounds.

This interstellar material is not uniformly distributed through space; it consists of a complex tangle of cool, dense clouds pushed and twisted by currents of hot, low-density gas. Although the cool clouds contain only 10 to 1,000 atoms/cm³ (fewer than in any laboratory vacuum on Earth), astronomers refer to them as dense clouds in contrast with the hot, low-density gas that fills the spaces between clouds. That thin gas contains only about 0.1 atom/cm³.

How do you know what the properties of the interstellar medium are? As you just learned, the ISM is in some cases easily visible as clouds of gas and dust, as in, for example, the Great Nebula in Orion. Astronomers call such a cloud a **nebula** from the Latin word for "cloud." Such nebulae (plural) give clear evidence of an interstellar medium. Study the "Three Kinds of Nebulae" on pages 134–135 and notice three important points:

1. First, very hot stars can excite clouds of gas and dust to emit light, and this reveals that the clouds contain mostly hydrogen gas at very low densities.

2. Second, where dusty clouds reflect the light of stars you see evidence that the dust in the clouds is made up of very small particles.

3. The third thing to notice is that some dense clouds of gas and dust are dark and are detectable only where they are silhouetted against background regions filled with stars or bright nebulae.

If a cloud is less dense, the starlight may be able to penetrate it, and stars can be seen through the cloud; but the stars look dimmer because the dust in the cloud scatters some of the light. Because shorter wavelengths are scattered more easily than longer wavelengths, the redder photons are more likely to make it through the cloud, and the stars look redder than they should for their respective spectral types. This effect is called **interstellar reddening** and is illustrated in Figure 7.7 on page 133.

Interstellar medium (ISM): The gas and dust distributed between the stars.

Interstellar dust: Microscopic solid grains in the interstellar medium.

Nebula: A relatively dense cloud of interstellar gas and dust.

Interstellar reddening: The process in which dust scatters blue light out of starlight and makes the stars look redder.

The same physical process makes the setting sun look red. The fact that distant stars are dimmed and reddened by intervening gas and dust is clear evidence of an interstellar medium.

Although you have been imagining individual nebulae, the thin gas and dust of the interstellar medium also fills the spaces between the nebulae. You can see evidence of that because the gas forms interstellar absorption lines in the spectra of distant stars, as shown in Figure 7.8. As starlight travels through the interstellar medium, gas atoms of elements such as calcium and sodium absorb photons of certain wavelengths, producing absorption lines. You can be sure those lines originate in the interstellar medium because they appear in the spectra of O and B stars—stars too hot to have visible calcium and sodium absorption lines in their own atmospheres. Also, the narrowness of the interstellar lines indicates they could not have been formed in the hot atmospheres of stars—the line widths indicate a very small range of Doppler shifts, meaning the atoms are moving at speeds corresponding to temperatures of only 10 to 50 K.

Observations at nonvisible wavelengths provide valuable evidence about the interstellar medium. Infrared observations can directly detect dust in the interstellar medium. Although the dust grains are very small and very cold, there are huge numbers of grains in a cloud, and each grain emits infrared radiation at long wavelengths, as shown in Figure 7.9a on page 136. Furthermore, some molecules in the cold gas emit in the infrared, so infrared observations can detect very cold clouds of gas. Also, infrared light penetrates ISM dust better than shorter-wavelength radiation, so astronomers can use infrared cameras to look into and through interstellar clouds that are opaque to visible light. X-ray observations can detect regions of very hot gas apparently produced by exploding stars, like those in the constellation Cygnus. Radio observations reveal the emissions of specific molecules in the interstellar medium—the equivalent of emission lines in visible light. Such studies show that some of the atoms in space have linked together to form molecules.

There is no shortage of evidence concerning the interstellar medium. Clearly, space is not empty, and from the correlation between young stars and clouds of gas and dust you can suspect that stars form from these clouds.

The Formation of Stars from the Interstellar Medium

To study the formation of stars, you can continue comparing theory with evidence. The theory of gravity predicts that the combined gravitational attraction of the atoms in a cloud of gas will shrink the cloud, pulling every

Figure 7.6

Young stars are found in clouds of gas and dust from which they have been born. The nebula N44 is 170,000 ly from Earth in a nearby galaxy, and the Horsehead Nebula is only about 1,500 ly distant in our own galaxy. Gas in both nebulae is excited to glow by hot, young stars. Interstellar dust is visible as dark, twisting clouds seen against the bright background gas.

Nebula N44

Roughly 40 young stars are inflating a bubble of hot gas inside the nebula from which they formed.

100 ly

Visual-wavelength image

Horsehead Nebula

Dusty foreground gas silhouetted against glowing gas illuminated by hot, young stars.

Young star embedded in the nebula.

5 ly

Visual-wavelength image

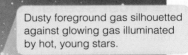

Figure 7.7

Interstellar reddening makes stars seen through a cloud of gas and dust look redder than they should because shorter wavelengths are more easily scattered. If the gas and dust is especially dense, no stars are visible through the cloud at visual wavelengths except near the edges. At longer near-infrared wavelengths, many stars can be detected behind the cloud.

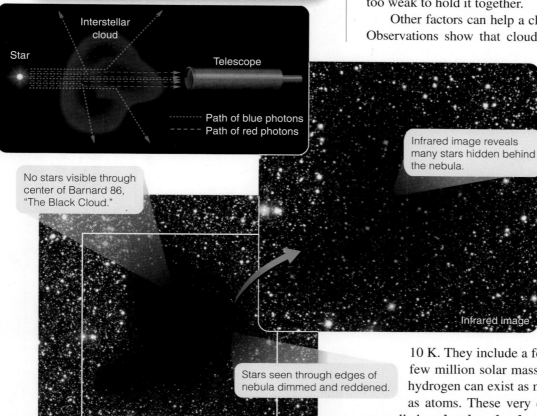

Star

Interstellar cloud

Telescope

······· Path of blue photons
- - - - Path of red photons

No stars visible through center of Barnard 86, "The Black Cloud."

Infrared image reveals many stars hidden behind the nebula.

Infrared image

Stars seen through edges of nebula dimmed and reddened.

Visual-wavelength image

© ESO

Figure 7.8

Interstellar absorption lines can be recognized easily because they are much narrower than the spectral lines produced in stellar atmospheres. The multiple interstellar lines in this spectrum are produced by separate interstellar clouds with slightly different radial velocities and are seen superimposed on a broad stellar line.

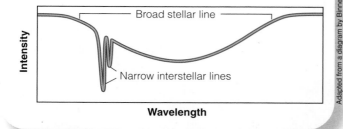

Adapted from a diagram by Binnendijk.

Broad stellar line

Intensity

Narrow interstellar lines

Wavelength

atom toward the center. That might lead you to expect that every cloud would eventually collapse and become a star; however, the thermal energy in the cloud resists collapse. Interstellar clouds are very cold; but, even at a temperature of only 10 K, the average hydrogen atom moves at about 0.33 km/s (740 mph). That much thermal motion would make the cloud drift apart if gravity were too weak to hold it together.

Other factors can help a cloud resist its own gravity. Observations show that clouds are turbulent with currents of gas pushing through and colliding with each other. Also, magnetic fields in clouds may resist being squeezed. The thermal motion of the atoms, turbulence in a cloud, and magnetic fields resist gravity, and only the densest clouds are likely to contract.

The densest interstellar clouds contain from 10^3 to 10^5 atoms/cm^3 and have temperatures as low as 10 K. They include a few hundred thousand to a few million solar masses. In such dense clouds, hydrogen can exist as molecules (H_2) rather than as atoms. These very densest parts of the ISM are called **molecular clouds**, and the largest are called giant molecular clouds. Although hydrogen molecules cannot be detected by radio telescopes, the clouds can be mapped by the emission lines of carbon monoxide molecules (CO) present in small amounts in the gas. By now you may have realized the main point: Stars can form inside molecular clouds when the densest parts of the clouds become unstable and contract under the influence of their own gravity.

Most clouds do not appear to be gravitationally unstable and will not contract to form stars on their own. However, a stable cloud colliding with a **shock wave** (the astronomical equivalent of a sonic boom) can be compressed and disrupted into

Molecular cloud: A dense interstellar gas cloud in which atoms are able to link together to form molecules such as H_2 and CO.

Shock wave: A sudden change in pressure that travels as an intense sound wave.

Three Kinds of Nebulae

1 **Emission nebulae** are produced when a hot star excites the gas near it to produce an emission spectrum. The star must be hotter than about B1 (25,000 K). Cooler stars do not emit enough ultraviolet radiation to ionize the gas. Emission nebulae have a distinctive pink color produced by the blending of the red, blue, and violet Balmer lines. Emission nebulae are also called HII regions, following the custom of naming gas with a roman numeral to show its state of ionization. HI is neutral hydrogen, and HII is ionized.

In an HII region, the ionized nuclei and free electrons are mixed. When a nucleus captures an electron, the electron falls down through the atomic energy levels, emitting photons at specific wavelengths. Spectra indicate that the nebulae have compositions much like that of the sun – mostly hydrogen. Emission nebulae have densities of 100 to 1000 atoms per cubic centimeter, better than the best vacuums produced in laboratories on Earth.

Visual-wavelength image

European Southern Observatory

2 A **reflection nebula** is produced when starlight scatters from a dusty nebula. Consequently, the spectrum of a reflection nebula is just the reflected absorption spectrum of starlight. Gas is surely present in a reflection nebula, but it is not excited to emit photons. See image below.

Reflection nebulae NGC 1973, 1975, and 1977 lie just north of the Orion Nebula. The pink tints are produced by ionized gases deep in the nebulae.

Visual-wavelength image

Anglo-Australian Observatory/David Malin Images

2a Reflection nebulae look blue for the same reason the sky looks blue. Short wavelengths scatter more easily than long wavelengths. See image below.

Sunlight enters Earth's atmosphere

Blue photons are scattered more easily than longer wavelengths, and blue photons enter your eyes from all directions, making the sky look blue.

2b The blue color of reflection nebulae at left shows that the dust particles must be very small in order to preferentially scatter the blue photons. Interstellar dust grains must have diameters ranging from 0.01 mm down to 100 nm or so.

2c In the star cluster called the Pleiades, the hottest stars are B6, not hot enough to ionize hydrogen in the interstellar medium. Instead, the brightest stars produce a reflection nebula as their light is scattered from interstellar dust.

Visual-wavelength image

Anglo-Australian Observation

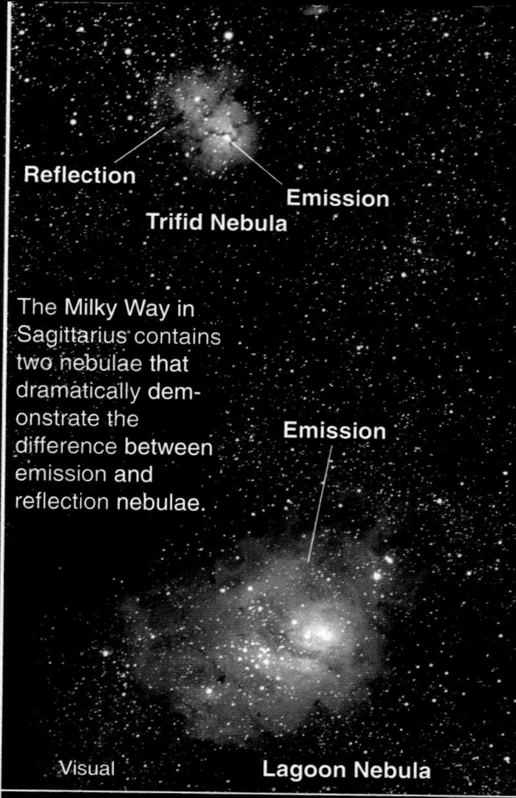

Anglo-Australian Observatory/
David Malin Images

Reflection

Emission

Trifid Nebula

The Milky Way in Sagittarius contains two nebulae that dramatically demonstrate the difference between emission and reflection nebulae.

Emission

Visual **Lagoon Nebula**

Daniel Good

Dark Nebula Barnard 86

Star Cluster NGC6520

Visual

Anglo-Australian Observatory/David Malin Images

3 **Dark nebulae** are dense clouds of gas and dust that obstruct the view of more distant stars. Some are generally round, but others are twisted and distorted, as shown at the left, suggesting that even when there are no nearby stars to ionize the gas or produce a reflection nebula, there are breezes and currents pushing through the interstellar medium.

Northern Coalsack

Cygnus

Milky Way

Great Rift

Twisted by intense light from nearby stars, this dark nebula is visible because it obscures more distant stars.

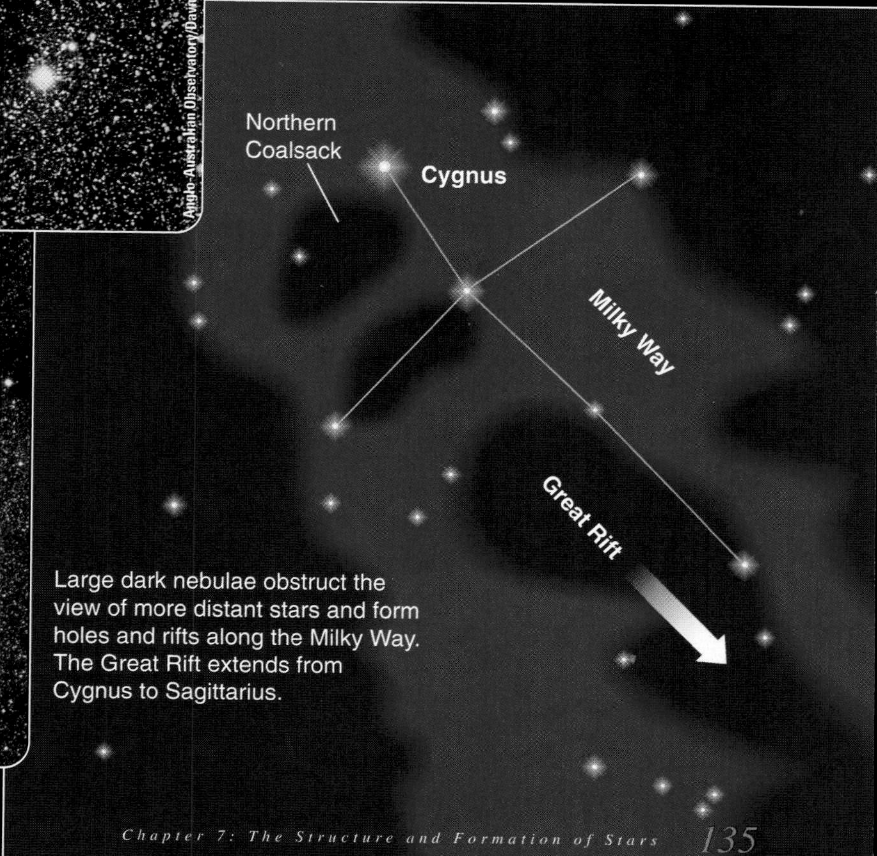

Visual-wavelength image

Large dark nebulae obstruct the view of more distant stars and form holes and rifts along the Milky Way. The Great Rift extends from Cygnus to Sagittarius.

fragments. Theoretical calculations show that some of these fragments can become dense enough to collapse under the influence of their own gravity and form stars, as in Figure 7.10.

Supernova explosions (exploding stars described in the next chapter) produce shock waves that compress the interstellar medium, and recent observations show young stars forming at the edges of such shock waves. Another source of shock waves may be the birth of very hot stars. A massive star is so luminous and hot that it emits vast amounts of ultraviolet photons. When such a star is born, the sudden blast of light, especially ultraviolet radiation, can ionize and drive away nearby gas, forming a shock wave that could compress nearby clouds and trigger further star formation. Even the collision of two interstellar clouds can produce a shock wave and trigger star formation. Some of these processes are shown in Figure 7.11 on page 138.

Although these are important sources of shock waves, the dominant trigger of star formation in our galaxy may be the spiral pattern itself. In Chapter 1, you learned that our galaxy contains spiral arms. As interstellar clouds encounter these spiral arms, the clouds are compressed, and star formation can be triggered (see Chapter 9).

Once begun, star formation can spread like a grass fire. Both high-mass and low-mass stars form together, but when the massive stars form, their intense radiation or eventual supernova explosions push back the surrounding gas and compress it. This compression in turn can trigger the formation of more stars, some of which will be massive. Thus, a few massive stars can drive a continuing cycle of star formation in a giant molecular cloud.

A collapsing cloud of gas does not form a single star; because of instabilities, the cloud breaks into fragments, producing perhaps thousands of stars. A stable group of stars that formed and are held together by their combined gravity is called a **star cluster**. In contrast, a **stellar association** is a group of stars that formed together but are not gravitationally bound to one another. The stars in an association drift away from each other in a few million years. The youngest associations are rich in young stars, including O and B stars. O and B stars have lifetimes so short in astronomical terms (Table 7.2) that they must be located close to their birthplaces, so these stars serve as brilliant signposts pointing to regions of star formation.

The Formation of Protostars

You might be wondering how the unimaginably cold gas of an interstellar cloud can heat up to form a star. The answer is gravity.

Once part of a cloud is triggered to collapse, gravity draws each atom toward the center. At first the atoms fall

Figure 7.9

Infrared and X-ray observations reveal different parts of the interstellar medium. (a) This far-infrared image shows the infrared cirrus, wispy clouds of very cold dust and gas spread all across the sky. (b) The Cygnus Superbubble, detectable in X rays, triggers star formation where it pushes into the surrounding gas.

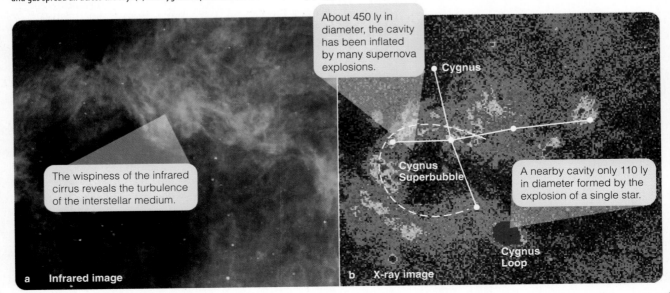

About 450 ly in diameter, the cavity has been inflated by many supernova explosions.

Cygnus

Cygnus Superbubble

A nearby cavity only 110 ly in diameter formed by the explosion of a single star.

Cygnus Loop

The wispiness of the infrared cirrus reveals the turbulence of the interstellar medium.

a **Infrared image**

b **X-ray image**

© NASA/IPAC/Courtesy Deborah Levine / Courtesy of Steve Snowden and Max-Planck Institute for Extraterrestrial Physics, Germany

unopposed; they hardly ever collide with each other. In this free-fall contraction, the atoms pick up speed as they fall until, by the time the gas becomes dense enough for the atoms to collide often, they are traveling very fast. Collisions convert the inward velocities of the atoms into random motions, and you remember that temperature is a measure of the random velocities of the atoms in a gas. Thus, the temperature of the gas goes up.

The initial collapse of the gas forms a dense core of gas, and as more gas falls in, a warm **protostar** develops, buried deep in the dusty gas cloud that continues to contract, although much more slowly than free-fall. A protostar is an object that will eventually become a star. Theory predicts that protostars are luminous red objects larger than main-sequence stars, with temperatures ranging from a few hundred to a few thousand degrees K.

Throughout its contraction, the protostar converts its gravitational energy into thermal energy. Half of this thermal energy radiates into space, but the remaining half raises the internal temperature. As the internal temperature climbs, the gas becomes ionized, changing into a mixture of positively charged atomic nuclei and free electrons. When the center gets hot enough, nuclear reactions begin generating enough energy to replace the radiation leaving the surface of the star. The protostar then halts its contraction and becomes a stable, main-sequence star.

The time a protostar takes to contract from a cool interstellar gas cloud to a main-sequence star depends on its mass. The more massive the star, the stronger its gravity and the faster it contracts. The sun took about 30 million years to reach the main sequence, but a 15-solar-mass star can contract in only 100,000 years. Conversely, a star of 0.2 solar mass takes 1 billion years to reach the main sequence.

Observations of Star Formation

By understanding what the interstellar medium is like and by understanding how the laws of physics work, astronomers have been able to work out the story of how stars must form. But you can't accept a scientific theory without testing it, and that means you must compare the theory with the evidence.

The protostar stage is predicted to be less than 0.1 percent of a star's total lifetime, so, although that is a long time in human terms, you cannot expect to find many stars in the protostar stage. Furthermore, protostars form deep inside clouds of dusty gas that absorb any light the protostar might emit. Only when the protostar is hot

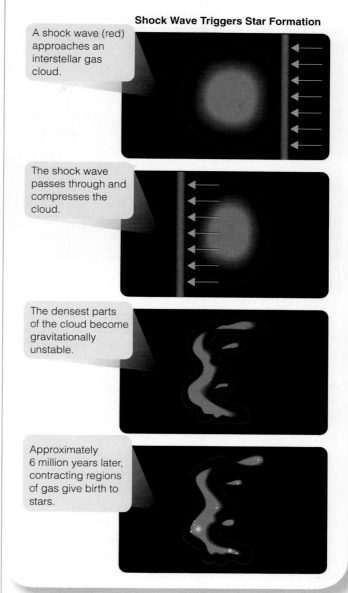

Figure 7.10

Shock Wave Triggers Star Formation

A shock wave (red) approaches an interstellar gas cloud.

The shock wave passes through and compresses the cloud.

The densest parts of the cloud become gravitationally unstable.

Approximately 6 million years later, contracting regions of gas give birth to stars.

enough to drive away its enveloping cloud of gas and dust does it become easy to observe at wavelengths your eye can see.

Although astronomers cannot easily observe protostars at visible wavelengths, protostars can be detected in the infrared. Protostars are cooler than stars and radiate as blackbodies predominantly at infrared wavelengths. The dust of the surrounding interstellar cloud absorbs light from the protostar and grows warm, and that warm dust radiates additional large amounts of infrared radiation. Infrared radiation

Protostar: A collapsing cloud of gas and dust destined to become a star.

These massive stars were triggered into formation by compression from the formation of earlier stars out of the image to the left.

New stars are forming in these dense clouds because of compression from the stars to the left.

a Visual-wavelength image

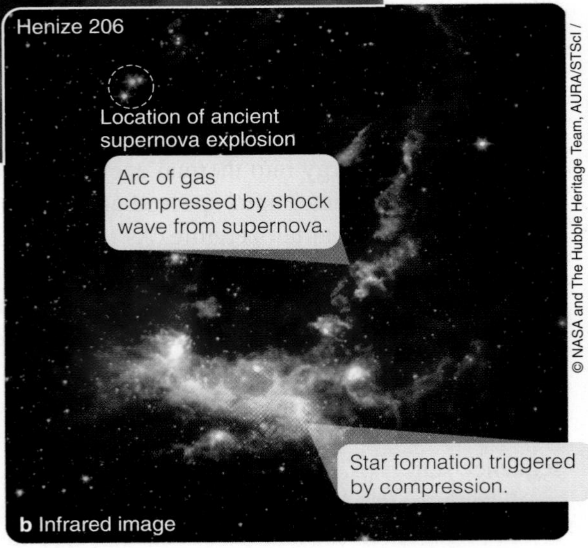

Henize 206

Location of ancient supernova explosion

Arc of gas compressed by shock wave from supernova.

Star formation triggered by compression.

b Infrared image

© NASA and The Hubble Heritage Team, AURA/STScI /
© NASA/JPL-CalTech/V. Gorjian and NOAO

Figure 7.11

(a) The blast of light and ultraviolet radiation from an earlier generation of massive stars has compressed neighboring gas and triggered the formation of more stars. Those stars are now triggering the birth of a third generation. (b) The explosion of a supernova a few million years ago produced a shock wave that is triggering star formation where it slams into nearby clouds of gas and dust.

also can penetrate the dust to reach us. Thus, infrared observations reveal many bright sources of radiation that are protostars buried in interstellar clouds.

You can be sure that star formation is going on right now because astronomers find regions containing massive stars with lifetimes so short they must still be in their birthplaces. Small, very dense zones of gas and dust called **Bok globules** within larger nebulae probably represent the very first stages of star formation. Observations of jets coming from hidden protostars show that protostars are often surrounded by disks of gas and dust, as in Figure 7.12. These jets appear to produce small flickering nebulae called **Herbig–Haro objects**. The same regions contain many stars with temperatures and luminosities that correspond to model predictions of the properties of very young stars. These include **T Tauri stars**, protostars with about the mass of the

Bok globule: Small, dark cloud only about 1 ly in diameter that contains 10 to 1,000 solar masses of gas and dust, thought to be related to star formation.

Herbig–Haro object: A small nebula that varies irregularly in brightness, evidently associated with star formation.

T Tauri star: A young star surrounded by gas and dust, understood to be contracting toward the main sequence.

sun that show signs of active chromospheres as you might expect from young rapidly rotating stars with strong magnetic dynamos (see Chapter 5). T Tauri stars are also often surrounded by thick disks of gas and dust that are very bright at infrared wavelengths.

These observations all confirm the theoretical models of contracting protostars. Although star formation still holds many puzzles, the general process seems clear. In at least some cases, interstellar gas clouds are compressed by passing shock waves, and the clouds' gravity, acting unopposed, draws the matter inward to form protostars. Now you know how to make a star: Just find some way to compress more than 0.08 solar masses of interstellar material into a small enough region of space, and gravity will do the rest of the job.

The Orion Nebula

On a clear winter night, you can see with your unaided eye the Great Nebula of Orion as a fuzzy wisp in Orion's sword. With binoculars or a small telescope it is striking, and through a large telescope it is breathtaking. At the

Figure 7.12

Visible light image of the Herbig-Haro object HH 47, a jet of gas about half a light year long. The wavy nature of the jet suggests that it is perturbed by a second body, perhaps a binary companion to the new star.

center lie four brilliant blue-white stars known as the Trapezium, the brightest of a cluster of a few hundred stars. Surrounding the stars are the glowing filaments of a nebula more than 8 pc across. Like a great thundercloud illuminated from within, the churning currents of gas and dust suggest immense power. The significance of the Orion Nebula lies hidden, figuratively and literally, beyond the visible nebula. The region is ripe with star formation.

Evidence of Young Stars

You should not be surprised to find star formation in Orion. The constellation is a brilliant landmark in the winter sky because it is marked by hot blue stars. These stars are bright not because they are nearby but because they are tremendously luminous. These O and B stars cannot live more than a few million years, so you can conclude that they must have been born astronomically recently, near where they are seen now. Furthermore, the constellation contains large numbers of T Tauri stars, which are known to be young. Orion is rich with young stars.

The history of star formation in the constellation of Orion is written in its stars. The stars at Orion's west shoulder are about 12 million years old, whereas the stars of Orion's belt are about 8 million years old. The stars of the Trapezium at the center of the Great Nebula are no older than 2 million years. Apparently, star formation began near the west shoulder, and the massive stars that formed there triggered the formation of the stars you see in Orion's belt. That star formation may have triggered the formation of the stars you see in the Great Nebula. Like a grass fire, star formation has swept across Orion from northwest to southeast.

Study "Star Formation in the Orion Nebula" on pages 140–141 and notice four points:

1. The nebula you see is only a small part of a vast, dusty molecular cloud. You see the nebula because the larger stars born within it have ionized the nearby gas and driven it outward, breaking out of the much larger molecular cloud.

2. A single very hot and short-lived O star is almost entirely responsible for ionizing the gas and making the nebula glow. This massive star has already become a main sequence star while its smaller siblings are still protostars.

3. Infrared observations reveal ongoing star formation deep in the molecular cloud behind the visible nebula.

4. Finally, notice that many stars visible in the Orion Nebula are surrounded by disks of gas and dust. Such disks do not last long and are clear evidence that the stars are very young.

Observations of stars in the Orion nebula show that some of the gas and dust in the cloud from which a star forms settles into a disk. Make special note of these disks, because planets may form in them. In Chapter 12, you will study how our solar system formed in the disk of gas and dust that encircled the young sun.

RECAP

Now you can really see how the universe works. Here you have put together observations and theories to understand how nature makes stars. In doing so, you should be able to answer six essential questions about stars:

1. What are the insides of stars like?

2. How do stars make energy?

3. What determines the properties of main-sequence stars?

4. How long can a star survive?

5. How are stars born?

6. How do you know that theories of star formation are correct?

4×5 FILM

220 EPC SSO

Star Formation in the Orion Nebula

1 The visible Orion Nebula shown below is a pocket of ionized gas on the near side of a vast, dusty molecular cloud that fills much of the southern part of the constellation Orion. The molecular cloud can be mapped by radio telescopes. To scale, the cloud would be many times larger than this page. As the stars of the Trapezium were born in the cloud, their radiation has ionized the gas and pushed it away. Where the expanding nebula pushes into the larger molecular cloud, it is compressing the gas (see diagram at right) and may be triggering the formation of the protostars that can be detected at infrared wavelengths within the molecular cloud.

Hundreds of stars lie within the nebula, but only the four brightest, those in the Trapezium, are easy to see with a small telescope. A fifth star, at the narrow end of the Trapezium, may be visible on nights of good seeing.

The cluster of stars in the nebula is less than 2 million years old. This must mean the nebula is similarly young.

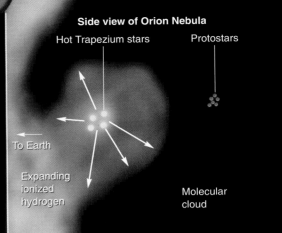

Side view of Orion Nebula

Hot Trapezium stars
Protostars
To Earth
Expanding ionized hydrogen
Molecular cloud

Infrared

The near-infrared image above reveals more than 50 low-mass, cool protostars.

Trapezium

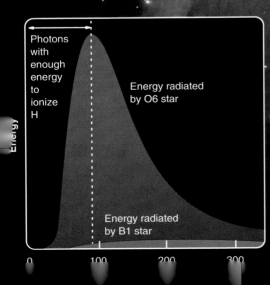

Visual

Small dark clouds called **Bok globules**, named after astronomer Bart Bok, are found in and near star-forming regions. The one pictured above is part of nebula NGC 1999 near the Orion Nebula. Typically about 1 light-year in diameter, they contain from 10 to 1000 solar masses.

Visual-wavelength image
Credit: NASA, ESA, M. Robberto, STScI and the Hubble Space Telescope Orion Treasury Project Team

Photons with enough energy to ionize H

Energy radiated by O6 star

Energy radiated by B1 star

Energy

0   100   200   300

2 Of all the stars in the Orion Nebula, only one is hot enough to ionize the gas. Only photons with wavelengths shorter than 91.2 nm can ionize hydrogen. The second-hottest stars in the nebula are B1 stars, and they emit little of this ionizing radiation. The hottest star, however, is an O6 star 30 times the mass of the sun. At a temperature of 40,000 K, it emits plenty of photons with wavelengths short enough to ionize hydrogen. Remove that one star, and the nebula would turn off its emission.

3 Below, a far-infrared image has been combined with an ultraviolet and visible image to reveal extensive nebulosity surrounding the visible Orion Nebula. Red and orange show the location of cold, carbon-rich gas molecules. Green areas outline hot, ionized gas around young stars. The infrared image reveals protostars buried in the gas cloud behind the visible nebula.

In this near-infrared image, known among some astronomers as the "Hand of God" image, fingers of gas rush away from the region of the infrared protostars.

Infrared

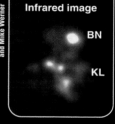

Infrared image

BN

KL

The Becklin-Neugebauer object (BN) is a hot B star just reaching the main sequence. It is not detectable at visual wavelengths. The Kleinmann-Low nebula (KL) is a cluster of cool young protostars also detectable only in the infrared.

The spectral types of the Trapezium stars are shown here. The gas looks green because of filters used to record the image.

Trapezium cluster

B3

B1

B1

O6

Visual-wavelength image

500 AU

4 As many as 85 percent of the stars in the Orion Nebula are surrounded by disks of gas and dust. One such disk is seen at the upper right of this Hubble Space Telescope image, magnified in inset. Radiation from the nearby hot Trapezium stars is evaporating gas from the disk and driving it away to form an elongated nebula.

Visual

Ultraviolet + visual + infrared image

Chapter 8

The Deaths
of Stars

Gravity is patient—so patient it can kill stars. Astronomers occasionally see what seems to be a new star in the sky that grows brighter, then fades away after a few weeks or a year. You will discover that a **nova**, an apparently new star in the sky, is in fact produced by an eruption around a stellar remnant, and that a **supernova**, a particularly luminous and long-lasting nova, is caused by the violent explosive death of a massive star. Modern astronomers find a few novae (plural of *nova*) each year, but supernovae (plural) are so rare that only one or two happen each century in our galaxy. In the previous chapter, you saw that stars resist their own gravity by generating energy through nuclear fusion. The energy keeps their interiors hot, and the resulting high pressure balances gravity and prevents the star from collapsing. Stars, however, have limited fuel. When they exhaust their fuel, gravity wins, and the stars die.

The mass of a star is critical in determining its fate. Massive stars use up their nuclear fuel at a furious rate and die after only a few million years. In contrast, the lowest-mass stars use their fuel sparingly and may be able to live hundreds of billions of years. Stars with different masses lead dramatically different lives and die in different ways.

8.1 Giant Stars

Nova: From the Latin, meaning "new," a sudden and temporary brightening of a star making it appear as a new star in the sky, evidently caused by an explosion of nuclear fuel on the surface of a white dwarf.

Supernova: A "new star" in the sky that is roughly 4000 times more luminous than a normal nova and longer lasting, evidently the result of an explosion of a star.

A main-sequence star generates its energy by nuclear fusion reactions that combine hydrogen to make helium. A star remains on the main sequence for a time span equal to 90 percent of its total existence as an energy-generating star. When the hydrogen is exhausted, however, the star begins to evolve rapidly. It swells into a giant star and then begins to fuse helium into heavier elements. A star can remain in this giant stage for only about 10 percent of its total lifetime; then it must die. The giant-star stage is the first step in the death of a star.

looking back

You are beginning to understand stars. The preceding chapter described how stars are born from clouds of gas and dust in the interstellar medium and how nuclear fusion maintains high temperatures and pressures inside stars, which keeps them from collapsing under the influence of their own gravity. Stars are the lighting system of the universe, but they can last only as long as their fuel supplies.

looking ahead

This chapter ends the story of individual stars, but it does not end the story of stars. In the next chapter, you will begin exploring the giant communities in which stars live—the galaxies.

Matter ejected repeatedly from the dying star at its center has formed the nebula known as the Cat's Eye.

astro

Expansion into a Giant

The nuclear reactions in a main-sequence star's core produce helium. Helium can fuse into heavier elements only at temperatures higher than 100,000,000 K, and no main-sequence star has a core that hot, so helium accumulates at the star's center like ashes in a fireplace. Initially, this helium ash has little effect on the star, but as hydrogen is exhausted and the stellar core becomes almost pure helium, the star's ability to generate nuclear energy is reduced. Because the energy generated at the center is what opposes gravity and supports the star, the core begins to contract as soon as the energy generation starts to decline.

Although the core of helium ash cannot generate nuclear energy, it can grow hotter because it contracts and converts gravitational energy into thermal energy. The rising temperature heats the unprocessed hydrogen just outside the core—hydrogen that was never previously hot enough to fuse. When the temperature of the surrounding hydrogen becomes high enough, hydrogen fusion begins in a spherical layer, called a shell, surrounding the exhausted core of the star. Like a grass fire burning outward from an exhausted campfire, the hydrogen-fusion shell creeps outward, leaving helium ash behind and increasing the mass of the helium core.

Figure 8.1 illustrates how the flood of energy produced by the hydrogen-fusion shell pushes toward the surface, heating the outer layers of the star and forcing them to expand dramatically. Stars like the sun become **giant stars** 10 to 100 times the present diameter of the sun, and the most massive stars become **supergiant stars** as much as 1,000 times larger than the sun.

The expansion of a star to giant or supergiant size cools the star's outer layers, and so the stars move toward the upper right in the H-R diagram. Look at Figure 6.8 and notice that some of the most familiar stars, such as Aldebaran, Betelgeuse, and Polaris, are giants or supergiants.

Although the energy output of the hydrogen-fusion shell can force the envelope of the star to expand, it cannot stop the contraction of the helium core. Because the core is not hot enough to fuse helium, gravity squeezes it to a relatively tiny size. If you represented a typical giant star as being the size of a baseball stadium, its helium core would be only about the size of a baseball, yet would contain about 10 percent of the star's mass.

Helium Fusion

As a star becomes a giant, fusing hydrogen in a shell, the inert core of helium ash contracts and grows hotter. When the core finally reaches a temperature of 100,000,000 K, helium nuclei can begin fusing to make carbon nuclei.

The ignition of helium in the core changes the structure of the star. The star now makes energy in two locations by two different processes, helium fusion in the core and hydrogen fusion in the surrounding shell. The energy flowing outward from the core halts the contraction of the core, and at the same time the star's envelope

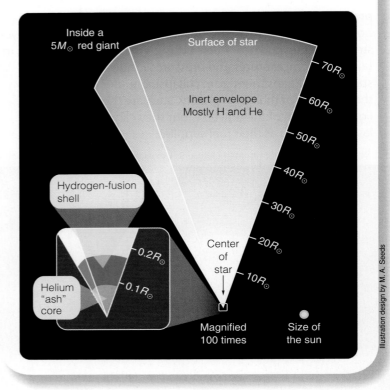

Figure 8.1

When a star runs out of hydrogen at its center, the core contracts to a small size, becomes very hot, and begins nuclear fusion in a shell (blue). The outer layers of the star expand and cool. The red giant star shown here has an average density much lower than the air at Earth's surface. Here $M_\odot$ stands for the mass of the sun, and $R_\odot$ stands for the radius of the sun.

Inside a $5M_\odot$ red giant

Surface of star

Inert envelope Mostly H and He

70$R_\odot$
60$R_\odot$
50$R_\odot$
40$R_\odot$
30$R_\odot$
20$R_\odot$
10$R_\odot$

Hydrogen-fusion shell

$0.2R_\odot$
$0.1R_\odot$

Helium "ash" core

Center of star

Magnified 100 times

Size of the sun

Illustration design by M. A. Seeds

contracts and grows hotter. Consequently, the point that represents the star in the H-R diagram moves downward, corresponding to lower luminosities, and to the left, corresponding to higher surface temperatures, to a region above the main sequence called the **horizontal branch**. Astronomers sometimes refer to stars with those characteristics as "yellow giants."

Helium fusion produces carbon and oxygen that accumulate in an inert core. Once again, the core contracts and heats up, and a helium-fusion shell ignites below the hydrogen-fusion shell. The star now makes energy in two fusion shells, so it quickly expands and its surface cools once again. The point that represents the star in the H-R diagram moves back to the right, completing a loop to become a red giant again. The approximate rule is that if the core of a post-main sequence star is "dead"

(has no nuclear reactions), the star is a red giant, and if the core is "alive" (has fusion reactions), the star is a yellow giant.

Horizontal branch: The location in the H-R diagram of giant stars fusing helium.

Now you can understand why giant stars are relatively rare (see page 117). A star spends about 90 percent of its lifetime on the main sequence and only 10 percent as a giant star. At any moment you look, only a fraction of the visible stars will be passing through the red and yellow giant stages.

What happens to a star after helium fusion depends on its mass, but in any event it cannot survive long: It must eventually collapse and end its career as a star. The remainder of this chapter will trace the details of this process of stellar death, but before you begin that story, you must ask the most important question in science: What is the evidence? What evidence shows that stars actually evolve as theory predicts? You can find the answers in observations of star clusters.

Star Clusters: Evidence of Evolution

Astronomers look at star clusters and say, "Aha! Evidence to solve a mystery." The stars in a star cluster all formed at about the same time and from the same cloud of gas, so they must be about the same age and composition. Each star cluster freeze-frames and makes visible a moment in stellar evolution. The differences you see among stars in one cluster must arise from differences in their masses.

Study "Star Cluster H-R Diagrams" on pages 148–149 and notice three important points:

1. There are two kinds of star clusters, but they are similar in the way their stars evolve. You will learn more about these clusters in the next chapter.

2. You can estimate the age of a star cluster by observing the distribution of the points that represent its stars in the H-R diagram.

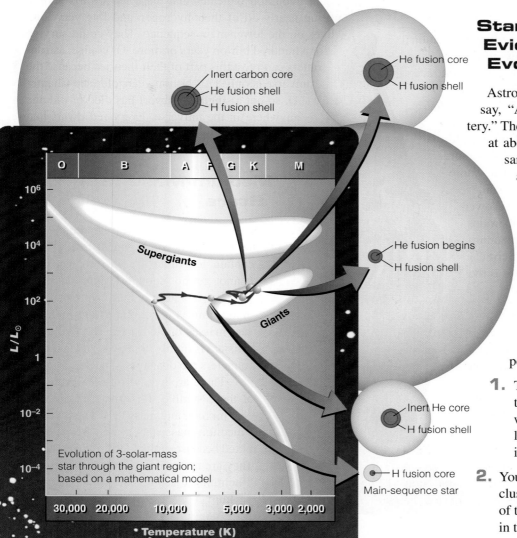

Inert carbon core
He fusion shell
H fusion shell

He fusion core
H fusion shell

Supergiants

He fusion begins
H fusion shell

Giants

Inert He core
H fusion shell

H fusion core
Main-sequence star

Evolution of 3-solar-mass star through the giant region; based on a mathematical model

Temperature (K)

3. Finally, a star cluster's H-R diagram shape is governed by the evolutionary path the stars take. By comparing clusters of different ages, you can visualize how stars evolve almost as if you were watching a film of a star cluster changing over billions of years.

Were it not for star clusters, astronomers would have little confidence in the theories of stellar evolution. Star clusters make that evolution visible and assure astronomers that they really do understand how stars are born, live, and die.

8.2 The Deaths of Low-Mass Stars

Contracting stars heat up by converting gravitational energy into thermal energy. Low-mass stars have little gravitational energy, so when they contract, they can't get very hot. This limits the nuclear fuels they can ignite. In Chapter 7, you saw that objects less massive than 0.08 solar mass cannot sustain hydrogen fusion. Consequently, this section will concentrate on stars more massive than 0.08 solar mass but no more than about 4 times the mass of the sun.

Structural differences divide the lower-main-sequence stars into two subgroups, very-low-mass red dwarfs and medium-mass stars such as the sun. The critical difference between the two groups is the extent of interior convection.

Red Dwarfs

Stars less massive than about 0.4 solar mass—the red dwarfs—have two important differences from more massive stars. First, because they have very small masses, they have very little weight to support. Their pressure–temperature thermostats are set low, and they consume their hydrogen fuel very slowly. The discussion of the life expectancies of

stars in Chapter 7 implies that red dwarfs have very long lives.

The red dwarfs have a second advantage because they are completely convective. That means they are stirred by circulating currents of hot gas rising from the interior and cool gas sinking inward, extending all the way from their cores to their surfaces. The interiors of these stars are mixed like a pot of soup that is constantly stirred as it cooks. Hydrogen is consumed and helium accumulates uniformly throughout the star, which means the star is not limited to using only the fuel in its core. It can use all of its hydrogen to prolong its life on the main sequence.

Because a red dwarf is mixed by convection, it won't develop an inert helium core surrounded by unprocessed hydrogen; therefore it never ignites a hydrogen shell and cannot become a giant star. Rather, nuclear fusion continues to convert hydrogen into helium, but the helium does not fuse into heavier elements because the star can never get hot enough. What astronomers know about stellar evolution indicates that these red dwarfs should use up nearly all of their hydrogen and live very long lives on the lower main sequence. They could survive for a hundred billion years or more. Of course, astronomers can't test this part of their theories because the universe is only 13.7 billion years old (see Chapter 11), so not a single red dwarf has died of old age anywhere in the universe. Every red dwarf that has ever been born is still glowing today.

Medium-Mass (Sunlike) Stars

Stars like the sun can ignite hydrogen and helium and become giants; but, if they contain less than 4 solar masses, they cannot get hot enough to ignite carbon, the next fuel after helium. Note that this mass limit is uncertain, as are many of the masses quoted here. The evolution of stars is highly complex, and such parameters are difficult to specify.

There are two keys to the evolution of these sunlike stars, the lack of complete mixing, and mass loss. The interiors of medium-mass stars are not completely mixed because, unlike the red dwarf stars, they are not completely convective. The helium ash accumulates in an inert helium core surrounded by unprocessed

© Hubble Heritage Team, STScI/AURA/NASA/ © NASA

Figure 8.2

(a) The Ring Nebula in the constellation Lyra is visible even in small telescopes. Note the hot blue star at its center and the radial texture in the gas that suggests outward motion. (b) Some planetary nebulae such as M2-9 are highly elongated, and it has been suggested that the Ring Nebula in (a) actually has a tubular shape like M2-9 that happens to be pointed approximately toward Earth.

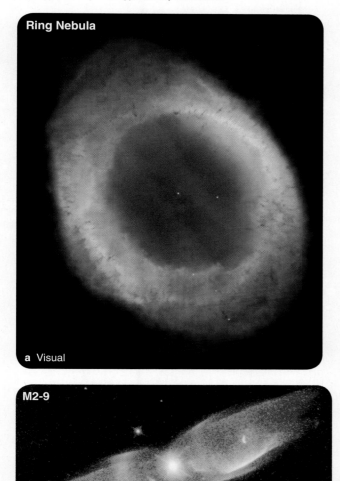

Ring Nebula

a Visual

M2-9

b Visual

oxygen core is the dead end for these stars.

All of this discussion is based on theoretical models of stars and a general understanding of how stars evolve. Does it really happen? Astronomers need observational evidence to confirm this theoretical discussion, and the gas that is expelled from these giant stars gives visible evidence that sun-like stars have gone through these stages of element building and do indeed die in this way.

> **Planetary nebula:** An expanding shell of gas ejected from a medium-mass star during the latter stages of its evolution.

Planetary Nebulae

When a medium-mass star like the sun becomes a distended giant, its atmosphere becomes cool and consequently more opaque. Light has to push against it to escape. At the same time, the fusion shells become thin and unstable, and they begin to flare, which pushes the atmosphere outward. An aging giant can expel its outer atmosphere in repeated surges to form one of the most beautiful objects in astronomy, a **planetary nebula**, so called because the first ones discovered looked like the greenish-blue disks of planets such as Uranus or Neptune. In fact, a planetary nebula has nothing to do with a planet. It is composed of ionized gases expelled by a dying star, like the nebulae in Figure 8.2.

You can understand what planetary nebulae are like by using simple observations and theoretical methods explained in Chapter 5 such as Kirchhoff's laws and the Doppler effect. Although real nebulae are quite complex, the simple model shown in Figure 8.3 on page 150 of a slow stellar wind followed by a fast wind explains their structures fairly well and provides a way to organize the observed phenomena. The complexities and asymmetries seen in planetary nebulae may be due to repeated expulsions of expanding shells and oppositely directed jets, much like bipolar flows observed coming from protostars. Observations and stellar evolution models indicate the central star of a planetary nebula star finally must contract and become a white dwarf.

White Dwarfs

As you have just learned, medium-mass stars die by ejecting gas into space and contracting into white dwarfs. In Chapter 6, you surveyed the stars and learned that

hydrogen. As described earlier, when this core contracts, the unprocessed hydrogen ignites in a shell and swells the star into a red giant. During red giant stages, stars tend to lose mass due to strong stellar winds. When the helium core finally ignites, the star becomes a yellow giant. When the core fills up with carbon–oxygen ash, that is the end of fusion for stars in the medium-mass range because they have masses too low to make their cores hot enough to ignite carbon or oxygen fusion. The carbon–

1 An **open cluster** is a collection of 10 to 1000 stars in a region about 25 pc in diameter. Some open clusters are quite small and some are large, but they all have an open, transparent appearance because the stars are not crowded together.

In a star cluster, each star follows its orbit around the center of mass of the cluster.

AURA/NOAO/NSF

Visual-wavelength image

Open Cluster
The Jewel Box

1a A **globular cluster** can contain 10^5 to 10^6 stars in a region only 10 to 30 pc in diameter. The term "globular cluster" comes from the word "globe," although globular cluster is pronounced like "glob of butter." These clusters are nearly spherical, and the stars are much closer together than the stars in an open cluster.

Globular Cluster
47 Tucanae

Anglo-Australian Telescope Board

Astronomers can construct an H-R diagram for a star cluster by plotting a point to represent the luminosity and temperature of each star.

Visual-wavelength image

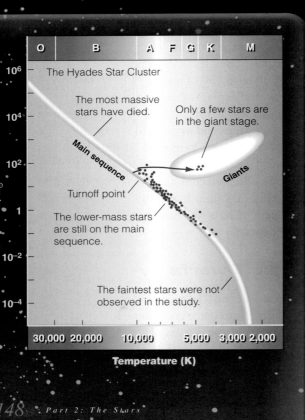

The Hyades Star Cluster

The most massive stars have died.

Only a few stars are in the giant stage.

Main sequence

Turnoff point

Giants

The lower-mass stars are still on the main sequence.

The faintest stars were not observed in the study.

O   B   A F G K   M

10^6

10^4

10^2

$L/L_\odot$

1

10^{-2}

10^{-4}

30,000 20,000   10,000   5,000 3,000 2,000

Temperature (K)

2 The H-R diagram of a star cluster can make the evolution of stars visible. The key is to remember that all of the stars in the star cluster have the same age but differ in mass. The H-R diagram of a star cluster provides a snapshot of the evolutionary state of the stars at the time you happen to be alive. The diagram here shows the 650-million-year-old star cluster called the Hyades. The upper main sequence is missing because the more massive stars have died, and our snapshot catches a few medium-mass stars leaving the main sequence to become giants.

As a star cluster ages, its main sequence grows shorter like a candle burning down. You can judge the age of a star cluster by looking at the turnoff point, the point on the main sequence where stars evolve to the right to become giants. Stars at the **turnoff point** have lived out their lives and are about to die. Consequently, the life expectancy of the stars at the turnoff point equals the age of the cluster.

From theoretical models of stars, you could construct a film to show how the H-R diagram of a star cluster changes as it ages. You can then compare y (left) with observation (right) to understand how stars evolve. Note that the time step for each frame in this film increases by a factor of 10.

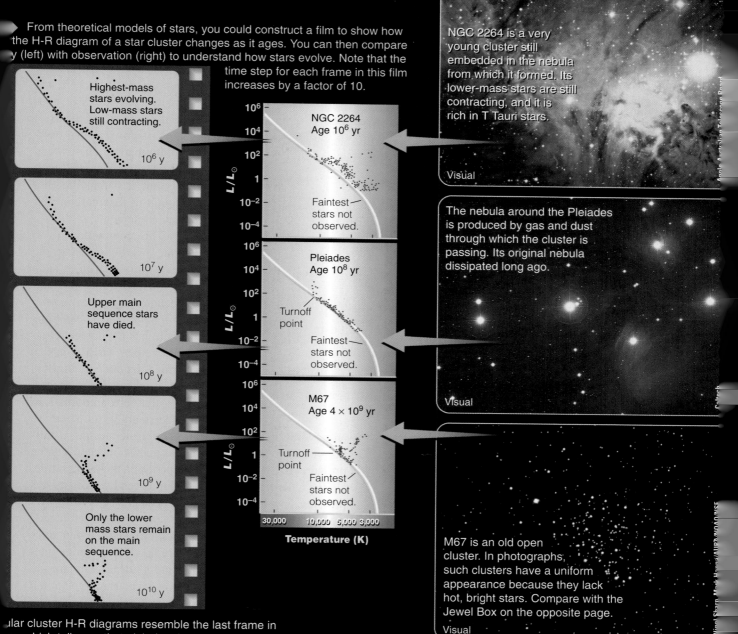

Highest-mass stars evolving. Low-mass stars still contracting.

10^6 y

10^7 y

Upper main sequence stars have died.

10^8 y

10^9 y

Only the lower mass stars remain on the main sequence.

10^{10} y

NGC 2264
Age 10^6 yr

Faintest stars not observed.

Pleiades
Age 10^8 yr

Turnoff point

Faintest stars not observed.

M67
Age 4×10^9 yr

Turnoff point

Faintest stars not observed.

$L/L_\odot$

30,000   10,000   5,000 3,000

Temperature (K)

NGC 2264 is a very young cluster still embedded in the nebula from which it formed. Its lower-mass stars are still contracting, and it is rich in T Tauri stars.

Visual

The nebula around the Pleiades is produced by gas and dust through which the cluster is passing. Its original nebula dissipated long ago.

Visual

M67 is an old open cluster. In photographs, such clusters have a uniform appearance because they lack hot, bright stars. Compare with the Jewel Box on the opposite page.

Visual

ular cluster H-R diagrams resemble the last frame in m, which tells you that globular clusters are very old.

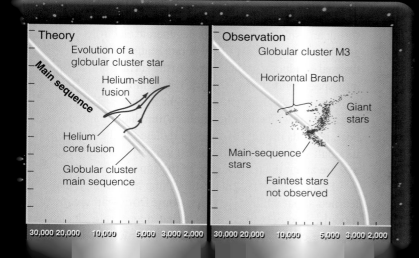

Theory

Evolution of a globular cluster star

Main sequence

Helium-shell fusion

Helium core fusion

Globular cluster main sequence

30,000 20,000   10,000   5,000 3,000 2,000

Observation

Globular cluster M3

Horizontal Branch

Giant stars

Main-sequence stars

Faintest stars not observed

30,000 20,000   10,000   5,000 3,000 2,000

3a The H-R diagrams of globular clusters have very faint turnoff points, which show that they are very old clusters. The best analysis suggests these clusters are about 11 billion years old.

The horizontal branch stars are giants fusing helium in their cores and then in shells. The shape of the horizontal branch outlines the evolution of these stars.

The main-sequence stars in globular clusters are fainter and bluer than the zero-age main sequence. Spectra reveal that globular cluster stars are poor in elements heavier than helium, and that means their gases are less opaque. Therefore energy can flow outward more easily, which makes the stars slightly smaller and hotter. Again the shapes of star clusters H-R diagrams illustrate principles of stellar evolution.

FIGURE 8.3 CREATING PLANETARY NEBULAE

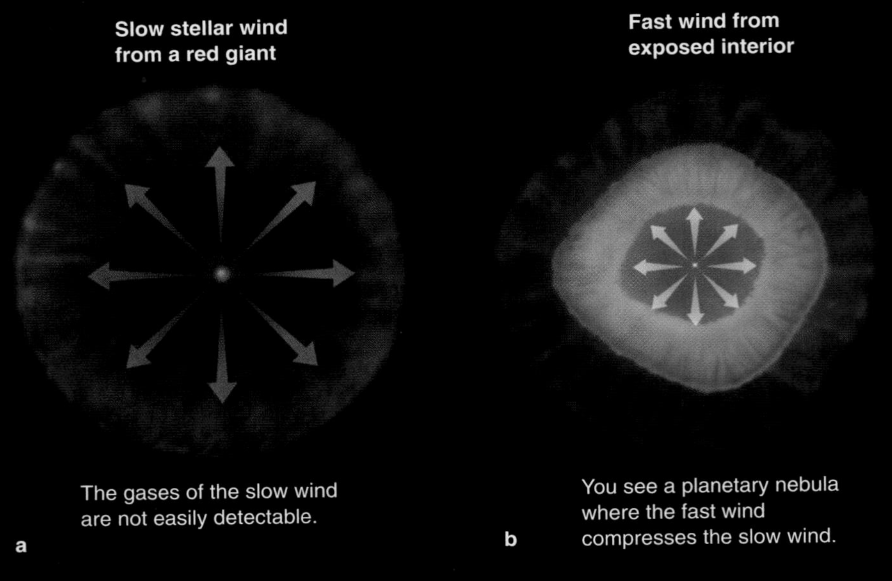

Slow stellar wind from a red giant

The gases of the slow wind are not easily detectable.

a

Fast wind from exposed interior

You see a planetary nebula where the fast wind compresses the slow wind.

b

fusion. It has exhausted its hydrogen and helium fuel and produced carbon and oxygen. As the star contracts into a white dwarf, it converts gravitational energy into thermal energy, and its interior becomes very hot, but it cannot get hot enough to fuse carbon into heavier elements.

The contraction of a white dwarf compresses the gases in its interior to such high densities that quantum mechanical laws become important and the electrons in the gas cannot get closer together. Such a gas is termed **degenerate matter**, and it takes on two properties that are important in understanding the structure and evolution of dying stars. A degenerate gas is millions of times harder to compress than solid steel, and the pressure in the gas no longer depends on the temperature. Unlike a normal star, which is supported by ordinary gas pressure, a white dwarf is supported against its own gravity by the resistance to compression of a degenerate gas.

Clearly, a white dwarf is not a true star. It generates no nuclear energy, is almost completely degenerate matter, and, except for a thin layer at its surface, contains no gas. Instead of calling a white dwarf a "star," you could call it a **compact object**. Later sections of this chapter discuss two other types of compact objects, neutron stars and black holes.

A white dwarf's future is bleak. As it radiates energy into space, its temperature gradually falls, but it cannot shrink any smaller because its degenerate electrons cannot get closer together. This degenerate matter is a very good thermal conductor, so heat flows to the surface and escapes into space, and the white dwarf gets fainter and cooler, moving downward and to the right in the H-R diagram. Because the white dwarf contains a tremendous amount of heat, it needs billions of years to radiate that heat through its small surface area. The coolest white dwarfs in our galaxy are about the temperature of the sun.

Perhaps the most interesting thing astronomers have learned about white dwarfs has come from mathematical models. The equations predict that degenerate electron pressure cannot support an object with more than about

white dwarfs are the second most common kind of star (see page 117). Only red dwarfs are more abundant. The billions of white dwarfs in our galaxy must be the remains of medium-mass stars.

The first white dwarf discovered was the faint companion to the well-known star Sirius, the brightest star in the sky. Sirius is a visual binary star, the most luminous member of which is Sirius A. The white dwarf, Sirius B, is 10,000 times fainter than Sirius A. The orbital motions of the stars (see Chapter 6) reveal that the white dwarf's mass is 0.98 solar mass, and its blue-white color tells you that its surface is hot, about 25,000 K. Although it is very hot, it has a very low luminosity, so it must have a small surface area. In fact, it is about the size of Earth. Dividing its mass by its volume reveals that it is very dense—about 2×10^6 g/cm^3. On Earth, a teaspoonful of Sirius B material would weigh more than 11 tons. These basic observations and simple physics lead to the conclusion that white dwarfs are astonishingly dense.

A normal star is supported by energy flowing outward from its core, but a white dwarf cannot generate energy by nuclear

Degenerate matter: Extremely high-density matter in which pressure no longer depends on temperature due to quantum mechanical effects.

Compact object: One of the three final states of stellar evolution, which generates no nuclear energy and is much smaller and denser than a normal star.

1.4 solar masses. A white dwarf with that mass would have such strong gravity that its radius would shrink to zero. This is called the **Chandrasekhar limit** after Subrahmanyan Chandrasekhar, the astronomer who calculated it. This seems to imply that a star more massive than 1.4 solar masses could not become a white dwarf unless it got rid of the extra mass in some way.

Can stars lose substantial amounts of mass? Observations provide clear evidence that young stars have strong stellar winds, and aging giants and supergiants also lose mass. This suggests that stars more massive than the Chandrasekhar limit can eventually end up as white dwarfs if they reduce their mass under the limit. Theoretical models show that stars that begin life with as much as 8 solar masses could lose mass fast enough to reduce their mass so low they can collapse to form white dwarfs with masses below 1.4 solar masses. With mass loss, a wide range of medium-mass stars can eventually die as white dwarfs.

> Natural laws have no pity.
> –Robert Heinlein
> The Notebooks of
> Lazarus Long

The Fate of the Sun and the End of Earth

Astronomy is about you. Although this chapter has been discussing the deaths of medium-mass stars, it has also been discussing the future of our Earth. The sun is a medium-mass star and must eventually die by becoming a giant, possibly producing a planetary nebula, and collapsing into a white dwarf. That will spell the end of Earth.

Evolutionary models of the sun suggest that it may survive for another 6 billion years or so. In about 5 billion years, it will exhaust the hydrogen in its core, begin burning hydrogen in a shell, and swell into a red giant star about 30 times its present radius. Later, helium fusion will ignite in the core and the sun will become a horizontal branch star. Once the helium fuel is exhausted in its core, helium fusion will begin in a shell, and the sun will expand again. That second red giant version of the sun will be about as large as the orbit of Earth. Before that, the sun's increasing luminosity will certainly evaporate Earth's oceans, drive away the atmosphere, and even vaporize much of Earth's crust. Astronomers are still uncertain about some of the details, but computer models that include tidal effects predict that the expanding sun eventually will engulf and destroy Mercury, Venus, and Earth.

While it is a giant star, the sun will have a strong wind and lose a substantial fraction of its mass into space. The atoms that were once in Earth will be part of the expanding nebula around the sun. Your atoms will be part of that nebula. If the white dwarf remnant sun becomes hot enough, it will ionize the expelled gas and light it (and you) up as a planetary nebula.

Models of the sun's evolution are not precise enough to predict whether its white dwarf remnant will become hot enough soon enough to light up its expelled gas and create a planetary nebula before that gas disperses. Whether the expelled gas lights up or not, it would include atoms that were once part of Earth. Some research also suggests that a star needs a close binary companion to speed up its spin in order to create a planetary nebula. The sun, of course, has no close stellar companion. This is an area of active research, and there are as yet no firm conclusions.

There is no danger that the sun will explode as a nova; it has no binary companion (see the next section of this chapter). And, as you will see, the sun is definitely not massive enough to die the violent supernova death of the most massive stars.

Chandrasekhar limit: The maximum mass of a white dwarf, about 1.4 solar masses. A white dwarf of greater mass cannot support itself and will collapse.

8.3 The Evolution of Binary Systems

Stars in binary systems can evolve independently of each other if they orbit at a large distance from each other. In this situation, one of the stars can swell into a giant and collapse without disturbing its companion. Other binary stars are as close to each other as 0.1 AU, and when one of those stars begins to swell into a giant, its companion can be disturbed in surprising ways.

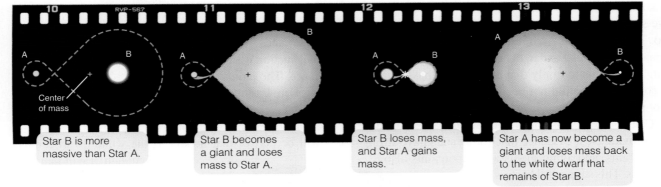

The Evolution of a Binary System

Star B is more massive than Star A.

Star B becomes a giant and loses mass to Star A.

Star B loses mass, and Star A gains mass.

Star A has now become a giant and loses mass back to the white dwarf that remains of Star B.

© iStockphoto.com

Mass Transfer and Accretion Disks

Binary stars can sometimes interact by transferring mass from one star to the other. Of course, the gravitational field of each star holds its mass together, but the gravitational fields of the two stars, combined with the rotation of the binary system, define a dumbbell-shaped volume called the **Roche lobes** around the pair of stars. Matter inside a star's Roche lobe is gravitationally bound to the star, but matter outside the lobe can be transferred to the other star or lost completely from the system. If the stars are close together, the Roche lobes are relatively small and can interfere with the evolution of the stars. When an evolving star in a close binary system expands so far that it fills its Roche lobe, matter can flow from that star's lobe into the other lobe and onto the other star, as you can see in the series above.

Matter flowing from one star to another cannot fall directly into the star. Rather, because of conservation of angular momentum, it must flow into a whirling disk around the star. **Angular momentum** refers to the tendency of a rotating object to continue rotating. All rotating objects possess some angular momentum, and in the absence of external forces, an object maintains (conserves) its total angular momentum. An ice dancer takes advantage of conservation of angular momentum by starting a spin slowly with arms extended and then drawing them in. As her mass becomes concentrated closer to her axis of rotation, she spins faster. The same effect causes the slowly circulating water in a bathtub to spin faster in a whirlpool as it approaches the drain.

Mass transferred from one star to another in a binary must conserve its angular momentum. Thus it must flow into a rapidly rotating whirlpool called an **accretion disk** around the second star, as in Figure 8.4. The gas in the disk grows hot due to friction and tidal forces (see Chapter 2) and eventually falls onto the second star. If that second star, the one receiving the matter lost from its companion, is a compact object like a white dwarf, the gas in the accretion disk can become very compressed. The gas temperature can exceed a million K, producing X rays. In addition, the matter accumulating on the white dwarf can eventually cause a violent explosion called a nova.

Novae

At the beginning of this chapter you saw that the word *nova* refers to an apparently new star that appears in the sky for a while and then fades away. Modern astronomers know that a nova is not a new star but an old star flaring

Roche lobe: The volume of space a star controls gravitationally within a binary system.

Angular momentum: A measure of the tendency of a rotating body to continue rotating. Mathematically, the product of mass, velocity, and radius.

Accretion disk: The rotating disk that forms in some situations as matter is drawn gravitationally toward a central body.

Figure 8.4

Matter falling into a compact object forms a whirling accretion disk. Friction and tidal forces can make the disk very hot.

© David A. Hardy/PPARC

up. After a nova fades, astronomers can photograph the spectrum of the remaining faint point of light. Invariably, they find a normal star and a white dwarf in a close binary system. A nova is evidently an explosion involving a white dwarf.

Observational evidence can tell you how nova explosions occur. As the explosion begins, spectra show blueshifted absorption lines that tell you the gas is dense and coming toward you at a few thousand kilometers per second. After a few days, the spectral lines change to emission lines, which tells you the gas has thinned, but the blueshifts remain, showing that a cloud of debris has been ejected into space.

Nova explosions occur when mass transfers from a normal star into an accretion disk around a white dwarf. As the matter loses its angular momentum in the accretion disk, it settles inward onto the surface of the white dwarf and forms a layer of unused nuclear fuel—mostly hydrogen. As the layer deepens, it becomes denser and hotter until the hydrogen fuses in a sudden explosion that blows the surface off the white dwarf. Although the expanding cloud of debris contains less than 0.0001 solar mass, it is hot, and its expanding surface area makes it very luminous. Nova explosions can become 100,000 times more luminous than the sun. As the debris cloud expands, cools, and thins over a period of weeks and months, the nova fades from view.

The explosion of its surface hardly disturbs the white dwarf and its companion star. Mass transfer quickly resumes, and a new layer of fuel begins to accumulate. How fast the fuel builds up depends on the rate of mass transfer. Accordingly, you can expect novae to repeat each time an explosive layer accumulates. Many novae take thousands of years to build an explosive layer, but some take only decades.

© iStockphoto.com

8.4 The Deaths of Massive Stars

You have seen that low- and medium-mass stars die relatively quietly as they exhaust their hydrogen and helium, and then some push away their surface layers to form planetary nebulae. In contrast, massive stars live spectacular lives, aspects of which can be seen in Figure 8.5 on the next page, and destroy themselves in violent explosions.

Nuclear Fusion in Massive Stars

Stars on the upper main sequence have too much mass to die as white dwarfs, but their evolution begins much like that of their lower-mass cousins. They consume the hydrogen in their cores, ignite hydrogen shells, and become giants or, for the most massive stars, supergiants. Their cores contract and fuse helium first in the core and then in a shell, producing a carbon–oxygen core.

Unlike medium-mass stars, the massive stars finally can get hot enough to ignite carbon fusion at a temperature of about 1 billion Kelvin. Carbon fusion produces more oxygen plus neon. As soon as the carbon is exhausted in the core, the core contracts, and carbon ignites in a shell. This pattern of core ignition and shell ignition continues with a series of heavier nuclei as fusion fuel, and the star develops a layered structure as shown in Figure 8.5, with a hydrogen-fusion shell surrounding a helium-fusion shell surrounding a carbon-fusion shell surrounding . . . and so on. At higher temperatures than carbon fusion, nuclei of oxygen, neon, and magnesium fuse to make silicon and sulfur, and at even higher temperatures silicon can fuse to make iron.

The fusion of the nuclear fuels in this series goes faster and faster as the massive star evolves rapidly. The amount of energy released per fusion reaction decreases as the mass of the types of atoms involved increases. To support its weight and remain stable, a star must fuse oxygen much faster than it fused hydrogen. Also, there are fewer nuclei in the core of the star by the time heavy nuclei begin to fuse. Four hydrogen nuclei made one helium nucleus, and three helium nuclei make one carbon, so there are 12 times fewer nuclei of carbon available for fusion than there were hydrogen nuclei. This means the heavy elements are used up, and fusion goes very

Table 8.1 Heavy-Element Fusion in a 25-*M* Star

Fuel	Time	Percentage of Lifetime
H	7,000,000 years	93.3
He	500,000 years	6.7
C	600 years	0.008
O	0.5 years	0.000007
Si	1 day	0.00000004

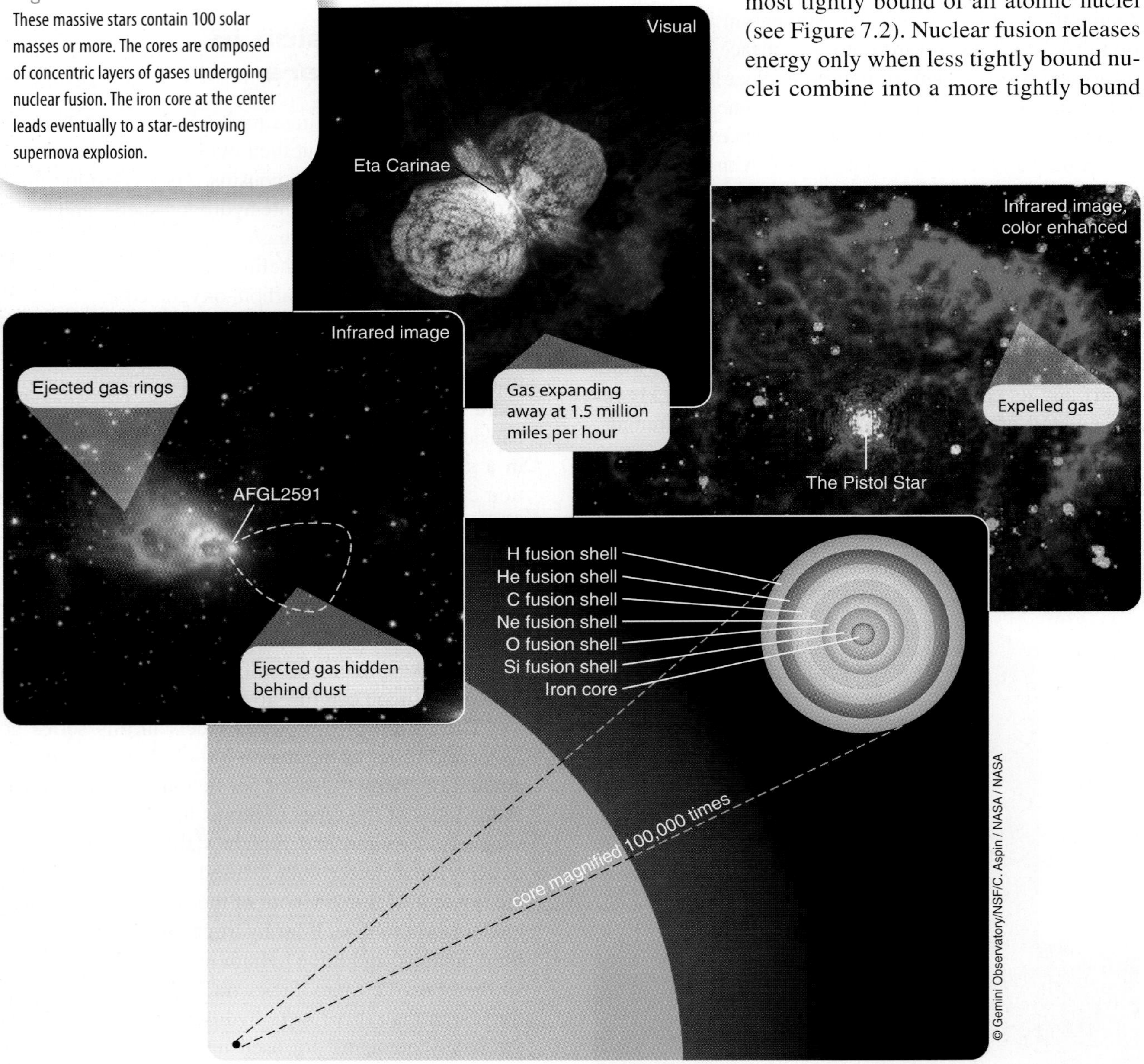

Figure 8.5

These massive stars contain 100 solar masses or more. The cores are composed of concentric layers of gases undergoing nuclear fusion. The iron core at the center leads eventually to a star-destroying supernova explosion.

Visual

Eta Carinae

Infrared image, color enhanced

Infrared image

Ejected gas rings

Gas expanding away at 1.5 million miles per hour

Expelled gas

AFGL2591

The Pistol Star

Ejected gas hidden behind dust

H fusion shell
He fusion shell
C fusion shell
Ne fusion shell
O fusion shell
Si fusion shell
Iron core

core magnified 100,000 times

© Gemini Observatory/NSF/C. Aspin / NASA / NASA

quickly in massive stars (Table 8.1). Hydrogen fusion can last 7 million years in a 25-solar-mass star, but that same star will fuse its oxygen in 6 months and its silicon in just one day.

Supernova Explosions of Massive Stars

Theoretical models of evolving stars combined with nuclear physics allow astronomers to describe what happens inside a massive star when the last nuclear fuels are exhausted. It begins with iron nuclei and ends in cosmic violence.

Silicon fusion produces iron, the most tightly bound of all atomic nuclei (see Figure 7.2). Nuclear fusion releases energy only when less tightly bound nuclei combine into a more tightly bound nucleus. Once the gas in the core of the star has been converted to iron, there are no further nuclear reactions that can release energy. The iron core is a dead end in the evolution of a massive star.

As a star develops an iron core, energy production begins to decline, and the core contracts. Nuclear reactions involving iron begin, but they remove energy from the core, causing it to contract even further. Once this process starts, the core of the star collapses inward in less than a tenth of a second.

The collapse of a giant star's core after iron fusion starts is calculated to happen so rapidly that the most powerful computers are unable to predict the details. Thus, models of supernova explosions contain many approximations. Nevertheless, the models predict exotic nuclear reactions in the collapsing core that should produce a flood of neutrinos (see Chapter 7). In fact, for a short time the core produces more energy per second than all of the stars in all of the visible galaxies in the universe, and 99 percent of that energy is in the form of neutrinos. This flood of neutrinos carries large amounts of energy out of the core, allowing the core to collapse further. The models also predict that the collapsing core of the star must quickly become a neutron star or a black hole, the subjects of the next sections of this chapter, while the envelope of the star is blasted outward.

To understand how the inward collapse of the core can produce an outward explosion, you can think about a traffic jam. The collapse of the innermost part of the degenerate core allows the rest of the core to fall inward, and this creates a tremendous traffic jam as all of the nuclei fall toward the center. The position of the traffic jam, called a shock wave, begins to move outward as more in-falling material encounters the jam. The torrent of neutrinos, as well as energy flowing out of the core in sudden violent convective turbulence, help drive the shock wave outward. Within a few hours, the shock wave bursts outward through the surface of the star and blasts it apart.

The supernova seen from Earth is the brightening of the star as its envelope is blasted outward by the shock wave. As months pass, the cloud of gas expands, thins, and fades, but the manner in which it fades tells astronomers more about the death throes of the star. The rate at which the supernova's brightness decreases matches the rate at which radioactive nickel and cobalt decay, so the explosion must produce great abundances of those atoms. The radioactive cobalt decays into iron, so destruction of iron in the core of the star is followed by the production of iron through nuclear reactions in the expanding outer layers.

> **Type I supernova:** A supernova explosion caused by the collapse of a white dwarf.
>
> **Type II supernova:** A supernova explosion caused by the collapse of a massive star.

Types of Supernovae

In studying supernovae in other galaxies, astronomers have noticed that there are a number of different types. **Type I supernovae** have no hydrogen lines in their spectra, and astronomers have thought of at least two ways a supernova could occur without involving much hydrogen (see below). **Type II supernovae**, in contrast, have spectra containing hydrogen lines and appear to be produced by the collapse and explosion of a massive star, the process discussed in the previous section.

A type Ia supernova is thought to occur when a white dwarf in a binary system receives enough mass to exceed the Chandrasekhar limit and collapse. The collapse of a white dwarf is different from the collapse of a massive star because the core of the white dwarf contains usable fuel. As the collapse begins, the temperature and density shoot up, and the carbon–oxygen core begins to fuse in violent nuclear reactions. In a few seconds, the carbon–oxygen interior is entirely consumed, and the outermost layers are blasted away in a violent explosion that, at its brightest, is about six times more luminous than a type II supernova. The white dwarf is entirely destroyed; no neutron star or black hole is left behind. This explains why no hydrogen lines are seen in the spectrum of a type Ia supernova explosion—white dwarfs contain very little hydrogen.

The less common type Ib supernova is understood to occur when a massive star in a binary system loses its hydrogen-rich outer layers to its companion star. The remains of the massive star could develop an iron core and collapse, as described in the previous section, producing a supernova explosion that lacked hydrogen lines in its spectrum.

Figure 8.6

The Crab Nebula is located in the constellation Taurus the Bull, just where Chinese astronomers saw a brilliant guest star in the year 1054 CE. High-speed electrons produced by the central neutron star spiral through magnetic fields and produce the foggy glow of synchrotron radiation that fills the nebula. The star just to the right of the neutron star lies much closer to Earth and is not in the Crab Nebula.

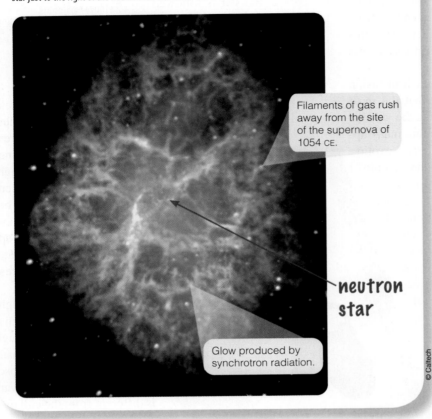

Filaments of gas rush away from the site of the supernova of 1054 CE.

neutron star

Glow produced by synchrotron radiation.

© Caltech

Astronomers working with the largest and fastest computers are using modern theory to try to understand supernova explosions. The companion to theory is observation, so you should ask what observational evidence supports this story of supernova explosions.

Observations of Supernovae

In the year 1054 CE, Chinese astronomers saw a "guest star" appear in the constellation known in the Western tradition as Taurus the Bull. The star quickly became so bright it was visible in the daytime, and then, after a month, it slowly faded, taking almost two years to vanish from sight. When modern astronomers turned their telescopes to the location of the guest star, they found a peculiar nebula now known as the Crab Nebula for its

Synchotron radiation: Radiation emitted when high-speed electrons move through a magnetic field.

many-legged shape, which you can see in Figure 8.6. In fact, the legs of the Crab Nebula are filaments of gas that are moving away from the site of the explosion at about 1,400 km/s. Comparing the radius of the nebula, 1.35 pc, with its velocity of expansion reveals that the nebula began expanding nine or ten centuries ago, just when the guest star made its appearance. The Crab Nebula is clearly the remains of the supernova seen in 1054 CE. In the next section, you will meet the neutron star found at the center of the Crab Nebula.

The blue glow of the Crab Nebula is produced by **synchrotron radiation**. This form of electromagnetic radiation, unlike blackbody radiation, is produced by rapidly moving electrons spiraling through magnetic fields and is common in the nebulae produced by supernovae. In the case of the Crab Nebula, the electrons travel so fast they emit visual wavelengths. In most such nebulae, the electrons move slower, and the synchrotron radiation is at radio wavelengths.

Supernovae are rare. Only a few have been seen with the naked eye in recorded history. Arab astronomers saw one in 1006 CE, and the Chinese saw one in 1054 CE. European astronomers observed two—one in 1572 CE (Tycho's supernova) and one in 1604 CE (Kepler's supernova). In addition, the guest stars of 185, 386, 393, and 1181 CE may have been supernovae.

In the centuries following the invention of the astronomical telescope in 1609, no supernova was bright enough to be visible to the naked eye. Then, in the early morning hours of February 24, 1987, astronomers around the world were startled by the discovery of a naked-eye supernova still growing brighter in the southern sky, which you can see in Figure 8.7. The supernova, known officially as SN 1987A, is 53,000 pc away in the Large Magellanic Cloud, a small satellite galaxy to our own Milky Way Galaxy. This first naked-eye supernova in 383 years has given astronomers a ringside seat for the most spectacular event in stellar evolution.

One observation of SN 1987A is critical in that it confirms the theory of core collapse. At 2:35 a.m. EST on February 23, 1987, nearly 4 hours before the supernova was first seen, a blast of neutrinos swept through

Earth. Instruments buried in a salt mine under Lake Erie and in a lead mine in Japan, though designed for another purpose, recorded 19 neutrinos in less than 15 seconds. Neutrinos are so difficult to detect that the 19 neutrinos actually detected mean that some 10^{17} neutrinos must have passed through the detectors in those 15 seconds. Furthermore, the neutrinos were arriving from the direction of the supernova. Thus, astronomers conclude that the burst of neutrinos was released when the iron core collapsed, and the supernova was first seen at visual wavelengths hours later when the shock wave blasted the star's surface into space.

Most supernovae are seen in distant galaxies, and careful observations allow astronomers to compare types, which can be done with charts like the one in Figure 8.8. Type Ia supernovae, caused by the collapse of white dwarfs, are more luminous at maximum brightness and decline rapidly at first and then more slowly. Type II supernovae, produced by the collapse of massive stars,

are not as bright at maximum, and they decline in a more irregular way.

SN 1987A was a type II supernova, although its light curve shown in Figure 8.8 is not typical. Models indicate that most type II supernovae are caused by the collapse of red supergiants, but SN 1987A was produced by the explosion of a hot, blue supergiant. Astronomers hypothesize that this star was once a red supergiant but later contracted and heated up slightly, becoming bluer before it exploded.

Although the supernova explosion fades to obscurity in a year or two, an expanding shell of gas marks the site of the explosion. The gas, originally expelled at 10,000 to 20,000 km/s, may carry away a fifth of the mass of the star. The collision of that expanding gas with

Figure 8.8

Type I supernovae decline in brightness rapidly at first and then more slowly, but type II supernovae pause for about 100 days before beginning a steep decline. Supernova 1987A was odd in that it did not rise directly to maximum brightness. These light curves have been adjusted to the same maximum brightness.

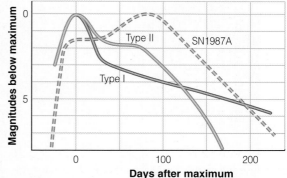

FIGURE 8.7

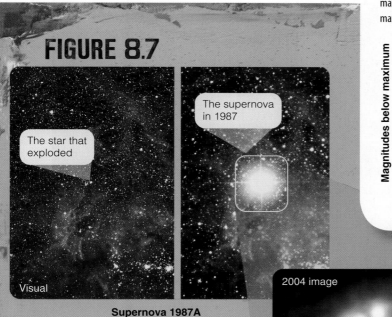

The star that exploded

The supernova in 1987

Visual

Supernova 1987A

The star that exploded as supernova 1987A was about 20 times the mass of the sun. The interaction of matter previously lost by the star with gas recently ejected plus the burst of light from the explosion have produced rings around the central glow. A shock wave from the explosion is now expanding into a 1-ly-diameter ring of gas ejected roughly 20,000 years before the explosion. As that ring is excited, it will light up the region and reveal how the star shed mass before it collapsed.

2004 image

Remains of the star

Shockwave expanding into ring of gas

Visual-wavelength image

Supernova remnant: The expanding shell of gas and dust marking the site of a supernova explosion.

the surrounding interstellar medium can sweep up even more gas and excite it to produce a **supernova remnant**, the nebulous remains of a supernova explosion. Figure 8.9 show images of 4 supernova remnants; images made at other than visible wavelengths are displayed in false color (see Chapter 4).

Supernova remnants look quite delicate and do not survive very long—a few tens of thousands of years—before they gradually mix with the interstellar medium and vanish. The Crab Nebula is a young remnant, only about 950 years old and about 8.8 ly in diameter. Older remnants can be larger. Some supernova remnants are visible only at radio and X-ray wavelengths. They have become too tenuous to emit detectable light, but the collision of the expanding hot gas with the interstellar medium can generate radio and X-ray radiation. You learned in Chapter 7 that the compression of the interstellar medium by expanding supernova remnants can also trigger star formation.

Figure 8.9

A supernova remnant is an expanding bubble of hot gas created by a supernova explosion. As the remnant expands and pushes into neighboring gas, it can emit radiation at many wavelengths.

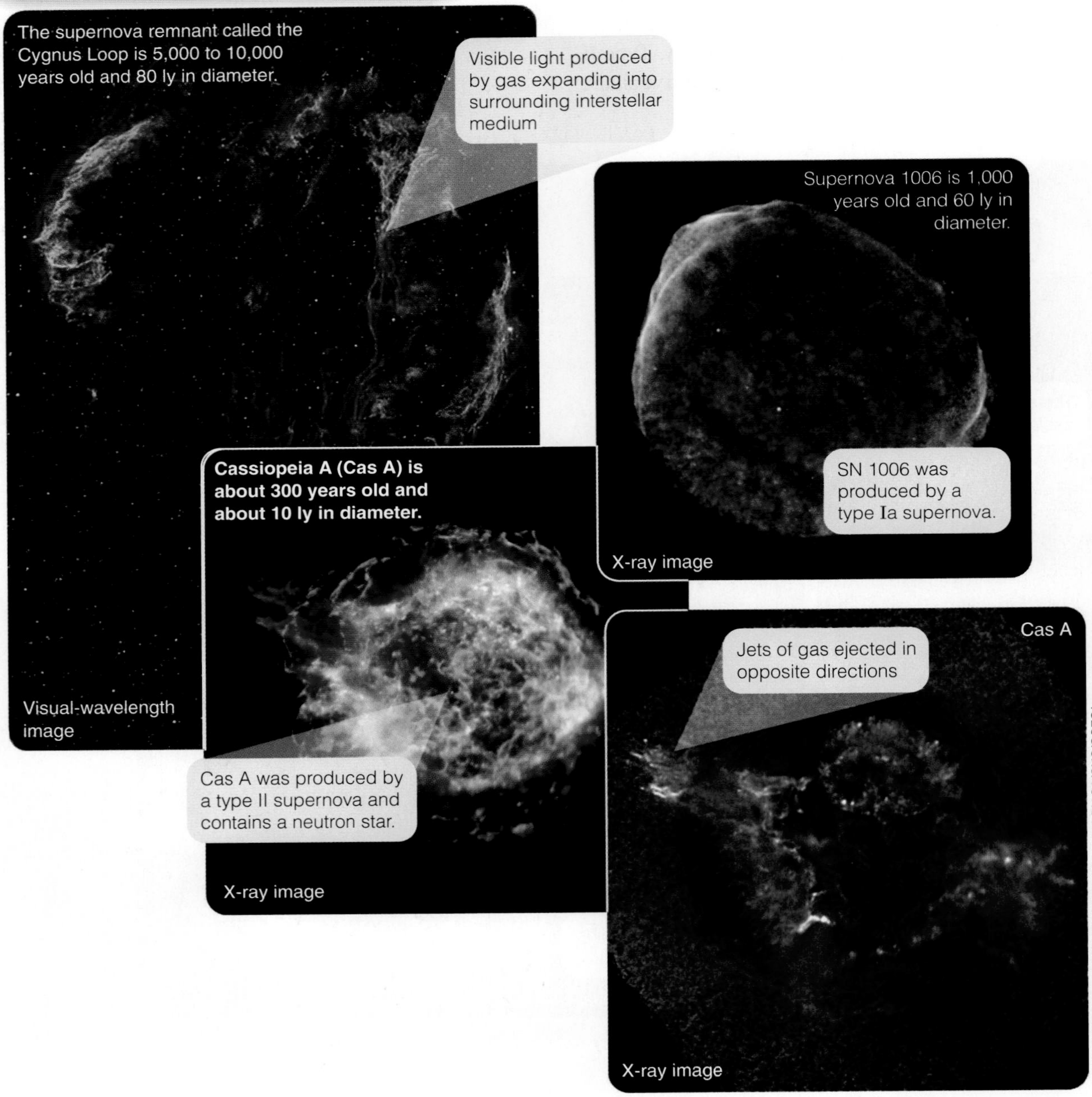
The supernova remnant called the Cygnus Loop is 5,000 to 10,000 years old and 80 ly in diameter.
Visible light produced by gas expanding into surrounding interstellar medium
Visual-wavelength image
Supernova 1006 is 1,000 years old and 60 ly in diameter.
SN 1006 was produced by a type Ia supernova.
X-ray image
Cassiopeia A (Cas A) is about 300 years old and about 10 ly in diameter.
Cas A was produced by a type II supernova and contains a neutron star.
X-ray image
Cas A
Jets of gas ejected in opposite directions
X-ray image

Mikael Svalgaard/NASA/CXC/J. Hughes et al./NASA/CXC/GSFC/U. Hwang et. al. / NASA/CXC/GSFC/U. Hwang et al.

Gravity always wins. However a star lives, theory predicts it must eventually die and leave behind one of three types of final remains—a white dwarf, neutron star, or black hole. These compact objects are small monuments to the power of gravity. Almost all of the energy available has been squeezed out of compact objects, and you find them in their final, high-density states.

8.5 Neutron Stars

A **neutron star** contains a little over 1 solar mass compressed to a radius of about 10 km. Its density is so high that the matter is stable only as a fluid of pure neutrons. Two questions should occur to you immediately. First, how could any theory predict such a strange object? And second, how do you know they exist?

Theoretical Prediction of Neutron Stars

The subatomic particles called neutrons were discovered in a laboratory in 1932, and physicists quickly realized that because neutrons spin much like electrons, a gas of neutrons could become degenerate and therefore nearly incompressible. Just two years later, in 1934, two astronomers, Walter Baade and Fritz Zwicky, suggested that some of the most luminous novae in the historical record were not regular novae but were caused by the collapse plus explosion of a massive star in a cataclysm they named a "super-nova." If the collapsing core is more massive than the Chandrasekhar limit of 1.4 solar masses, then the weight is too great to be supported by degenerate electron pressure, and the core cannot become a stable white dwarf. The collapse would force protons to combine with electrons and become neutrons. The envelope of the star would be blasted away in a supernova explosion, and the core of the star would be left behind as a small, tremendously dense sphere of neutrons that Zwicky called a "neutron star."

Mathematical models predict that a neutron star will be only 10 or so kilometers in radius (Figure 8.10) and will have a density of about 10^{14} g/cm^3. That is roughly the density of atomic nuclei, and you can think of a neutron star as matter with all of the empty space squeezed out of it. On Earth, a sugar-cube-sized lump of this material would weigh 100 million tons, the mass of a small mountain.

Simple physics, the physics you have used in previous chapters to understand normal stars, predicts that neutron stars should (1) spin rapidly, perhaps 100 to 1,000 rotations per second; (2) be hot, with surface temperatures of millions of degrees K; and (3) have strong magnetic fields, up to a trillion times stronger than the sun's or Earth's magnetic fields. For example, the collapse of a massive star's core would greatly increase its spin rate by conservation of angular momentum. Other processes during core collapse should create high temperature and magnetic field strength. Despite their high temperature, neutron stars should be difficult to detect because of their tiny size.

Neutron star: A small, highly dense star, with radius about 10 km, composed almost entirely of tightly packed neutrons.

Figure 8.10

A tennis ball and a road map illustrate the relative size of a neutron star. Such an object, containing slightly more than the mass of the sun, would fit with room to spare inside the beltway around Washington, D.C.

Photo by M. Seeds

Pulsar: A source of short, precisely timed radio bursts, understood to be spinning neutron stars.

Lighthouse model: The explanation of a pulsar as a spinning neutron star sweeping beams of electromagnetic radiation around the sky.

What is the maximum mass for a stable neutron star? In other words, is there an upper limit to the mass of neutron stars like the Chandrasekhar limit that defines the maximum mass of a white dwarf star? That is difficult to answer, because physicists don't know enough about the properties of pure neutron material. It can't be made in a laboratory, and theoretical calculations in this case are very difficult. The most widely accepted results suggest that a neutron star can't be more massive than 2 to 3 solar masses. An object more massive than that can't be supported by degenerate neutron pressure, so it would collapse, presumably becoming a black hole.

What size stars will end their lives with supernova explosions that leave behind neutron star corpses? Theoretical calculations suggest that stars that begin life on the main sequence with 8 to about 15 solar masses will end up as neutron stars. Stars more massive than about 15 solar masses are expected to form black holes when they die.

The Discovery of Pulsars

In November 1967, Jocelyn Bell, a graduate student at Cambridge University in England, found a peculiar pattern in the data from a radio telescope. Unlike other radio signals from celestial bodies, this was a series of regular pulses, graphed in Figure 8.11. At first she and the leader of the project, Anthony Hewish, thought the signal was interference, but they found it day after day at the same celestial latitude and longitude. Clearly, it was cosmic in origin.

Figure 8.11

The 1967 detection of regularly spaced pulses in the output of a radio telescope led to the discovery of pulsars. This record of the radio signal from the first pulsar, CP1919, contains regularly spaced pulses (marked by ticks). The period is 1.33730119 seconds.

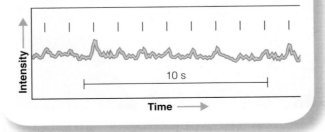

Another possibility, that it came from a distant civilization, led them to consider naming it LGM for Little Green Men. Within a few weeks, the team found three more objects in other parts of the sky pulsing with different periods. The objects were clearly natural, and the team dropped the name LGM in favor of **pulsar**—a contraction of "pulsing star." The pulsing radio source Bell had observed with her radio telescope was the first known pulsar. Hewish received the Nobel Prize in physics for this work, and Bell (now Bell Burnell) has been remarkably gracious about that.

As more pulsars were found, astronomers argued over their nature. The pulses, which typically last only about 0.001 second, gave astronomers an important clue. The pulse length places an upper limit on the size of the object producing the pulse. This is a very important principle in astronomy—an object cannot change its brightness significantly in an interval shorter than the time light takes to cross its diameter. If pulses from pulsars are no longer than 0.001 second, then the objects cannot be larger than 0.001 light second, or 300 km (190 mi) in diameter, smaller than white dwarfs, which makes neutron stars the only reasonable explanation.

The missing link between pulsars and neutron stars was found in late 1968, when astronomers discovered a pulsar at the heart of the Crab Nebula (see Figure 8.6). The Crab Nebula is a supernova remnant, which agrees nicely with Zwicky and Baade's prediction that some supernovae should produce a neutron star. The short pulses and the discovery of the pulsar in the Crab Nebula are strong evidence that pulsars are neutron stars.

The modern model of a pulsar has been called the **lighthouse model**, and is illustrated in Figure 8.12. The name *pulsar* is, in a sense, inaccurate: A pulsar does not pulse (vibrate) but rather emits beams of radiation that sweep around the sky as the neutron star rotates, like a rotating lighthouse light. The mechanism that produces the beams involves extremely high energies and strong electric and magnetic fields and is not fully understood. Over 1,000 pulsars are now known. There may be many more which are undetected because their beams never point toward Earth.

The Evolution of Pulsars

Neutron stars are not simple objects, and modern astronomers need knowledge of frontier physics to understand them. Nevertheless, the life story of pulsars can be worked out to some extent. When a pulsar first forms, it may be spinning as many as 100 times a second. The energy it radiates into space ultimately comes from its energy of rotation, so as it blasts beams of radiation outward, its rotation slows. Judging from their pulse periods and rates at which they slow down, the average pulsar is apparently only a few million years old and the oldest has an age of about 10 million years. Presumably, neutron stars older than that rotate too slowly to generate detectable radio beams. Supernova remnants last only about 50,000 years before they mix into the interstellar medium and disappear, so most pulsars have long outlived the remnants in which they were originally embedded.

You can expect that a young neutron star should emit especially strong beams of radiation powered by its rapid rotation. The Crab Nebula provides an example of such a system. Only about 950 years old and spinning 30 times per second, the Crab pulsar is so powerful that astronomers can detect photons of radio, infrared, visible, X-ray, and gamma-ray wavelengths from it.

The explosion of Supernova 1987A in February 1987 probably formed a neutron star. You can draw this conclusion because a burst of neutrinos was detected passing through Earth, and theory predicts that the collapse of a massive star's core into a ball of neutrons would produce such a burst of neutrinos. The neutron star initially should be hidden at the center of the expanding shells of gas ejected into space by the supernova explosion, but as the gas continues to expand and become thinner, you can expect that astronomers might eventually be able to detect it. As of this writing, no neutron star has been detected in the SN 1987A remnant, but astronomers continue to watch the site hoping to find the youngest pulsar known.

Binary Pulsars

One reason pulsars are so fascinating is the extreme conditions found in spinning neutron stars. To see even more extreme natural processes, you have only to look at the pulsars that are members of binary systems. These pulsars are of special interest because astronomers can learn more about the neutron star and about the behavior of matter in unusual circumstances by studying them.

Binary pulsars can be sites of tremendous violence because of the strength of gravity at the surface of a neutron star. Matter falling onto a neutron star can release titanic amounts of energy. If you dropped a single marshmallow onto the surface of a neutron

star from a distance of 1 AU, it would hit with an impact equivalent to a 3-megaton nuclear warhead. Even a small amount of matter flowing from a companion star to a neutron star can generate high temperatures and release X rays and gamma rays.

Figure 8.12

A neutron star contains a powerful magnetic field and spins very rapidly. The spinning magnetic field generates a tremendously powerful electric field, and the field causes the creation of electron-positron pairs that accelerate through the magnetic field, emit photons in the directions of their motion, and thereby produce powerful beams of electromagnetic radiation emerging from the neutron star's magnetic poles.

Neutron Star Rotation with Beams

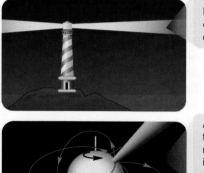

The pulsing of a lighthouse is actually caused by the rotation of beams of light.

As in the case of Earth, the magnetic axis of a neutron star could be inclined to its rotational axis.

The rotation of the neutron star will sweep its beams around like beams from a lighthouse.

While a beam points roughly toward Earth, observers detect a pulse.

While neither beam is pointed toward Earth, observers detect no energy.

As an example of such an active system, examine Hercules X-1 in Figure 8.13. It emits pulses of X rays with a period of about 1.2 seconds, but every 1.7 days the pulses vanish for a few hours. Hercules X-1 seems to contain a 2-solar-mass star and a neutron star that orbit each other with a period of 1.7 days. Matter flowing from the normal star into an accretion disk around the neutron star reaches temperatures of millions of degrees and emits a powerful X-ray glow, some of which is in beams that sweep around with the rotating neutron star. Earth receives a pulse of X rays every time a beam points this way. The X rays shut off completely every 1.7 days when the neutron star is eclipsed behind the normal star. Hercules X-1 is an intricate system with many different high-energy processes going on simultaneously, and this quick analysis only serves to illustrate how complex and powerful such binary systems are during mass transfer.

A binary system in which both objects are neutron stars was discovered in 1974 when radio astronomers Joseph Taylor and Russell Hulse noticed that the pulse period of the pulsar PSR 1913+16 grew longer and then shorter in a cycle that takes 7.75 hours. Taylor and Hulse realized that must be the binary orbital period of the pulsar. They analyzed the system with the same techniques used to study spectroscopic binary stars (see Chapter 6) to find that PSR 1913+16 consists of two neutron stars separated by a distance roughly equal to the radius of our sun. The masses of the two neutron stars are each about 1.4 solar masses, in good agreement with models of neutron stars and how they are created.

A nice surprise was hidden in the motion of PSR 1913+16. In 1916 Einstein published his **general theory of relativity** that described gravity as a curvature of space-time. Einstein realized that any rapid change in a gravitational field should spread outward at the speed of light as **gravitational radiation**. Gravity waves themselves have not been detected yet, but Taylor and Hulse were able to show that the orbital period of the binary pulsar is slowly growing shorter because the stars are gradually spiraling toward each other at the rate expected if they radiate orbital energy away as gravitational radiation. Taylor and Hulse won the Nobel Prize in 1993 for their work with binary pulsars.

The Fastest Pulsars

This discussion of pulsars suggests that newborn pulsars should blink rapidly, and old pulsars should blink slowly, but a few that blink the fastest may be quite old. A number of **millisecond pulsars** have been found, so called because their pulse periods are almost as short as a millisecond (0.001 s). The energy stored in a neutron star rotating at this rate is equal to the total energy of a supernova explosion, and these pulsars generally have weak magnetic fields consistent with advanced age, so it seemed difficult at first to understand their rapid rotation. Astronomers hypothesized that an old neutron star could gain rotational energy from a companion star in a binary system. Some of the millisecond pulsars are caught in the act of receiving matter from companions in a fashion that should speed the pulsar's rotation to the observed high rates, so the hypothesis seems to be confirmed.

Figure 8.13

In Hercules X-1, matter flows from a star into an accretion disk around a neutron star producing X rays, which heat the near side of the star to 20,000 K compared with only 7,000 K on the far side. X rays turn off when the neutron star is eclipsed behind the star.

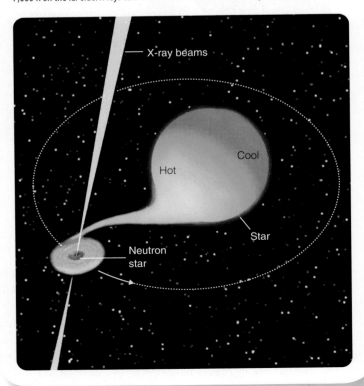

X-ray beams

Cool

Hot

Star

Neutron star

no proof!
(or the impossibility of proof in science)

No scientific theory or hypothesis can be proved correct. You can test a hypothesis over and over by performing experiments or making observations, but you can never prove that the hypothesis is absolutely true. It is always possible that you have misunderstood the hypothesis or the evidence, and the next observation you make might disprove the hypothesis.

For example, you might propose the hypothesis that the sun is mostly calcium and iron vapor because those elements are known to be abundant on Earth. You might test the hypothesis by looking at the calcium and iron lines in the solar spectrum, and the strength of the lines would suggest your hypothesis is right. Although your observation has confirmed your hypothesis, it has not proven the hypothesis is right. You might confirm the hypothesis many times in many ways before you realize that at the temperature of the sun, iron and calcium atoms absorb photons much more efficiently than hydrogen. Although the hydrogen lines are weak in the sun's spectrum, a careful analysis shows that most of the atoms in the sun are hydrogen and not iron or calcium.

The nature of scientific thinking can lead to two common misconceptions. Sometimes nonscientists will say, "You scientists just want to tear everything down—you don't believe in anything." Scientists test a hypothesis over and over to test its worth. If a hy-

pothesis survives many tests, scientists begin to have confidence it is true, and the hypothesis "graduates" to being considered a theory.

The second common misconception arises when nonscientists say, "You scientists are never sure of anything." Again, the scientist knows that no hypothesis can be proved correct. That the sun will rise tomorrow is very likely, and scientists have great confidence in that theory. But it is still only a theory. People will say of an idea they dislike, "That is only a theory," as if a theory were simply a random guess. In fact, a theory is a hypothesis that has been well tested many times and so has become a truth in which all scientists have great confidence. Yet you can never prove that any hypothesis or theory is absolutely true.

"Show me," say scientists; and, in the case of neutron stars, the evidence seems very strong. Of course, you can never prove a scientific hypothesis or theory is absolutely true, but the evidence for neutron stars is so strong that astronomers have great confidence that they really do exist. Other theories that describe how they emit beams of radiation and how they form and evolve are less certain, but continuing observations at many wavelengths are expanding the understanding of these last embers of massive stars. In fact, precise observations have turned up objects no one expected.

Pulsar Planets

Because a pulsar's period is so precise, astronomers can detect tiny variations by comparison with atomic clocks. When astronomers checked pulsar PSR 1257+12, they found variations in the period of pulsation (Figure 8.14a) analogous to the variations caused by the orbital motion of the binary pulsar but much smaller. When these variations were interpreted as Doppler shifts, it became evident that the pulsar was being orbited by at least two objects with planet like masses of about 4 and 3 Earth masses. The gravitational tugs of the planets make the pulsar wobble about the center of mass of the system by about 800 km, and that produces the tiny changes in period that are observed (Figure 8.14b).

Astronomers greeted this discovery with both enthusiasm and skepticism. As usual, they looked for ways to test the hypothesis. Simple gravitational theory predicts that the planets should interact and slightly modify each other's orbit. When the data were analyzed, that interaction was found, further confirming the hypothesis that the variations in the period of the pulsar are caused by planets. In fact, more observations and analyses revealed the presence of a third planet of about twice the mass of Earth's moon, and possibly a fourth planet of only 3 percent the mass of Earth's moon. This illustrates the astonishing precision of studies based on pulsar timing.

Astronomers wonder how a neutron star can have planets. The inner three planets that orbit PSR 1257+12 are closer to the pulsar than Venus is to the sun. Any planets that orbit a star would be lost or vaporized when

the star exploded. Furthermore, a star about to explode as a supernova would be a large giant or a supergiant, and planets only a few AU distant would be inside such a large star and could not survive. It seems more likely that these planets are the remains of a stellar companion that was devoured by the neutron star. In fact, PSR1257+12 is very fast (162 pulses per second), suggesting that it was spun up in a binary system.

8.6 Black Holes

You have studied white dwarfs and neutron stars, two of the three end states of dying stars. Now it's time to think about the third end state—black holes.

Although the physics of black holes is difficult to discuss without sophisticated mathematics, simple logic is sufficient to predict that they should exist. The problem is to use their predicted properties and attempt to confirm that they exist. What objects observed in the heavens could be real black holes? More difficult than the search for neutron stars, the quest for black holes has nevertheless met with success.

To begin, you can consider a simple question. How fast must an object travel to escape from the surface of a celestial body? The answer will lead to black holes.

Escape Velocity

In Chapter 3 you learned that the escape velocity is the initial velocity an object needs to escape from a celestial body. Whether you are discussing a baseball leaving Earth or a photon leaving a collapsing star, the escape velocity depends on two things, the mass of the celestial body and the distance from the center of mass to the escapee's starting point. If the celestial body has a large mass, its gravity is strong, and you need a high velocity to escape; but if you begin your journey farther from the center of mass, the velocity needed is less. For example, to escape from Earth, a spaceship would have to leave Earth's surface at 11 km/s (25,000 mph), but if you could launch spaceships from the top of a tower

Figure 8.14

(a) The dots in this graph are observations showing that the period of pulsar PSR 1257+12 varies from its average value by a fraction of a billionth of a second. The blue line shows the variation that would be produced by planets orbiting the pulsar. (b) As the planets orbit the pulsar, they cause it to wobble by less than 800 km, a distance that is invisibly small in this diagram.

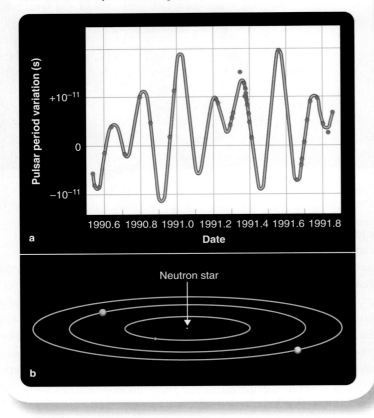

© NASA-JPL

smaller than a proton. Astronomically, it seems to make little difference.

If the contracting core of a star becomes small enough, the escape velocity in a region around it is so large that no light can escape. You can receive no information about the object or about the region of space near it. Such a region is called a **black hole**. Note that the term *black hole* refers to a volume of space, not just the singularity at the region's center. If the core of an exploding star collapsed to create a black hole, the expanding outer layers of the star could produce a supernova remnant, but the core would vanish without a trace.

1,000 miles high, the escape velocity would be only 10 km/s (22,000 mph).

An object massive enough and/or small enough could have an escape velocity greater than the speed of light. Relativity says that nothing can travel faster than the speed of light, so even photons would be unable to escape. Such a small, massive object could never be seen because light could not leave it. This was first noted by British astronomer Reverend John Mitchell in 1783, long before Einstein and relativity.

Schwarzschild Black Holes

If the core of a star collapses and contains more than about 3 solar masses, no force can stop the collapse. When the object reaches the size of a white dwarf, the collapse continues because degenerate electrons cannot support that much weight. It also cannot stop when it reaches the even smaller size of a neutron star because degenerate neutrons also cannot support that weight. No force remains to stop the object from collapsing to zero radius.

As an object collapses, its density and the strength of its surface gravity increase; and if an object collapses to zero radius, its density and gravity become infinite. Mathematicians call such a point a **singularity**, but in physical terms it is difficult to imagine an object of zero radius. Some theorists think that a singularity is impossible and that the laws of quantum physics must somehow halt the collapse at some subatomic radius roughly 10^{20} times

Albert Einstein's general theory of relativity treats space and time as a single entity—space-time. His equations showed that gravity could be described as a curvature of space-time, and almost immediately astronomer Karl Schwarzschild found a way to solve the equations to describe the gravitational field around a single, nonrotating, electrically neutral lump of matter. That solution contained the first general relativistic description of a black hole, and nonrotating, electrically neutral black holes are now known as Schwarzschild black holes. In recent decades, theorists such as Roy Kerr and Stephen Hawking have found ways to apply the sophisticated mathematical equations of the general theory of relativity and quantum mechanics to black holes that are rotating and have electrical charges. For this discussion the differences are minor, and you may proceed as if all black holes were Schwarzschild black holes.

Schwarzschild's solution shows that if matter is packed into a small enough volume, then space-time curves back on itself. Objects can still follow paths that lead into the black hole, but no path leads out, so nothing can escape, not even light. Consequently, the inside of the black hole is totally beyond the view of an outside observer. The

Singularity: An object of zero radius and infinite density.

Black hole: A mass that has collapsed to such a small volume that its gravity prevents the escape of all radiation. Also, the volume of space from which radiation may not escape.

event horizon is the boundary between the isolated volume of space-time and the rest of the universe, and the radius of the event horizon is called the **Schwarzschild radius, R_s**—the radius within which an object must shrink to become a black hole, as in Figure 8.15, and the point of no return for any object falling in later.

Although Schwarzschild's work was highly mathematical, his conclusions were quite simple. The size of a black hole, its Schwarzschild radius, is simply proportional to its mass. A 3-solar-mass black hole will have a Schwarzschild radius of about 9 km, a 10-solar-mass black hole will have a Schwarzschild radius of 30 km, and so on. Note that even a very massive black hole would not be very large—just a few miles across.

It is a common misconception to think of black holes as giant vacuum cleaners that will suck up everything in the universe. A black hole is just a gravitational field, and at a large distance its gravity is no greater than that of a normal object of similar mass. If the sun were replaced by a 1-solar-mass black hole, the orbits of the planets would not change at all. The gravity of a black hole becomes extreme only when you approach close to it. Figure 8.16 illustrates this by representing gravitational fields as curvature of the fabric of space-time. Physicists like to graph the strength of gravity around a black hole as curvature in a flat sheet. The graphs look like funnels in which the depth of the funnel indicates the strength of the gravitational field, but black holes themselves are not shaped like funnels; they are spheres or spheroids. In Figure 8.16, you should note that the strength of the gravitational field around the black hole becomes extreme only if you venture too close.

This chapter discusses black holes that might originate from the deaths of massive stars. In later chapters, you will encounter black holes whose masses might exceed a million solar masses, located in the centers of galaxies.

Figure 8.15

A black hole forms when an object collapses to a small size (perhaps to a singularity) and the escape velocity becomes so great light cannot escape. The boundary of the black hole is called the event horizon because any event that occurs inside is invisible to outside observers. The radius of the black hole R_s is the Schwarzschild radius.

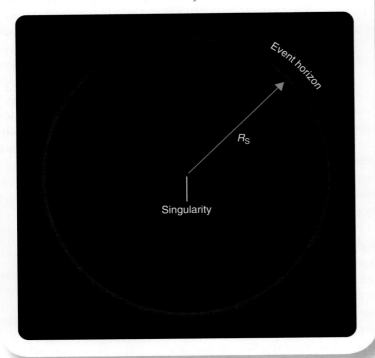

Leaping into a Black Hole

Before you can search for real black holes, you must understand what theory predicts about the behavior of a black hole. To explore that idea, you can imagine that you leap, feet-first, into a Schwarzschild black hole.

If you were to leap into a black hole of a few solar masses from a distance of an astronomical unit, the gravitational pull would not be very large, and you would fall slowly at first. Of course, the longer you fell and the closer you came to the center, the faster you would travel. Your wristwatch would tell you that you fell for about two months before you reached the event horizon.

Your friends who stayed behind would see something different. They would see you falling more slowly as you came closer to the event horizon because, as described by general relativity, time slows down in curved space-time. This is known as **time dilation**. In fact, your friends would never actually see you cross the event horizon. To them you would fall more and more slowly until you seemed hardly to move. Generations later, your descendants could focus their telescopes on you and see you still inching closer

Figure 8.16

If you fell into the gravitational field of a star, you would hit the star's surface before you fell very far. Because a black hole is so small, you could fall much deeper into its gravitational field and eventually cross the event horizon. At a distance, the two gravitational fields are the same.

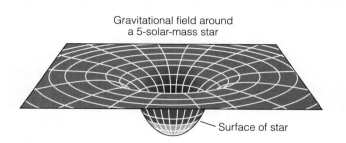

Gravitational field around a 5-solar-mass star

Surface of star

Gravitational field around a 5-solar-mass black hole

To the event horizon

Figure 8.17

Leaping feet-first into a black hole. A person of normal proportions (left) would be distorted by tidal forces (right) long before reaching the event horizon around a typical black hole of stellar mass. Tidal forces would stretch the body lengthwise while compressing it laterally.

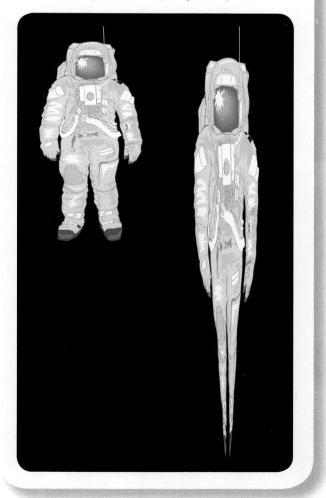

to the event horizon. You, however, would have sensed no slowdown and would conclude that you had crossed the event horizon after about two months.

Another relativistic effect would make it difficult to see you with normal telescopes. As light travels out of a gravitational field, it loses energy, and its wavelength grows longer. This is known as the **gravitational redshift**. Although you would notice no effect as you fell toward the black hole, your friends would need to observe at longer and longer wavelengths in order to detect you.

While these relativistic effects seem merely peculiar, other effects would be quite unpleasant. Imagine again that you are falling feet-first toward the event horizon of a black hole. You would feel your feet, which would be closer to the black hole, being pulled in more strongly than your head. This is a tidal force, and at first it would be minor. As you fell closer, however, the tidal force would become very large. Another tidal force would compress you as your left side and your right side both fell toward the center of the black hole. For any black hole with a mass like that of a star, the tidal forces would crush you laterally and stretch you longitudinally long before you reached the event horizon (Figure 8.17). Needless to say, this would render you inoperative as a thoughtful observer.

Gravitational redshift: The lengthening of the wavelength of a photon as it escapes from a gravitational field.

Science fiction books, movies, and TV shows sometimes suggest that you can travel through the universe by jumping into a black hole in one place and popping out of another somewhere far across space. That might make for good science fiction, but tidal forces would make it an unpopular form of transportation even if it worked. You would certainly lose your luggage.

Your imaginary leap into a black hole is not frivolous. You now know how to find a black hole: Look for a strong source of X rays that may be from matter compressed and stressed just before disappearing as it approaches the event horizon.

The Search for Black Holes

An isolated black hole is totally invisible because nothing can escape from the event horizon. But a black hole into which matter is flowing would be a source of X rays. Of course, X rays can't escape from *inside* the event horizon, but X rays emitted by the heated matter flowing into the black hole can escape if the X rays are emitted *before* the matter crosses the event horizon. An isolated black hole in space will not have much matter flowing into it, but a black hole in a binary system might receive a steady flow of matter transferred from its companion star. This suggests you can search for black holes by searching among X-ray binaries.

You have learned about X-ray binaries such as Hercules X-1 that contain a neutron star, and they emit X rays much as a binary containing a black hole should. You can tell the difference in two ways. If the compact object emits pulses, you know it is a neutron star. Otherwise, you might check the mass of the object. If the mass of the compact object is greater than about 3 solar masses, the object cannot be a neutron star, and you can conclude that it must be a black hole.

The first X-ray binary suspected of harboring a black hole was Cygnus X-1. It contains a supergiant B0 star and a compact object orbiting each other with a period of 5.6 days. Matter flows from the B0 star as a strong stellar wind, and some of that matter enters a hot accretion disk around the compact object (Figure 8.18). The accretion disk is about 5 times larger in diameter than the orbit of Earth's moon, and the inner few hundred kilometers of the disk have a temperature of about 2 million Kelvin—hot enough to radiate X rays. The compact object is invisible, but Doppler shifts in the spectrum reveal the motion of the B0 star around the center of mass of the binary. From the geometry of the orbit, astronomers were able to calculate the mass of the compact object—at least 3.8 solar masses, well above the maximum for a neutron star.

As X-ray telescopes have found more X-ray objects, the list of black hole candidates has grown to a few

Figure 8.18

The X-ray source Cygnus X-1 consists of a supergiant B star and a compact object orbiting each other. Gas from the B star's stellar wind flows into the hot accretion disk around the compact object, and astronomers detect X rays from the disk.

dozen. Some of these objects are shown in Table 8.2. Each candidate is a compact object surrounded by a hot accretion disk in a close X-ray binary system without regular pulsations. A few of the binary systems are easier to analyze than others, but, in the end, it has become clear that some of these objects are too massive to be neutron stars. Scientists cannot have absolute proof, but the evidence is now overwhelming: Black holes really do exist.

Another way to confirm that black holes are real is to search for evidence of their distinguishing characteristic—event horizons—and that search also has been successful. In one study, astronomers selected twelve X-ray binary systems, six of which seemed to contain neutron stars and six of which were expected to contain black holes. Using X-ray telescopes, the astronomers could see flares of energy as blobs of matter fell into the accretion disks and spiraled inward. In the six systems thought to contain neutron stars, the astronomers could also detect bursts of energy when the blobs of matter finally fell onto the surfaces of the neutron stars. In the six systems thought to contain black holes, however, the blobs of matter spiraled inward through the accretion disks and vanished without final bursts of energy. Evidently, those blobs of matter had approached the event horizons and become undetectable due to time dilation and gravitational redshift. This is dramatic evidence that event horizons are real.

Energy from Compact Objects—Jets

As you can see, it is a common misconception to think that it is impossible to get any energy out of a black hole. In a later chapter, you will meet galaxies radiating vast amounts of energy from massive black holes, so you should pause here to see how a compact object and an accretion disk can eject powerful jets and beams of energy.

Whether a compact object is a black hole or a neutron star, it has a strong gravitational field. Any matter

Table 8.2 **Nine Black Hole Candidates**

Object	Location	Companion Star	Orbital Period	Mass of Compact Object
Cygnus X-1	Cygnus	B0 supergiant	5.6 days	$>3.8\ M_\odot$
LMC X-3	Dorado	B3 main-sequence	1.7 days	$\sim 10\ M_\odot$
A0620-00	Monocerotis	K main-sequence	7.75 hours	$10 \pm 5\ M_\odot$
V404 Cygni	Cygnus	K main-sequence	6.47 days	$12 \pm 2\ M_\odot$
GRO J1655-40	Scorpius	F–G main-sequence	2.61 days	$6.9 \pm 1\ M_\odot$
QZ Vul	Vulpecula	K main-sequence	8 hours	$10 \pm 4\ M_\odot$
4U 1543-47	Lupus	A main-sequence	1.12 days	$2.7–7.5\ M_\odot$
V4641 Sgr	Sagittarius	B supergiant	2.82 days	$8.7–11.7\ M_\odot$
XTEJ1118+480	Ursa Major	K main-sequence	0.170 days	$>6\ M_\odot$

flowing into that field is accelerated inward; and, because it must conserve angular momentum, it flows into an accretion disk made so hot by friction that the inner regions can emit X rays and gamma rays. Somehow the spinning accretion disk can emit powerful beams of gas and radiation along its axis of rotation. The process isn't well understood, but it seems to involve magnetic fields that get caught in the accretion disk and are twisted into tightly wound tubes that squirt gas and radiation out of the disk and confine it in narrow beams.

This process is similar to the bipolar outflows ejected by protostars (see Chapter 7), but it is much more powerful. One example of this process is an X-ray binary called SS 433. Its optical spectrum shows sets of spectral lines that are Doppler-shifted by about one-fourth the speed of light, with one set shifted to the red and one set shifted to the blue. Apparently, SS 433 is a binary system in which a compact object (probably a black hole) pulls matter from its companion star and forms an extremely hot accretion disk. Jets of high-temperature gas blast away in beams aimed in opposite directions. SS 433 is a prototype that illustrates how the gravitational field around a compact object can produce powerful beams of radiation and matter.

Energy from Compact Objects—Gamma-Ray Bursts

During the 1960s the United States put a series of satellites in orbit to watch for bursts of gamma rays coming from Earth indicating nuclear weapons tests that would be violations of an international treaty. The experts were startled when the satellites detected about one **gamma-ray burst** coming from space per day. The Compton Gamma Ray Observatory launched in 1991 discovered that gamma-ray bursts were occurring all over the sky and not from any particular region. Starting in 1997, new satellites in orbit were able to detect gamma-ray bursts, determine their location in the sky, and immediately alert astronomers on the ground. When telescopes swiveled to image the locations of the bursts, they detected fading glows that resembled supernovae, and that has led to the conclusion that some relatively long gamma-ray bursts are produced by a kind of supernova explosion called a **hypernova** (Figure 8.19).

Theoretical calculations indicate that a star more massive than some threshold around 15 or 20 solar masses will collapse and become a black hole when its nuclear fuel is exhausted. Models show that the collapsing star would conserve angular momentum and spin very rapidly, and that would slow the collapse of the equatorial parts of the star. The poles of the star would fall in quickly, and that would focus beams of intense radiation and ejected gas blasting out along the axis of rotation, resulting in a hypernova. If either of those beams happens to point in the right direction, Earth would receive a powerful gamma-ray burst. The evidence seems clear that at least some of the gamma-ray bursts are produced by hypernovae. Massive stars explode as hypernovae only rarely in any one galaxy, but the gamma-ray bursts they produce are so powerful that astronomers can detect these explosions among a vast number of galaxies. Other gamma-ray bursts may be produced by the merger of two neutron stars or a neutron star and a black hole, and yet others by sudden shifts in the crusts of highly magnetized neutron stars.

Incidentally, if a gamma-ray burst occurred only 1,000 ly from Earth, the gamma rays would shower Earth with radiation equivalent to a 10,000-megaton nuclear blast. (The largest bombs ever made were a few tens of megatons in size.) The gamma rays could create enough nitric oxide in the atmosphere to produce intense acid rain and would destroy the ozone layer, exposing

Figure 8.19

A Hypernova Explosion

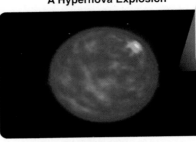

The collapsing core of a massive star drives its energy along the axis of rotation because. . .

the rotation of the star slows the collapse of the equatorial regions.

Within seconds, the remaining parts of the star fall in.

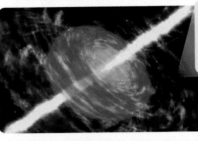

Beams of gas and radiation strike surrounding gas and generate beams of gamma rays.

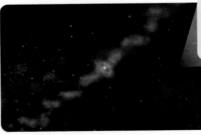

The gamma-ray burst fades in seconds, and a hot accretion disk is left around the black hole.

© NASA/Skyworks Digital

life on Earth to deadly levels of solar ultraviolet radiation. Gamma-ray bursts can occur relatively near the Earth as often as every few 100 million years and could be one of the causes of the mass extinctions that show up in the fossil record.

RECAP

The life and death stories of stars are important because Earth depends on one star, the sun. Perhaps even more important, the lives and deaths of previous generations of stars created the atomic elements of which Earth and you are made. If those stars hadn't lived and died, you would not exist. This chapter explored details of the life and death stories of stars and how the types of remains stars produce depends on the mass of the individual star. Stars with masses like that of the sun end as white dwarfs, but more massive stars leave behind the strangest beasts in the cosmic zoo, neutron stars and black holes. Some matter from dying stars escapes back into the interstellar medium and is incorporated into new stars and the planets that form with them. The deaths of stars are part of a great cycle of stellar birth and death that includes the sun, Earth, and you, and by understanding the deaths of stars, you can better understand your role in the evolution of the universe.

Test your knowledge by answering these questions in as much detail as you can.

1. What happens to a star when it uses up the last of the hydrogen fuel in its core?

2. How will the sun and stars smaller than the sun die?

3. What happens if an evolving star is in a binary system?

4. How do massive stars die?

5. What are neutron stars, and how do you know they really exist?

6. What are black holes, and how do you know they really exist?

Part 3

Chapter 9

The Milky Way
Galaxy

The Stars Are Yours was the title of a popular astronomy book written by James S. Pickering in 1948. The point of the title is that the stars belong to everyone equally, and you can enjoy the stars as if you owned them. You live inside one of the largest of the star systems that fill the universe. Our Milky Way Galaxy is over 80,000 ly in diameter and contains over 100 billion stars. As you read this chapter, you will be learning about your home galaxy, but you will also be learning how the stars of the galaxy have cooked up the atoms heavier than helium. You know already how the cores of massive stars make atoms heavier than helium and how supernovae blast those atoms back into space and add even heavier atoms made during the supernova explosion. In this chapter you will see how the stars in the Milky Way Galaxy have, generation after generation, made the atoms now in your body.

9.1 The Discovery of the Galaxy

It seems odd to say astronomers discovered something that is all around you. However, until the early 20th century, no one knew what the Milky Way was.

It isn't obvious that you live in a galaxy. You are inside, and you see nearby stars scattered in all directions over the sky, whereas the more distant clouds of stars in the galaxy make a faint band of light circling the sky, as the artwork in Figure 9.1 on page 174 demonstrates. The ancient Greeks named that band *galaxies kuklos*, the "milky circle." The Romans changed the name to *via lactea*, "milky road" or "milky way." It was not until early in the 20th century that astronomers understood that humans live inside a great wheel of stars and that the universe is filled with other such star systems. Drawing on the Greek word for milk, astronomers called these star systems *galaxies*.

looking back

In previous chapters, you learned how stars are born from clouds of gas and dust, how they evolve as they consume their nuclear fuels, and how they die. You discovered that most stars live long, quiet lives but that the most massive stars die in violent supernova explosions, leaving behind neutron stars or black holes as their corpses.

looking ahead

In the chapter that follows, you will meet some of the billions of galaxies that fill the depths of the universe. Understanding the Milky Way Galaxy, big as it is, is only a step in understanding the universe as a whole.

Figure 9.1

Nearby stars look bright, and cultures around the world group them into constellations. Nevertheless, the vast majority of the stars in our galaxy merge into a faintly luminous path that circles the sky, the Milky Way. This artwork shows the location of a portion of the Milky Way near a few bright winter constellations. (See Figure 9.5a and the star charts at the front and back of this book to further locate the Milky Way in your sky.)

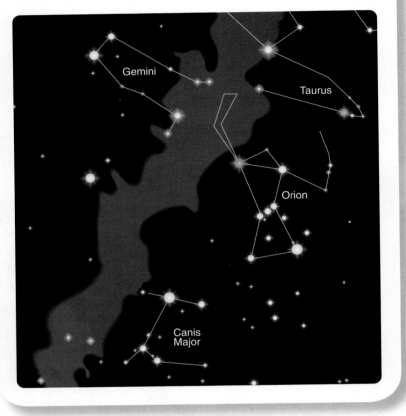

The Great Star System

Galileo's telescope revealed that the glowing Milky Way is made up of stars, and later astronomers realized that the great cloud of stars in which the sun is located, which they called the star system, must be wheel shaped. If the star system were spherical, for example, you would see stars scattered more or less uniformly in all directions over the sky. Only a wheel or disk shape would look, from the inside, like the Milky Way band encircling the sky.

The English astronomers Sir William Herschel (1738–1822) and his sister Caroline Herschel (1750–1848) attempted to gauge the true shape of the star system by counting stars in 683 different directions in the sky. In directions where they saw more stars, they assumed, the star system extended further into space. As you can see in Figure 9.2, which was published by the Herschels in

Cepheid variable stars: Variable stars with pulsation periods of 1 to 60 days whose period of variation is related to their luminosity.

1785, they concluded that the star system has a disk shape with some noticeable "holes" lacking stars around its edges. Modern astronomers know that those apparent holes are caused by dense clouds of gas and dust blocking the view of more distant stars. The Herschels counted similar numbers of stars in most directions around the Milky Way and concluded that the sun and Earth are near the center of the star system.

As the 20th century began, astronomers still believed that the sun was located near the center of a wheel-shaped star system that they estimated was about 15,000 ly in diameter. How humanity realized the truth about the size of the galaxy, Earth's location in it, and the galaxy's location in a much larger universe of other galaxies, is an adventure that begins with a woman studying stars that pulsate and leads to a man studying distant star clusters. The next section will follow that story because it is an important historical moment in human history and because it illustrates how scientists build on the work of their predecessors and step-by-step refine their ideas about the natural world.

The Size of the Milky Way

It is a common misconception of poets that the stars are eternal and unchanging, but astronomers have known for centuries that some stars change in brightness. Of course, novae and supernovae burst into view, grow brighter, and then fade, but many other variable stars actually pulsate like beating hearts. The period of pulsation is the time it takes a star to complete a cycle from bright to faint to bright again. You will learn next how the properties of some types of variable stars allowed astronomers to measure the size of the galaxy.

In 1912, Henrietta Leavitt (1868–1921) was studying a star cloud in the southern sky known as the Small Magellanic Cloud. On her photographic plates she found many variable stars, and she noticed that the brightest had the longest periods. Because all of the variables were in the same cloud at nearly the same distance, she concluded that there was a relationship between the pulsation periods and intrinsic brightness (luminosities, i.e. true total power output) of those variable stars.

The stars Leavitt saw, **Cepheid variable stars,** are named after the first such star discovered, Delta Cephei.

Figure 9.2

William Herschel's diagram showing the star system as a thick disk seen edge-on, with the sun located near the center of the star system.

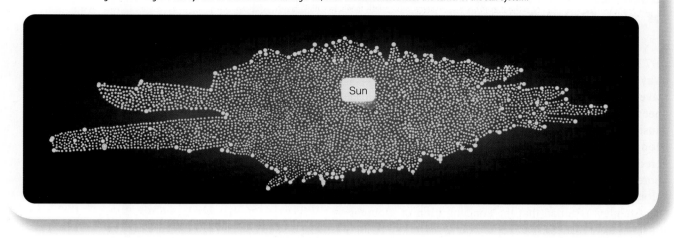

Sun

They are giant and supergiant stars with pulsation periods of 1 to 60 days with properties that lie in a region of the H-R diagram known as the **instability strip** (Figure 9.3). As stars evolve, and the points that represent their temperatures and luminosities move in the H-R diagram, they can cross into the instability strip and start pulsating; they stop pulsating when they evolve out of the strip. Massive stars are larger and pulsate slower, just as large bells vibrate slower and have deeper tones. Lower-mass stars are less luminous and, being smaller, pulsate faster.

This explains why, as first noticed by Leavitt, the long-period Cepheids are more luminous than the short-period Cepheids. That is now known as the **period–luminosity relation**, shown graphically in Figure 9.4. You may be interested to learn that the North Star, Polaris, is a Cepheid variable with a pulsation period of 4 days.

> **Instability strip:** The region of the H-R diagram in which stars are unstable to pulsation. A star evolving through this strip becomes a variable star.
>
> **Period–luminosity relation:** The relation between period of pulsation and intrinsic brightness among Cepheid variable stars.

Figure 9.3

The more massive a star is, the more luminous it is. Because both luminosity and period depend on mass, there is a relationship between period of pulsation and luminosity.

| O | B | A | F | G | K | M |

Stars evolving through the instability strip become unstable and pulsate.

$9M_\odot$

$5M_\odot$

$3M_\odot$

Sun

More massive stars are more luminous and larger, so they pulsate slower.

Stars evolving through the instability strip become unstable and pulsate as variable stars.

$L/L_\odot$

10^6

10^4

10^2

1

10^{-2}

10^{-4}

30,000 20,000   10,000   5,000   3,000

Temperature (K)

5.36634 days

3.5

4.0

4.5

Brightness of Delta Cephei

Visual magnitude

0   1   2   3   4   5   6   7   8

Time (days)

> Observations of Delta Cephei show the shape of its light curve.

Star Clusters and the Center of the Galaxy

A young astronomer named Harlow Shapley (1885–1972) began the discovery of the true nature of the Milky Way when he noticed that different kinds of star clusters have different distributions in the sky. In Chapter 8 you met two types of star clusters, open clusters and globular clusters (see page 148). Open clusters are concentrated along the Milky Way. Globular clusters are widely scattered, but Shapley noticed that the globular clusters were more common toward the constellations Sagittarius and Scorpius, as illustrated in Figure 9.5a.

Shapley assumed that the concentration of globular clusters is controlled by the combined gravitational field of all the stars in the galaxy. In that case, he realized, he could study the size and extent of the galaxy by studying the globular clusters. To do that, he needed to measure the distances to as many globular clusters as possible.

Globular clusters are much too far away to have measurable parallaxes, but they do contain variable stars. Shapley knew of Leavitt's work on these stars, which depended on their relative, rather than true, luminosities. Cepheids are rare, and there are none near enough to have measurable parallaxes, so their true luminosities were not then known. Shapley realized that he could calculate the distance to the globular clusters if he found the true luminosities of the Cepheid variable stars in the clusters.

Through a statistical process involving measurements of position shifts due to their motions through space, called **proper motions**, Shapley was able to find the average distances of a few of the nearest Cepheids, and from that their average luminosities. That meant he could replace Leavitt's apparent magnitudes with absolute magnitudes on the period–luminosity diagram (as shown in Figure 9.4). Astronomers say that Shapley **calibrated** the variable stars for distance.

Finally, having calibrated the Cepheids, Shapley could find the distance to the globular star clusters. He could identify the variable stars he found in the clusters and find their apparent magnitudes from his photographs. Comparison of apparent and absolute magnitudes gave him the distance to the star cluster based on the inverse square law for light (see Chapter 6).

Proper motion: The rate at which a star moves across the sky, measured in arc seconds per year.

Calibrate: To make observations of reference objects, checks on instrument performance, calculations of units conversions, and so on, needed to completely understand measurements of unknown quantities.

Figure 9.4

The period–luminosity diagram is a graph of the luminosity of variable stars versus their periods of pulsation. Because the diagram is used for comparison with apparent magnitudes for distance calculations, it is most convenient with absolute magnitudes on the vertical axis to express luminosity. Modern astronomers know that there are two types of Cepheids plus other similar types of variable stars, something that astronomers in the early 20th century could not recognize in their limited data, causing some errors in the first determinations of the size of the Milky Way.

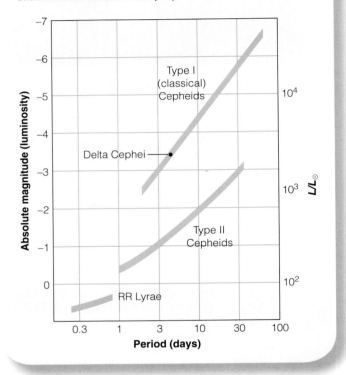

Shapley finally plotted the direction and distance to the globular clusters and found that they form a great swarm that is not centered on the sun. Instead, the center of the galaxy's cloud of clusters lies many thousands of light-years in the direction of Sagittarius. By analogy, if you see a bunch of tall buildings, and they are all together in one direction away from you, you might conclude that downtown is over there, and you are in the suburbs. Evidently the center of the star system is in Sagittarius and far away. The star system is much bigger than anyone had suspected (Figure 9.5b). You live not near the center of a small star system but in the suburbs of a very big wheel of stars, a galaxy.

Why did astronomers before Shapley think humanity lives near the center of a small star system? The answer: Space is filled with gas and dust that dims the

Figure 9.5

(a) Nearly half of cataloged globular clusters (red dots) are located in or near Sagittarius and Scorpius. A few of the brighter globular clusters, labeled with their catalog designations, are visible with binoculars or small telescopes. Constellations are shown as they appear above the southern horizon on a summer night as seen from latitude 40° N, typical for most of the United States. (b) A "side view" of the Milky Way Galaxy's globular clusters. Shapley's study of the clusters and their distances showed that they were not centered on the sun, which is located at the origin of this graph, but rather formed a great cloud centered far away in the direction of Sagittarius. Distances are given in thousands of parsecs and correspond to Shapley's original calibration that produced values more than 2× larger than the modern calibration.

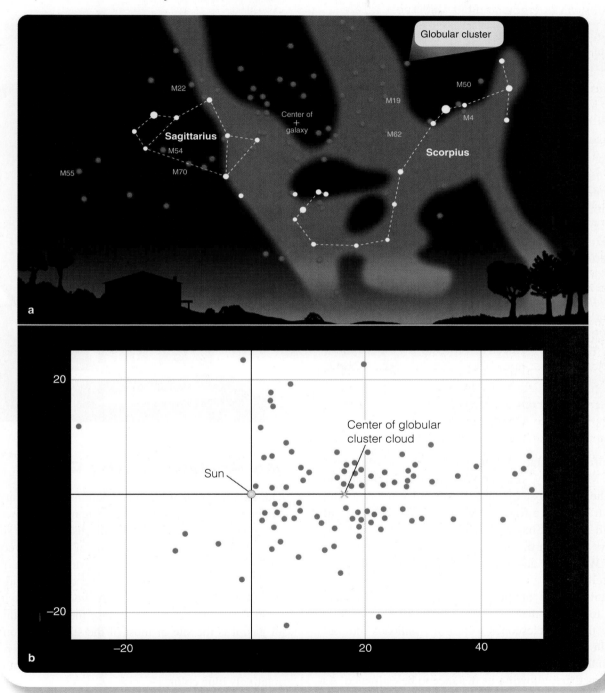

The Calibration of One Parameter to Find Another

Astronomers often say that Shapley "calibrated" the Cepheids for the determination of distance, meaning that he did all the detailed background work so that the Cepheids could be used to find distances. Other astronomers could then use the Cepheids without repeating the calibration.

Calibration is actually very common in science. Chemists, for instance, have carefully calibrated the colors of certain compounds against acidity. They can quickly measure the acidity of a solution by dipping into it a slip of paper containing the indicator compound and looking at the color. They don't have to repeat the careful calibration of the paper slips every time they measure acidity.

Engineers in steel mills have calibrated the color of molten steel against its temperature. They can use a handheld device that measures the color of the blackbody radiation from a ladle of molten steel and then looks up the temperature from a calibrated table. They don't have to repeat the calibration every time. Astronomers have made the same kind of color–temperature calibration for stars.

As you read about any science, notice how calibrations are used to simplify common measurements. But notice, too, how important it is to get the calibration right. An error in calibration can throw off every measurement made with that calibration. Some of the biggest errors in science have been errors of calibration.

© iStockphoto.com /
© Duff Bassett/iStockphoto.com

---

view of distant stars. When you look toward the band of the Milky Way, you can see only the neighborhood near the sun. Most of the star system is hidden and, like a traveler in a fog, you seem to be at the center of a small region. Shapley was able to see the globular clusters at greater distances because they lie outside the Milky Way plane and are not dimmed very much by the interstellar dust.

Building on Shapley's work, other astronomers began to suspect that some of the faint patches of light visible through telescopes were other galaxies like our own. In 1923, Edwin Hubble photographed individual stars in the Andromeda Galaxy, and in 1924 he identified Cepheids there, allowing its distance to be estimated. As a result, it became clear that our galaxy is just one in a universe filled with galaxies.

An Analysis of the Galaxy

Our galaxy, like many others, contains two primary components—a disk and a sphere. Figure 9.6 shows these components and other features discussed in this section.

The **disk component** consists of all matter confined to the plane of rotation—that is, everything in the disk itself. This includes stars, open star clusters, and nearly all of the galaxy's gas and dust. As you will learn in more detail in the next section, because the disk contains lots of gas, it is the site of most of the star formation in the galaxy. Consequently, the disk is illuminated by recently formed brilliant, blue, massive stars and has an overall relatively blue color.

The diameter of the disk and the position of the sun are difficult to determine accurately. Interstellar dust

Figure 9.6

An artist's conception of our Milky Way Galaxy, seen face-on and edge-on. Note the position of the sun and the distribution of globular clusters in the halo. Hot blue stars light up the spiral arms. Only the inner halo is shown here. At this scale, the entire halo would be larger than a dinner plate.

80,000 ly

Sun   Central bulge   Disk

Globular cluster   Halo

© NASA/JPL-Caltech/R. Hurt, SSC

blocks the view in the plane of the galaxy, so astronomers cannot see to the center or to the edge easily, and the outer edge of the disk is not well defined. Most recent studies suggest the sun is about 8.0 kpc from the center, where 1 kpc is a **kiloparsec**, or 1,000 pc. The sun and Earth seem to be about two-thirds of the way from the center to the edge, so the diameter of our galaxy appears to be about 25 kpc or about 80,000 ly, but that isn't known to better than 10 percent precision. This is the diameter of the luminous part of our galaxy, the part you would see from a distance. You will learn later that strong evidence shows that our galaxy is much larger than that, but the outer parts are not luminous.

Observations made at other than visual wavelengths can help astronomers peer through the dust and gas. Infrared and radio photons have wavelengths long enough to be unaffected by the dust. Thus, a map of the sky at long infrared wavelengths reveals the disk of our galaxy, which you can see in Figure 9.7.

The most striking features of the disk component are the **spiral arms**—long curves of bright stars, star clusters, gas, and dust. Such spiral arms are easily visible in other galaxies, and you will see later how astronomers found that our own galaxy has a spiral pattern.

The second component of our galaxy is the **spherical component**, which includes all matter in our galaxy scattered in a roughly spherical distribution around the center. This includes a large halo and the central bulge.

The **halo** is a spherical cloud of thinly scattered stars and globular star clusters. It contains only about 2 percent as many stars as the disk of the galaxy and has very little

*Figure 9.7

In these infrared images, the entire sky has been projected onto ovals with the center of the galaxy at the center of each oval. The Milky Way extends from left to right. In the near-infrared image, the central bulge is prominent, and dust clouds block the view along the Milky Way. At far-infrared wavelengths, the dust emits significant blackbody radiation and glows brightly.

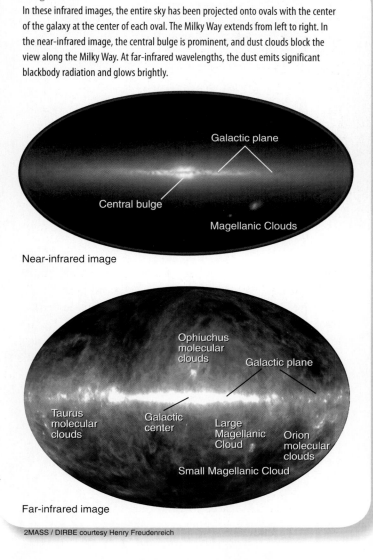

Galactic plane

Central bulge

Magellanic Clouds

Near-infrared image

Ophiuchus molecular clouds

Galactic plane

Taurus molecular clouds

Galactic center

Large Magellanic Cloud

Orion molecular clouds

Small Magellanic Cloud

Far-infrared image

2MASS / DIRBE courtesy Henry Freudenreich

gas and dust. Thus, with no raw material available, no new stars are forming in the halo. In fact, the halo stars are mostly old, cool giants or dim lower-main-sequence stars plus, as revealed by recent careful studies, old white dwarfs that are difficult to detect. Astronomers can map the halo of our galaxy by studying the more easily detected giant stars.

The **central bulge** is the dense cloud of stars that surrounds the center of our galaxy. It has a radius of about 2 kpc and is slightly flattened. It is hard to observe because thick dust in the disk scatters and absorbs radiation of visible wavelengths, but observations at longer wavelengths can penetrate the dust. The bulge seems to contain little gas and dust, and there is not much star

DESCRIBING OUR GALAXY

DISK COMPONENT: All material confined to the plane of the galaxy.

KILOPARSEC: A unit of distance equal to 1,000 pc or 3,260 ly.

SPIRAL ARMS: Long spiral pattern of bright stars, star clusters, gas, and dust. Spiral arms extend from the center to the edge of the disk of spiral galaxies.

SPHERICAL COMPONENT: The part of the galaxy including all matter in a spherical distribution around the center (the halo and central bulge).

HALO: The spherical region of a spiral galaxy, containing a thin scattering of stars, star clusters, and small amounts of gas.

NUCLEAR BULGE: The spherical cloud of stars that lies at the center of spiral galaxies.

ROTATION CURVE: A graph of orbital velocity versus radius in the disk of a galaxy.

DARK HALO: The low-density extension of the halo of our galaxy, believed to be composed of dark matter.

DARK MATTER: Nonluminous matter that is detected only by its gravitational influence.

formation there. Most of the stars in the central bulge are old, cool stars like those in the halo.

The Mass of the Galaxy

The vast numbers of stars in the disk, halo, and central bulge lead to a basic question: How massive is the galaxy?

When you needed to find the masses of stars, you studied the orbital motions of the pairs of stars in binary systems. To find the mass of the galaxy, you must look at the orbital motions of the stars within the galaxy. Every star in the galaxy follows an orbit around the center of mass of the galaxy. In the disk, the stars follow parallel circular orbits, and astronomers say the disk of the galaxy rotates. That rotation can allow an estimate of the mass of the galaxy. Consequently, any discussion of the mass of the galaxy is also a discussion of the rotation of the galaxy and the orbits of the stars within the galaxy.

Astronomers can find the orbits of stars by finding how they move. Of course, the Doppler effect reveals a star's radial velocity (see Chapter 5). In addition, if astronomers can measure the distance to a star and its proper motion, they can find the velocity of the star perpendicular to the radial direction. Combining all of this information, astronomers can find the shape of the star's orbit.

As you can see in Figure 9.8, the orbital motions of the stars in the halo are strikingly different from those in the disk. In the halo, each star and globular cluster follows its own randomly tipped elliptical orbit. These orbits carry the stars and clusters far out into the spherical halo, where they move slowly, but when they fall back into the inner part of the galaxy, their velocities increase. Motions in the halo do not resemble a general rotation but are more like the random motions of a swarm of bees. In contrast, the stars in the disk of the galaxy move in the same direction in nearly circular orbits that lie in the plane of the galaxy. The sun is a disk star and follows a nearly circular orbit around the galaxy that always remains within the disk.

You can use the orbital motion of the sun to find the mass of the galaxy inside the sun's orbit. By observing the radial velocity of distant galaxies in various directions around the sky, astronomers can tell that the sun moves about 220 km/s in the direction of Cygnus, carrying Earth and the other planets of our solar system along with it. Because its orbit is a circle with a radius of 8.0 kpc, you can divide the circumference of the orbit by

the velocity and find that the sun completes a single orbit in about 220 million years.

If you think of the sun and the center of mass of our galaxy as two objects orbiting each other, you now have enough information to determine the mass of the galaxy. The Milky Way Galaxy must have a mass of about 100 billion solar masses. This estimate is uncertain for a number of reasons. First, you don't know the radius of the sun's orbit with great certainty. Astronomers estimate the radius as 8.0 kpc, but they could be wrong by 10 percent or more, and the radius gets cubed in the calculation, which has a large effect. Second, this estimate of the mass includes only the mass inside the sun's orbit. Mass spread uniformly outside the sun's orbit will not affect its orbital motion. Thus 100 billion solar masses is a lower limit for the mass of the galaxy, but no one knows exactly how much to increase the estimate to include the rest of the mass in the galaxy that lies outside the sun's orbit.

The measured motion of the stars shows that the disk does not rotate as a solid body. Each star follows its own orbit, and stars in some regions have shorter or longer orbital periods than the sun. This is called differential rotation. (Recall that you met the term *differential rotation* in Chapter 5 where it was applied to the sun.) The graph in Figure 9.9 is an example of the orbital velocity of stars at various orbital radii in the galaxy, called a **rotation curve**. If all of the mass in the galaxy were concentrated at its center, then orbital velocity would be high near the center and would decline away from the center. This kind

Figure 9.8

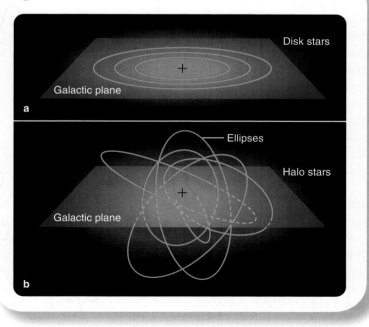

Figure 9.9

The rotation curve of our galaxy is plotted here as orbital speed versus radius. Data points show measurements made by radio telescopes. Observations outside the orbit of the sun are much more uncertain, and the data points scatter widely. Orbital speeds do not decline outside the orbit of the sun, as you would expect if most of the mass of the galaxy were concentrated toward the center. Rather, the curve is approximately flat at great distances, suggesting that the galaxy contains significant mass outside the orbit of the sun.

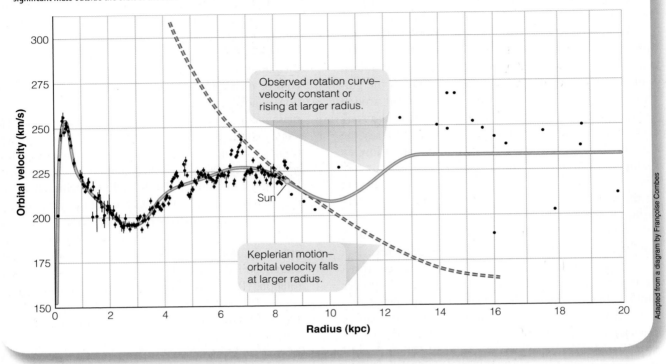

Observed rotation curve—velocity constant or rising at larger radius.

Keplerian motion—orbital velocity falls at larger radius.

Adapted from a diagram by Françoise Combes

of motion has been called *Keplerian motion* because it follows Kepler's third law, as in the case of our solar system, where nearly all of the mass is in the sun.

Of course, you already know the galaxy's mass is not all concentrated at its center, but if most of the mass were inside the orbit of the sun, then orbital velocities should decline at greater distances. Many observations confirm, however, that velocities do not decline and may actually increase at greater distance; this observation shows that larger and larger orbits enclose more and more mass. Although it is difficult to determine a precise edge to the visible galaxy, it seems clear that large amounts of matter are located beyond what seems to be the limit of the galaxy's luminous matter.

The evidence is clear that extra mass lies in an extended halo sometimes called a **dark halo**. It may extend up to 10 times farther than the edge of the visible disk and could contain up to two trillion solar masses. Some small fraction of this mass is made up of low-luminosity stars and white dwarfs, but most of the matter is not producing any light. Astronomers call it **dark matter** and conclude that it must be some as yet unknown form of matter. The following two chapters will return to the question of dark matter. It is one of the fundamental problems of modern astronomy.

9.2 Spiral Arms and Star Formation

The most striking feature of galaxies like the Milky Way is the system of spiral arms that wind outward through the disk and contain swarms of hot, blue stars; clouds of dust and gas; and young star clusters. These young objects hint that the spiral arms involve star formation. As you try to understand the spiral arms, you face two questions. First, how can you be sure our galaxy has spiral arms when Earth is embedded inside the galaxy and your view is obscured by gas and dust? Second, why doesn't the differential rotation of the galaxy destroy the arms? The answers to both questions involve star formation.

Tracing the Spiral Arms

Studies of other galaxies show that spiral arms contain hot, blue stars. Thus, one way to study the spiral arms of our own galaxy is to locate these stars. Fortunately, this is not difficult, since O and B stars are often found in associations and, being very luminous, are easy to detect across great distances. Unfortunately, at these great distances their parallax is too small to measure, so their distances must be found by other means, usually by spectroscopic parallax (see Chapter 6).

O and B associations in the sky are not located randomly but reveal parts of three spiral arms near the sun, which have been named for the prominent constellations through which they pass (Figure 9.10). Objects used to map spiral arms are called **spiral tracers**. O and B associations are good spiral tracers because they are bright and easy to see at great distances. Other tracers include young open clusters, clouds of hydrogen ionized by hot stars (emission nebulae), and certain higher-mass variable stars.

Notice that all spiral tracers are young objects. O stars, for example, live for only a few million years. If their orbital velocity is about 200 km/s, they cannot have moved more than about 500 pc since they formed. This is less than the width of a spiral arm. Because they don't live long enough to move away from the spiral arms, they must have formed there. The youth of spiral tracers provides an important clue about spiral arms. Somehow they are associated with star formation.

Radio telescopes can detect emission from clouds of cool neutral hydrogen gas. Radio maps of the galaxy disk, combined with optical and infrared data, allow astronomers to deduce the spiral pattern of our galaxy. The segments near the sun are part of a spiral pattern that continues throughout the disk. However, the maps show that the spiral arms are rather irregular and interrupted by branches, spurs, and gaps. The stars in Orion, for example, appear to be a detached segment of a spiral arm, a spur. There are significant sources of error in the mapping methods, but many of the irregularities along the arms seem real, and images of nearby spiral galaxies show similar features (Figure 9.10a). Studies comparing all of the available data on our galaxy's spiral pattern with patterns seen in other galaxies do not necessarily agree, although the newest models suggest that the central bulge in our galaxy is elongated into a bar (Figure 9.11). You will see in the next chapter that such bars are common in spiral galaxies.

Spiral tracers show that the arms contain young objects, and that suggests active star formation. The

Spiral tracer: Object used to map the spiral arms—for example, O and B associations, open clusters, clouds of ionized hydrogen, and some types of variable stars.

Figure 9.10

(a) Many of the galaxies in the sky are disk shaped, and most of those galaxies have spiral arms. Images of other galaxies show that spiral arms are marked by hot, luminous stars that must be very young; and this should make you suspect that spiral arms are related to star formation. (b) Gas and dust block your view of most of the disk of our galaxy, but near the sun, young O and B stars fall along bands that appear to be segments of spiral arms.

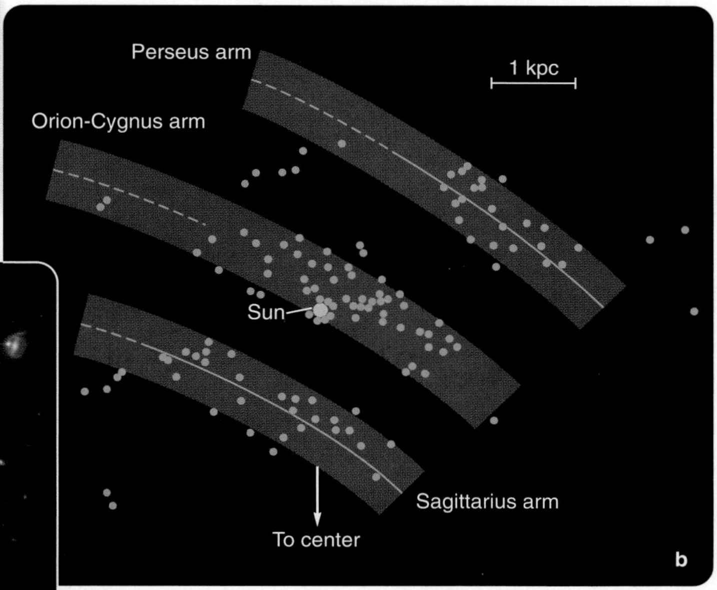

Visual-wavelength image

Spiral galaxy NGC 3370 contains many spiral arms.

NASA/Hubble Heritage Team

a

Perseus arm

Orion-Cygnus arm

1 kpc

Sun

To center

Sagittarius arm

b

Figure 9.11

This artist's impression of a two-armed model is based on observations with the Spitzer infrared space telescope. Notice the large central bar.

NASA/JPL-Caltech

The Spiral Density Wave

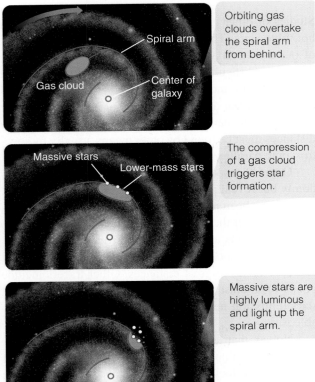

Orbiting gas clouds overtake the spiral arm from behind.

The compression of a gas cloud triggers star formation.

Massive stars are highly luminous and light up the spiral arm.

The most massive stars die quickly.

Low-mass stars live long lives but are not highly luminous.

radio maps confirm this suspicion by showing that the material needed to make stars, hydrogen gas, is abundant in spiral arms.

Star Formation in Spiral Arms

Having mapped the spiral pattern, you can ask, "What are spiral arms?" You can be sure that they are not physically connected structures. Like a kite string caught on a spinning hubcap, such arms would be wound up and pulled apart by differential rotation within a few tens of millions of years. Yet spiral arms are common in disk-shaped galaxies and must be reasonably permanent features.

The most prominent theory about spiral arms is called the **density wave theory**, which proposes that the arms are waves of compression that move around the galaxy triggering star formation. The density wave is a bit like a traffic jam behind a truck moving slowly along a highway. Seen from an airplane overhead, the jam seems a permanent, though slow-moving, feature. But individual cars overtake the jam from behind, slow down, move up through the jam, wait their turn, pass the truck, and re-

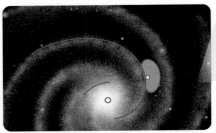

Density wave theory: Theory proposed to account for spiral arms as compressions of the interstellar medium in the disk of the galaxy.

sume speed along the highway. Similarly, clouds of gas overtake the spiral density wave, become compressed in the "traffic jam," and eventually move out in front of the arm, leaving the slower-moving density wave behind.

Of course, star formation will occur where the gas clouds are compressed. Stars pass through the spiral arms unaffected, like bullets passing through a wisp of fog, but large clouds of gas slam into the spiral density wave from behind and are suddenly compressed. The Spiral Density Wave series of images on page 184 demonstrates how this movement occurs. You saw in Chapter 7 that sudden compression could trigger the formation of stars in a gas cloud. Thus, new stars should form along the spiral arms. The spiral arms are not wound up by differential rotation because they are patterns, not physically connected structures.

The brightest stars, the O and B stars, live such short lives that they never travel far from their birthplace and are found only along the arms. Their presence is what makes the spiral arms glow so brightly, due to both their own light and the emission from clouds of gas excited by UV radiation from the stars, as in Figure 9.12a. Lower-mass stars, like the sun, live longer and have time to move out of the arms and continue their journey around the galaxy. The sun may have formed in a star cluster almost 5 billion years ago when a gas cloud smashed into a spiral arm. Since that time, the sun has escaped from its birth cluster and made about 20 trips around the galaxy, passing through spiral arms many times.

The density wave theory is very successful in explaining general properties of spiral galaxies, but it has two problems. First, what stimulates the formation of the spiral pattern in the first place? Theorists calculate that minor fluctuations in the galaxy's disk shape or gravitational interactions with passing galaxies may be able to start a density wave. Second, the density wave theory does not account for the branches and spurs in the spiral arms of our own and other galaxies. The solution to this second problem may lie in a process that sustains star formation once it begins.

Figure 9.12

(a) Some galaxies are dominated by two spiral arms; but, even in these galaxies, minor spurs and branches are common. The spiral density wave can generate the two-armed, grand-design pattern, but another process called self-sustained star formation may be responsible for the irregularities. (b) Many spiral galaxies do not appear to have two dominant spiral arms. Multiple spurs and branches suggest that star formation is proceeding rapidly in such galaxies.

Visual-wavelength images

a M100

b NGC 300

Star formation can control the shape of spiral patterns if the birth of stars in a cloud of gas can cause the birth of more new stars. Massive stars evolve so quickly that their lifetimes are only an instant in the history of a galaxy, and then they explode as supernovae. You learned in Chapter 7 that bursts of luminosity and jets from newborn stars and the expanding gases of supernova explosions can compress neighboring clouds of gas and trigger more star formation. This process is known as **self-sustaining star formation**. The Orion complex, consisting of the Great Nebula in Orion plus the protostars buried deep in the dark interstellar clouds behind the nebula (see pages 140–141), seems to be a region of self-sustaining star formation.

Astronomers have calculated models indicating that the differential rotation of the galaxy can drag the inner edge of a star-forming region ahead and let the outer edge lag behind to produce a cloud of star formation shaped like a segment of a spiral arm. Self-sustaining star formation plus differential rotation may produce the branches and spurs so prominent in some galaxies (Figure 9.12b), including our own, but only the spiral density wave can generate the beautiful two-armed grand spiral patterns.

9.3 The Origin and History of the Milky Way

Just as paleontologists reconstruct the history of life on Earth from the fossil record, astronomers try to reconstruct our galaxy's past from the fossil it left behind as it formed and evolved. That fossil is the spherical component of the galaxy. The stars in the halo formed when the galaxy was young. The chemical composition and the distribution of these stars can provide clues to how our galaxy formed.

The Age of the Milky Way

To begin, you should ask yourself how old our galaxy is. That question is easy to answer because you already know how to find the age of star clusters. But there are uncertainties that make that easy answer hard to interpret.

The oldest open clusters (see page 149) have ages of about 9 to 10 billion years. These ages are determined by analyzing the turnoff point in the cluster H-R diagram (see Chapter 8) and are somewhat uncertain. Nevertheless, from open clusters you can get a rough age for the disk of our galaxy of at least 9 billion years.

Globular clusters have faint turnoff points in their H-R diagrams and are clearly old, but finding their ages accurately is difficult. Clusters differ slightly in chemical composition, which must be accounted for in calculating the stellar models from which ages are determined. Also, to find the age of a cluster, astronomers must know the distance to the cluster. Precise parallaxes from the Hipparcos satellite have allowed astronomers to increase the precision of the Cepheid variable stars' calibration, and careful studies with the newest large telescopes have refined the cluster H-R diagrams. Globular cluster ages seem to average about 11 billion years, with the oldest being a bit over 13 billion years old. The halo of our galaxy therefore seems to be at least 13 billion years old, older than the disk.

Stellar Populations

In the 1940s, astronomers realized that there were two types of stars in the galaxy. The stars they were most accustomed to studying were located in the disk, such as the stars near the sun. These they called **population I stars**. The second type, called **population II stars**, are usually found in the halo, in globular clusters, or in the central bulge. In other words, the two stellar populations are associated respectively with the disk and spherical components of the galaxy.

The stars of the two populations fuse nuclear fuels and evolve in nearly identical ways. They differ only in their abundance of atoms heavier than helium, atoms that astronomers refer to collectively as **metals**. (Note that this is definitely not the way the word *metal* is commonly used by non-astronomers.) Population I stars are relatively metal rich, containing 2 to 3 percent metals,

Table 9.1 Stellar Populations

	Population I		Population II	
	Extreme	Intermediate	Intermediate	Extreme
Location	Spiral arms	Disk	Central bulge	Halo
Metals (%)	3	1.6	0.8	Less than 0.8
Shape of orbit	Circular	Slightly elliptical	Moderately elliptical	Highly elliptical
Average age (yr)	0.2 billion and younger	0.2–10 billion	2–10 billion	10–13 billion

while population II stars are metal poor, containing only about 0.1 percent metals or less. The metal content of a star defines its population.

Population I stars, sometimes called *disk population stars*, have circular orbits in the plane of the galaxy and are relatively young stars that formed within the last few billion years. The sun is a population I star. Population II stars belong to the spherical component of the galaxy and are sometimes called *halo population stars*. These stars have randomly tipped orbits with a wide range of shapes. A few follow circular orbits, but most follow elliptical orbits. The population II stars are all lower-mass main-sequence stars or giants. They are old stars. The metal-poor globular clusters are part of the halo population. Since the discovery of stellar populations, astronomers have realized that there is a gradation between populations, illustrated in Table 9.1.

Why do the disk and halo stars have different metal abundances? The two types of star must have formed at different stages in the life of the galaxy, at times when the chemical composition of the galaxy differed. This is a clue to the history of our galaxy, but to use the clue, you must understand the cycle of element building.

The Element-Building Cycle

The atoms of which you are made were created in a process that spanned a number of generations of stars. The process that built the chemical elements over the history of our galaxy led to the possibility of Earth and life on Earth.

You saw in Chapter 8 how elements heavier than helium but lighter than iron are built up by nuclear reactions inside evolving stars. More massive atoms are made only by short-lived nuclear reactions that occur during a supernova explosion. This explains why lower-mass atoms like carbon, nitrogen, and oxygen are more com-

mon and why more massive atoms—such as gold, silver, platinum, and uranium—are so rare and, often, valuable. Figure 9.13a shows the abundance of the chemical elements, but notice that the graph has an exponential scale. To get a feeling for the true abundance of the elements, you should draw this graph using a linear scale as in Figure 9.13b. With that scale you see how rare the elements heavier than helium really are.

Most of the matter in stars is hydrogen and helium. Other elements, including carbon, nitrogen, oxygen, and the rest of what astronomers call metals, were cooked up inside stars. When the galaxy first formed, there should have been no metals because stars had not yet manufactured any. Judging from the composition of stars in the oldest clusters, the gas from which the galaxy originally condensed must have contained about 90 percent hydrogen atoms and 10 percent helium atoms. (The hydrogen and helium came from the big bang that began the universe, which you will study in Chapter 11.)

The first stars to form from this gas were metal poor, and now, 13 billion years later, their spectra still show few metal lines. Of course, they may have manufactured many atoms heavier than helium in their cores, but because the stars' interiors are not mixed, those heavy atoms stay trapped at the centers of the stars where they were produced and do not affect the spectra (Figure 9.14). The population II stars in the halo are the survivors of an earlier generation of stars that formed in the galaxy.

Most of the first stars evolved and died, and various types of star death throes, including supernovae, enriched the interstellar gas with metals. Succeeding generations of stars formed from gas clouds that were more enriched, and each generation added to the enrichment with its death. By the time the sun formed, roughly 5 billion years ago, the element-building process had added about 1.6 percent metals. Since then, the metal abundance has increased further, and stars forming at the present time incorporate 2 to 3 percent metals and become extreme population I stars. Thus metal abundance varies between populations because of the production of heavy atoms in successive generations of stars as the galaxy aged.

The oxygen atoms you are breathing, the carbon atoms in your flesh and the calcium atoms in your bones

Figure 9.13

The abundance of the elements in the universe. (a) When the elements are plotted on an exponential scale, you see that elements heavier than iron are about a million times less common than iron and that all elements heavier than helium (referred to by astronomers as "metals") are quite rare. (b) The same data plotted on a linear scale provide a more realistic impression of how rare the metals are. Carbon, nitrogen, and oxygen make small peaks near atomic mass 15, and iron is just visible in the graph.

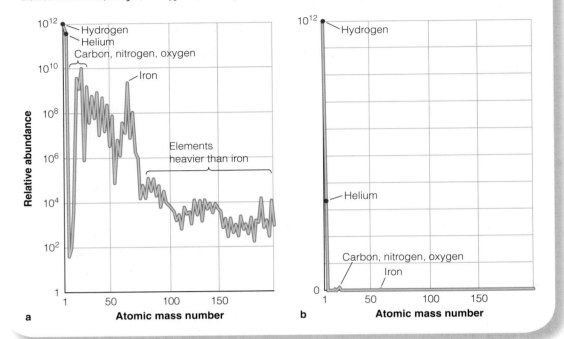

Figure 9.14

The difference between spectra of population I stars and population II stars is dramatic. (a) A graph of a population I star's spectrum reveals overlapping absorption lines of metals completely blanketing the continuum. (b) The lower spectrum is that of an extremely metal-poor star with only a few, weak metal lines of iron (Fe) and nickel (Ni). This population II star has about 10,000 times less metal content than the sun.

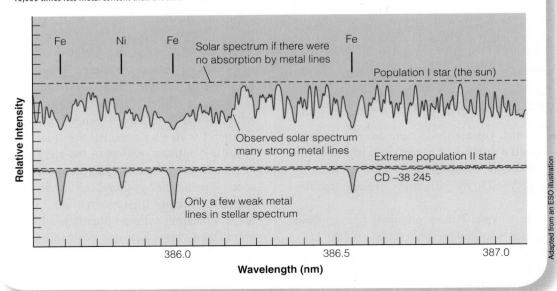

were all created in the interiors of red and yellow giant stars hot enough to sustain fusion reactions beyond helium. The gold and silver atoms in your jewelry and dental fillings, and the iodine atoms in your thyroid gland were all created in a few moments during supernova explosions, because those are the only places in the universe hot enough to make these types of atoms. This idea is perhaps the most significant one to take away from an introductory astronomy course and textbook—you, and everything on Earth, are made of stardust.

The History of the Milky Way Galaxy

The lack of metals in the spherical component of the galaxy tells you it is very old, a fossil left behind by the galaxy when it was young and drastically different from its present disk shape. The study of element building and stellar populations leads to the fundamental question, "How did our galaxy form?"

In the 1950s, astronomers began to develop a hypothesis, sometimes called the **monolithic collapse model**, to explain the formation of our galaxy. Recent observations are forcing a major reevaluation and revision of that traditional hypothesis.

The traditional hypothesis says that the galaxy formed from a single large cloud of turbulent gas over 13 billion years ago. Stars and star clusters that formed from this material went into randomly shaped and randomly tipped orbits. These first stars were metal poor because no stars had existed earlier to enrich the gas with metals. In this way, the initial contraction of a large, turbulent gas cloud produced the spherical component of the galaxy.

The second stage of this hypothesis accounts for the disk component. The turbulent motions would eventually have canceled out, leaving the cloud with uniform rotation. A rotating, low-density cloud of gas cannot remain spherical. A star is spherical because its high internal pressure balances its gravity, but in a low-density cloud, the pressure cannot support the weight. Like a blob of pizza dough spun in the air, such a cloud must flatten into a disk, as the figure at right demonstrates.

This contraction into a disk took billions of years, with the metal abundance gradually increasing as generations of stars were born from the gradually flattening gas cloud. The stars and globular clusters that formed first in the halo would not have been affected by the motions of the gas and would have been left behind by the cloud as it collapsed and flattened. Later generations of stars formed in flatter distributions. The gas distribution

Monolithic Collapse Model

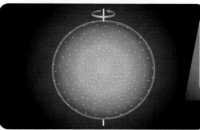

A spherical cloud of turbulent gas gives birth to the first stars and star clusters.

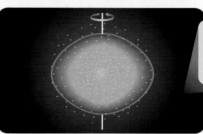

The rotating cloud of gas begins to contract toward its equatorial plane.

Stars and clusters are left behind in the halo as the gas cloud flattens.

New generations of stars have flatter distributions.

The disk of the galaxy is now very thin.

in the galaxy now is so flat that the youngest stars are confined in a thin disk only about 100 parsecs thick. These stars are metal rich and have nearly circular orbits.

Monolithic collapse model: An early hypothesis that says that the galaxy formed from the collapse of a single large cloud of turbulent gas over 13 billion years ago.

Figure 9.15

Looking toward Sagittarius, you see nothing to suggest this is the center of the galaxy. Gas and dust block your view. Only the distribution of globular clusters told Shapley the sun lies far from the center of the star system.

Center of galaxy

Visual-wavelength image

© Daniel Good

This monolithic collapse hypothesis accounts for many of the Milky Way's properties. Advances in technology, however, have improved astronomical observation; and, beginning in the 1980s, contradictions between theory and observations arose. For example, not all globular clusters have the same age, but, surprisingly, some of the younger clusters seem to be in the outer halo. In contrast, the monolithic collapse hypothesis says that the halo formed first and predicts that the clusters within it should either have a uniform age or the most distant ones should be slightly older. Another problem is that the oldest stars are observed to be metal poor but not completely metal free. There must have been at least a few massive stars to create these metals during a generation before the formation of the oldest stars now seen in the halo.

Can the original hypothesis be modified to explain these observations? Perhaps the galaxy began with the contraction of a gas cloud to form the central bulge, and the halo later accumulated from gas clouds that had been slightly enriched in metals by an early generation of massive stars. That first generation of stars would have formed from almost pure hydrogen and helium gas, which mathematical models indicate would form only stars with very high masses. Those stars would have lived very short lives, made metals, and died in supernova ex-

plosions; and none would be left today. This would explain the metals in the oldest stars surviving today.

There is also evidence that entire small galaxies were captured by the growing Milky Way, and that fresh gas was added to the disk in several episodes over the life of the galaxy and not in a single gravitational collapse when the galaxy first formed. If our galaxy absorbed a few small but partially evolved galaxies, then some of the globular clusters in the halo may be hitchhikers that originally belonged to the captured galaxies. This could explain the range of globular cluster ages and compositions. (You will see in the next chapter that such galaxy mergers do occur.)

The problem of the formation of our galaxy is frustrating because the explanations are incomplete. The older monolithic collapse hypothesis has proven inadequate to explain all of the observations and you see astronomers attempting to refine the observations and devise new theories. The metal abundances and ages of the stars in our galaxy seem to be important clues, but metal abundance and age do not tell the whole story.

9.4 The Nucleus

The most mysterious region of our galaxy is its very center, the nucleus. At visual wavelengths, this region is totally hidden by gas and dust that dim the light by 30 magnitudes (Figure 9.15). If a trillion (10^{12}) photons of light left the center of the galaxy on a journey to Earth, only one would make it through the gas and dust. The longer-wavelength infrared photons are scattered much less often: One in every ten makes it to Earth. Consequently, visual-wavelength observations reveal nothing about the nucleus, but it can be observed at longer wavelengths such as in the infrared and radio parts of the spectrum.

Observations of the Galactic Nucleus

Harlow Shapley's study of globular clusters placed the center of our galaxy in Sagittarius, and the first radio maps of the sky showed a powerful radio source located in Sagittarius. The first infrared map of the central bulge made by Eric Becklin in 1968 showed the location of intense radiation where the stars are most crowded together, identifying the gravitational center of the galaxy. Later high-resolution radio maps of that stellar center revealed a complex collection of radio sources, with one, **Sagittarius A*** (abbreviated Sgr A* and pronounced "sadge A-star"), lying at the expected location of the galactic core.

Observations show that Sgr A* is only a few astronomical units in diameter but is a powerful source of radio energy. The tremendous amount of infrared radiation coming from the central area appears to be produced by crowded stars and by dust warmed by those stars. But what could be as relatively tiny as Sgr A* and produce so much radio energy?

Study "Sagittarius A*" on pages 192–193 and notice three important points:

1. First, observations at radio wavelengths reveal complex structures near Sgr A* caused by magnetic fields and by rapid star formation. Supernova remnants show that massive stars have formed there recently and died explosively.

2. Also, the center is very crowded. Tremendous numbers of stars plus radiation from Sgr A* heat the dust, producing strong infrared emission.

3. Finally, there is evidence that Sgr A* is a supermassive black hole into which gas is flowing.

Astronomers continue to test the hypothesis that the center of our galaxy contains a supermassive black hole. Such an object is sufficient to explain the observations, but is it necessary? Is there some other way to explain what is observed? For example, astronomers have suggested that gas flowing toward the center of the galaxy could trigger tremendous bursts of star formation. Such hypotheses have been considered and tested against the evidence, but none appears to be adequate to explain all the observations. So far, the only hypothesis that seems adequate is that our galaxy's nucleus is home to a supermassive black hole.

Meanwhile, observations are allowing astronomers to improve their models. For instance, Sgr A* is not as bright in X rays as it should be if it had a hot accretion disk with matter constantly flowing into the black hole. Observations of X-ray and infrared flares lasting only a few hours suggest that mountain-size blobs of matter may occasionally fall into the black hole and be heated and ripped apart by tidal forces (see Chapter 8). The black hole may be mostly dormant and lack a fully developed hot accretion disk because the rate of matter flow into it is relatively low at the present time.

Such a supermassive black hole could not be the remains of a single dead star. It contains much too much mass. It probably formed when the galaxy first formed over 13 billion years ago. In the next chapter you will learn that supermassive black holes are commonly found in the centers of galaxies.

Sagittarius A*: The powerful radio source located at the core of the Milky Way Galaxy.

recap Hang on tight. The sun, with Earth in its clutch, is ripping along at high velocity as it orbits the center of the Milky Way Galaxy, our "parent" galaxy. Except for hydrogen atoms, which have survived unchanged since the universe began, you and Earth are made of metals, atoms heavier than helium. All of these atoms were cooked up by stars during their life—or death—throughout the history of our galaxy. Each generation of stars has produced elements heavier than helium and spread them into the interstellar medium. The abundance of metals has grown slowly in the galaxy. About 4.6 billion years ago a cloud of gas enriched by heavy atoms slammed into a spiral arm and produced the sun, the Earth, and you.

With your expanded knowledge of your home galaxy, you should be able to answer the following questions:

1. How do you know you live in a galaxy?

2. How do you know ours is a spiral galaxy, and what are the spiral arms?

3. How did our galaxy form and evolve?

4. What lies at the very center of our galaxy?

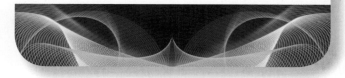

Sagittarius A*

1   There is so much interstellar dust in the plane of the Milky Way that you cannot observe the nucleus of our galaxy at visual wavelengths. The image below is a radio image of the innermost 300 pc. Many of the feature are supernova remnants (labeled SNR), and a few are star formation clouds. Peculiar features such as threads, the Arc, and the Snake may be gas trapped in magnetic fields. At the center of the image lies Sagittarius A, the location of the nucleus of our galaxy.

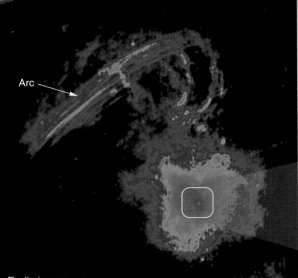

Arc

Radio image

NRAO/AUI/NSF

The radio map above shows Sgr A and the Arc filaments, 50 parsecs long. The image was made with the VLA radio telescope. The contents of the white box are shown on the opposite page.

Sgr D HII

Sgr D SNR

SNR 0.9 + 0.1

Sgr B2

Apparent angular size of the moon for comparison

Sgr B1

New SNR 0.3 + 0.0

Threads

The Cane

Arc

Background galaxy

Sgr A

Threads

Radio image

NRL

2   Infrared photons with wavelengths longer than 4 microns (4000 nm) are emitted almost entirely by warm interstellar dust. The radiation at those wavelengths coming from Sagittarius is intense, indicating that this region contains lots of dust, and is crowded with stars that warm the dust.

Sgr C

The Pelican

Coherent structure?

Snake

Sgr E

SNR 359.1 − 00.5

2MASS

Infrared image

1a This high-resolution radio image of Sgr A (within the white box in the small-scale map on the opposite page) reveals a spiral swirl of gas around an intense radio source known as Sgr A*. About 3 pc across, this spiral lies in a low-density cavity inside a larger disk of neutral gas. The arms of the spiral are thought to be streams of matter flowing into Sgr A* from the inner edge of the larger disk (drawing at right).

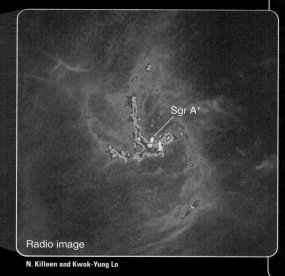

Radio image

N. Killeen and Kwok-Yung Lo

Evidence of a Black Hole in the Nucleus of Our Galaxy

3 Since the middle 1990s, astronomers have been able to use large infrared telescopes and active optics to follow the motions of stars orbiting around Sgr A*. A few of those orbits are shown here. The size and period of the orbit allows astronomers to calculate the mass of Sgr A* using Kepler's third law. The orbital period of the star SO-2, for example, is 15.2 years, and the semimajor axis of its orbit is 950 AU. The combined motions of the observed stars suggest that Sgr A* has a mass of 4 million solar masses.

The Chandra X-ray Observatory has imaged Sgr A* and detected over 2000 other X-ray sources in the area.

NASA/CXC/MIT/F.K. Baganoff et al.

Infrared image

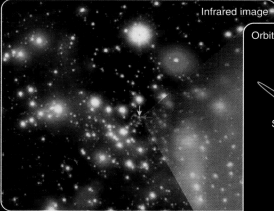

At its closest, SO-2 comes within 17 light-hours of Sgr A*. Alternative theories that Sgr A* is a cluster of stars, of neutron stars, or of stellar black holes are eliminated. Only a single black hole could contain so much mass in so small a region.

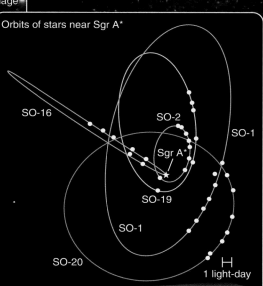

Orbits of stars near Sgr A*

SO-16
SO-2
SO-1
Sgr A*
SO-19
SO-1
SO-20

⊢—⊣
1 light-day

For comparison, the diameter of the planetary region of our solar system, defined by Neptune's orbit, is half a light-day.

3a A black hole with a mass of 4 million solar masses would have an event horizon with a size on the scale of this diagram smaller than the period at the end of this sentence. A slow dribble of only 0.0002 solar mass of gas per year flowing into the black hole could produce the observed energy. A sudden increase, such as when a star falls in, could produce a violent eruption.

The evidence of a massive black hole at the center of our galaxy seems conclusive. It is much too massive to be the remains of a dead star, however, and astronomers conclude that it probably formed as the galaxy first took shape.

Chapter 10

Galaxies

L ess than a century ago, astronomers did not even understand that galaxies exist. Nineteenth-century telescopes revealed faint nebulae scattered among the stars, some with spiral shapes. Astronomers argued about the nature of these nebulae, but it was not until the 1920s that astronomers understood that many of those nebulae are other galaxies much like our own, and it was not until recent decades that astronomical telescopes could reveal the variety, intricacy, and beauty of the galaxies. In this chapter, you will try to understand how galaxies form and evolve, and you will discover that the amount of gas and dust in a galaxy is a critical clue. You will also discover that interactions between galaxies can dramatically influence their structure and evolution.

Before you can build theories, however, you need to gather some basic data concerning galaxies. You can classify the different kinds of galaxies and discover their basic properties—diameter, luminosity, and mass. Once you know the typical properties of galaxies, you will be ready to theorize about their origin and evolution. You also will learn about some of the most energetic events in the universe: The energy pouring out of the nuclei of certain galaxies is enormously greater than mere supernova explosions.

sections

10.1 The Family of Galaxies

10.2 Measuring the Properties of Galaxies

10.3 The Evolution of Galaxies

10.4 Active Galaxies and Quasars

NASA-MSFC

10.1 The Family of Galaxies

Astronomers classify galaxies according to their shape using a system developed in the 1920s by Edwin Hubble (after whom the Hubble Space Telescope is named). Creating a system of classification is a fundamental technique in science.

Study "Galaxy Classification" on pages 198–199 and notice three important points:

1. Many galaxies have no disk, no spiral arms, and almost no gas and dust. These elliptical galaxies range in size from huge giants to small dwarfs.

looking back

In the preceding chapter, you learned about our Milky Way Galaxy, an important object to Earthlings but only one of billions of galaxies visible in the sky. The Milky Way Galaxy is a normal spiral galaxy, somewhat larger than average, with a history that is probably typical of other large galaxies. However, your understanding of galaxies from the study of a single example, the Milky Way, is no more complete than your understanding of humanity would be from knowing a single person.

looking ahead

In the next chapter, you will study the universe as a whole. Galaxies, active galaxies, and dark matter are the final puzzle pieces you need before you try to understand the birth and evolution of the entire universe.

Even at the speed of light, it would take you 29 million years to reach the Sombrero Galaxy, one of the closer galaxies to our Milky Way Galaxy. This is an infrared image of the same galaxy on the cover of this book.

astro

2. Disk-shaped galaxies usually have spiral arms and contain gas and dust, although some have very little. Many spiral galaxies have a barred structure.

3. Many spiral galaxies have a central region shaped like an elongated bar.

You might also wonder what proportion of the galaxies are elliptical, spiral, and irregular, but that is a difficult question to answer. In some catalogs of galaxies, about 70 percent are spiral, but that is the result of what scientists call a selection effect. Spiral galaxies contain hot, bright stars and are consequently very luminous and easy to see. From careful studies astronomers conclude that ellipticals are actually more common than spirals and that irregulars make up about 25 percent of all galaxies. Among spiral galaxies, about two-thirds are barred spirals.

Different kinds of galaxies have different colors, depending mostly on how much star formation is happening in them. Spirals and irregulars usually contain plenty of young stars, including massive, hot, luminous O and B stars. They produce most of the light and give spirals and irregulars a distinct blue tint. In contrast, elliptical galaxies usually have few young stars. The most luminous stars in ellipticals are red giants, which give those galaxies a red tint.

How many galaxies are there? A research effort called GOODS (Great Observatories Origins Deep Survey) has used the Hubble Space Telescope, the Chandra X-ray Observatory, the Spitzer Space Telescope, and the XMM-Newton X-ray Telescope, plus the largest ground-

M83 12 million ly

Dust clouds glow red in this infrared image.

Infrared image

ESO/Hubble Heritage Team/AURA/STScI/NASA

ESO 510-G13 150 million ly

Dusty disk of galaxy warped by interaction with another galaxy

New stars forming in dust clouds.

Visual-wavelength image

A few Galaxies

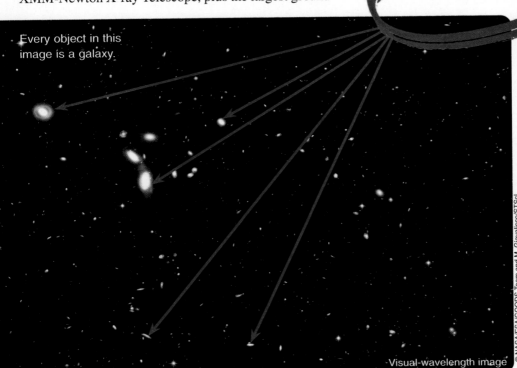

Every object in this image is a galaxy.

© NASA/ESA/GOODS Team and M. Giavalisco/STScI

Visual-wavelength image

based telescopes, to study two selected areas in the northern and southern skies. The GOODS images reveal tremendous numbers of galaxies, which are shown in the image on the left. There are good reasons to believe that the two regions of the sky chosen for study are typical, so evidently the entire sky is carpeted with galaxies. The result of the GOODS research is an estimate that at least 100 billion galaxies would be visible if today's telescopes were used for an all-sky census.

SELECTION EFFECTS

in Science

Many different kinds of science depend on selecting objects to study. Scientists studying insects in the rain forest, for example, must choose which ones to catch. They can't catch every insect they see, so they might decide on some way to select the ones they do catch. If they are not careful, a selection effect could bias their data and lead them to incorrect conclusions without their ever knowing it. The reason selection effects are dangerous in science is that they can have powerful influences without always being obvious. Scientists can avoid selection effects only by carefully designing a research project.

For example, suppose you decide to measure the speed of cars on a highway. There are too many cars to measure every one, so you reduce the workload and measure the speed of only red cars. It is quite possible that this selection criterion will mislead you because people who buy red cars may be more likely to be younger and drive faster. Should you measure only brown cars? No, because you might suspect that only older, more sedate people would buy a brown car. Should you measure the speed of any car in front of a truck? Perhaps you should pick any car following a truck? Again, you may be selecting cars

that are traveling a bit faster or a bit slower than normal. Only by very carefully designing your experiment can you be certain that the cars you measure are traveling at speeds truly representative of the entire population of cars.

Astronomers face the danger of selection effects quite often. Very luminous stars are easier to see at great distances than faint stars. Spiral galaxies are brighter, bluer, and more noticeable than elliptical galaxies. What astronomers see through a telescope depends on what they notice, and that is powerfully influenced by selection effects.

Scientists engaged in observation must spend considerable time designing their experiments. They must be careful to observe an unbiased sample if they expect to make logical deductions from their results. The scientists in the rain forest, for example, should not catch and study only the red insects. Often, the most brightly colored insects are poisonous (or at least taste bad) to predators. Catching only brightly colored insects could produce a highly biased sample of the insect population. Careful scientists must plan their work with great care and avoid any possible selection effects.

1   **Elliptical galaxies** are round or elliptical, contain no visible gas and dust, and have few or no bright stars. They are classified with a numerical index ranging from 1 to 7; E0s are round, and E7s are highly elliptical. The index is calculated from the largest and smallest diameter of the galaxy used in the following formula and rounded to the nearest integer.

$$\frac{10(a-b)}{a}$$

Outline of an E6 galaxy

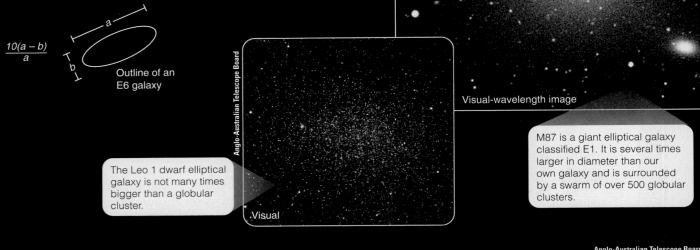

AURA/NOAO/NSF

Visual-wavelength image

The Leo 1 dwarf elliptical galaxy is not many times bigger than a globular cluster.

Anglo-Australian Telescope Board

Visual

M87 is a giant elliptical galaxy classified E1. It is several times larger in diameter than our own galaxy and is surrounded by a swarm of over 500 globular clusters.

2   **Spiral galaxies** contain a disk and spiral arms. Their halo stars are not visible, but presumably all spiral galaxies have halos. Spirals contain gas and dust and hot, bright O and B stars, as shown at right and below. The presence of short-lived O and B stars alerts us that star formation is occurring in these galaxies. Sa galaxies have larger nuclei, less gas and dust, and fewer hot, bright stars. Sc galaxies have small nuclei, lots of gas and dust, and many hot, bright stars. Sb galaxies are intermediate.

Anglo-Australian Telescope Board

Sa

Visual   NGC 3623

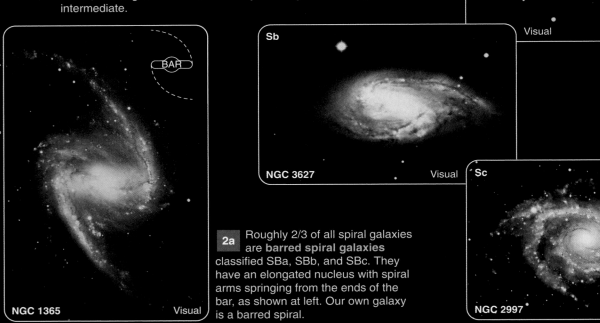

BAR

Sb

NGC 3627   Visual

NGC 1365   Visual

Sc

2a   Roughly 2/3 of all spiral galaxies are **barred spiral galaxies** classified SBa, SBb, and SBc. They have an elongated nucleus with spiral arms springing from the ends of the bar, as shown at left. Our own galaxy is a barred spiral.

NGC 2997   Visual

2b Some disk galaxies are rich in dust, which is concentrated along their spiral arms. NGC 4013, shown below, is a galaxy much like ours, but seen edge-on, its dust is dramatically apparent.

Visual

Dust visible in spiral arm crossing in front of more distant galaxy.

NGC 2207 and IC 2163

Visual

Dust in spiral galaxies is most common in the spiral arms. Here the spiral arms of one galaxy are silhouetted in front of a more distant galaxy.

The galaxy IC 4182 is a dwarf irregular galaxy only about 4 million parsecs from our galaxy.

2c Galaxies with an obvious disk and central bulge but no visible gas and dust and few or no hot bright stars are classified as S0 (pronounced "Ess Zero"). Compare this galaxy with the edge-on spiral above.

Visual

Visual

3 **Irregular galaxies** (classified Irr) are a chaotic mix of gas, dust, and stars with no obvious central bulge or spiral arms. The Large and Small Magellanic Clouds are visible to the unaided eye as hazy patches in the southern hemisphere sky. Telescopic images show that they are irregular galaxies that are interacting gravitationally with our own much larger galaxy. Star formation is dramatic in the Magellanic Clouds. The bright pink regions are emission nebulae excited by newborn O and B stars. The brightest nebula in the Large Magellanic Cloud is called the Tarantula Nebula.

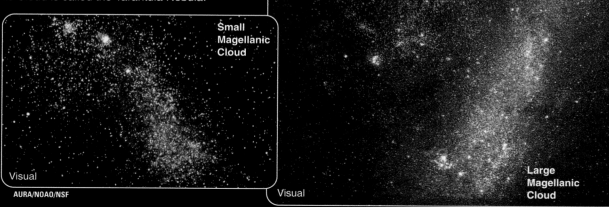

Tarantula Nebula

Small Magellanic Cloud

Visual

Visual

Large Magellanic Cloud

10.2 Measuring the Properties of Galaxies

Looking beyond the edge of our Milky Way Galaxy, astronomers find many billions of galaxies. What are the properties of these star systems? What are the diameters, luminosities, and masses of galaxies? Just as in your study of stellar characteristics (see Chapter 6), the first step in your study of galaxies is to find out how far away they are. Once you know a galaxy's distance, its size and luminosity are relatively easy to find. Later in this section, you will see that finding the masses of galaxies is more difficult, but the results are intriguing.

Distance

The distances to galaxies are so large that it is not convenient to express them in light-years, parsecs, or even kiloparsecs. Instead, astronomers use the unit **megaparsec (Mpc)**, or 1 million pc. One Mpc equals 3.26 million ly, or approximately 3×10^{19} km (2×10^{19} mi).

To find the distance to a galaxy, astronomers must search among its stars, nebulae, and star clusters for familiar objects whose luminosity they know. Such objects are called **standard candles**. If you can find a standard candle in a galaxy, you can judge its distance.

Cepheid variable stars are reliable standard candles because their period is related to their luminosity (see Figure 9.4). If you know the period of the star's variation, you can use the period–luminosity diagram to learn its absolute magnitude. Then, by comparing its absolute and apparent magnitudes, you can find its distance. Figure 10.1 shows a galaxy in which the Hubble Space Telescope detected Cepheids. Even with the Hubble Space Telescope, Cepheids are not detectable much beyond 100 million ly (30 Mpc), so astronomers must search for less common but brighter distance indicators and calibrate them using nearby galaxies containing detectable Cepheids.

When a supernova explodes in a distant galaxy, astronomers rush to observe it. Studies show that type Ia supernovae, those caused by the collapse of a white dwarf, all reach about the same luminosity at maximum (which make them more like "standard bombs" than standard candles). By searching for Cepheids and other distance indicators in nearby galaxies where type Ia supernovae

have occurred, astronomers have been able to calibrate these supernovae. As a result, when type Ia supernovae are seen in more distant galaxies, astronomers can measure the apparent brightness at maximum and compare that with the known luminosity of these supernovae to find their distances. Because type Ia supernovae are much brighter than Cepheids, they can be seen in galaxies at great distances. The drawback is that supernovae are rare, and none may occur during your lifetime in a galaxy you might be studying. You will see in Chapter 11 how these distance indicators have reshaped our understanding of the history of the universe.

Notice how astronomers use calibration to build a distance scale reaching from the nearest galaxies to the most distant visible galaxies. Often astronomers refer to this as the **distance ladder** because each step depends on the steps below it. The foundation of the distance scale rests on understanding the luminosities of stars, which

describing other galaxies

Megaparsec (Mpc): A unit of distance equal to 1,000,000 pc.

Standard candle: Object of known brightness that astronomers use to find distance—for example, Cepheid variable stars and supernovae.

Distance ladder: The calibration used to build a distance scale reaching from the size of Earth to the most distant visible galaxies.

Look-back time: The amount by which you look into the past when you look at a distant galaxy, a time equal to the distance to the galaxy in light-years.

Hubble law: The linear relation between the distances to galaxies and the apparent velocity of recession.

Hubble constant: A measure of the rate of expansion of the universe, the average value of the apparent velocity of recession divided by distance, about 70 km/s/Mpc.

Rotation curve method: A method of determining a galaxy's mass by observing the orbital velocity and orbital radius of stars in the galaxy.

ultimately rests on measurements of stellar parallax (see Chapter 6).

Telescopes as Time Machines

The most distant visible galaxies are a little over 10 billion ly (3,000 Mpc) away, and at such distances you see an effect like time travel. When you look at a galaxy millions of light-years away, you do not see it as it is now but as it was millions of years ago when its light began the journey toward Earth. When you look at a distant galaxy, you look into the past by an amount called the **look-back time**, a time in years equal to the distance to the galaxy in light-years.

The look-back time to nearby objects is usually not significant. The look-back time to the moon is only 1.3 seconds, to the sun 8 minutes, and to the nearest star about 4 years. The Andromeda Galaxy has a look-back time of about 2 million years, but that is a mere eye blink in the lifetime of a galaxy. When astronomers look at more distant galaxies, the look-back time becomes an appreciable part of the age of the universe. You will see evidence in Chapter 11 that the universe began about 13.7 billion years ago. When astronomers observe the most distant visible galaxies, they are looking back over 10 billion years to a time when the universe may have been significantly different.

The Hubble Law

Although astronomers find it difficult to measure the distance to a galaxy precisely, they often estimate such distances using a simple relationship. Early in the 20th century, astronomers noticed that the lines in galaxy spectra are generally shifted toward longer wavelengths—redshifted (see Chapter 5). These redshifts imply that the galaxies are receding from Earth.

In 1929, the American astronomer Edwin Hubble published a graph that plotted the apparent velocity of recession versus distance for a number of galaxies. The points in the graph fell along a straight line (Figure 10.2). This relation between apparent velocity of recession and distance is known as the **Hubble law**, and the slope of the line is known as the **Hubble constant**, symbolized by H.

The Hubble law is important in astronomy for two reasons. It is evidence that the universe is expanding, a subject to be discussed in detail in Chapter 11. The present chapter discusses the Hubble law because astronomers use it to estimate the distance to galaxies. The distance to a galaxy can be found by dividing its apparent velocity of recession by the Hubble constant. This is a very useful calculation, because it is usually possible to obtain a spectrum of a galaxy and measure its redshift even if it is too far away to have observable standard candles. Obviously, knowing the precise value of the Hubble constant is important.

Figure 10.1

The vast majority of spiral galaxies are too distant for Earth-based telescopes to detect Cepheid variable stars. The Hubble Space Telescope, however, can locate Cepheids in some of these galaxies (for example, in the bright spiral galaxy M100). From a series of images taken on different dates, astronomers can locate Cepheids (inset), determine the period of pulsation, and measure the average apparent brightness. They can then deduce the distance to the galaxy—16 Mpc (52 milion ly) for M100.

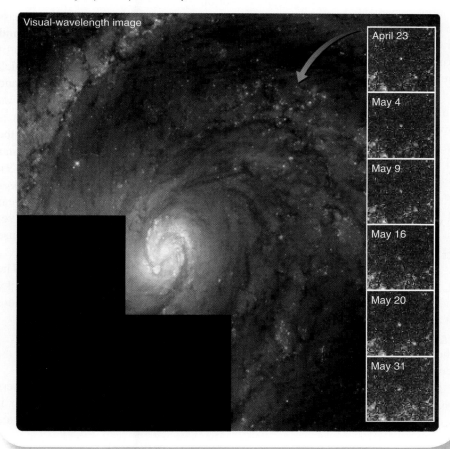

Visual-wavelength image

April 23
May 4
May 9
May 16
May 20
May 31

Figure 10.2

Edwin Hubble's first diagram of the apparent velocities of recession and distances of galaxies did not probe very deeply into space. Although Hubble's distance scale (horizontal axis) was later recalibrated, the diagram did show that the galaxies are receding at speeds proportional to their distances.

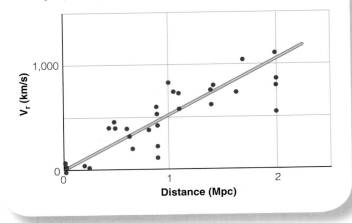

Edwin Hubble's original measurement of H was too large because of errors in his measurements of the distances to galaxies. Later astronomers have struggled to measure this important constant. The most precise measurements of the Hubble constant, made using the Hubble Space Telescope, yield a value for H of about 70 km/s/Mpc with an uncertainty of about 5 percent. This means that a galaxy at a distance of 1 Mpc from the Milky Way is receding from us at a rate of 70 km/s, a galaxy 2 Mpc away is receding at 140 km/s, and so on.

Note that the redshifts of galaxies are not really Doppler shifts, even though astronomers often express the red shifts in kilometers per second as if they were true velocities. In Chapter 11 you will discover a much more sophisticated way of understanding these redshifts. For the moment, you can just keep in mind that modern astronomers interpret redshifts of galaxies as caused by the expansion of the universe, and the Hubble law allows them to estimate the distance to a galaxy from its redshift.

Galaxy Diameters and Luminosities

The distance to a galaxy is the key to finding its diameter and its luminosity. With even a modest telescope and a CCD camera, you could photograph a galaxy and measure its angular diameter. If you know the distance to the galaxy, you can then find its linear diameter. Also, if you measure the apparent brightness of the galaxy, you can use the distance to find its luminosity, as you learned regarding stars in Chapter 6.

The results of such observations show that galaxies differ dramatically in size and luminosity. Irregular galaxies tend to be small, 1 to 25 percent the diameter of our galaxy, and of low luminosity. Although they are common, they are easy to overlook. Our Milky Way Galaxy is large and luminous compared with most spiral galaxies, though astronomers know of a few spiral galaxies that are even larger and more luminous. Elliptical galaxies cover a wide range of diameters and luminosities. The largest, called giant ellipticals, are five or more times the diameter of our Milky Way Galaxy, but many so-called dwarf elliptical galaxies are only 1 percent the diameter of our galaxy.

Clearly, the diameter and luminosity of a galaxy do not determine its type. Some small galaxies are irregular, and some are elliptical. Some large galaxies are spiral, and some are elliptical. Other factors must influence the evolution of galaxies.

Galaxy Masses

Although the mass of a galaxy is difficult to determine, it is an important quantity. It tells you how much matter the galaxy contains, which provides clues to the galaxy's origin and evolution.

The most precise method for measuring the mass of a galaxy is called the **rotation curve method** (Figure 10.3). It requires knowing: (1) the true sizes of the orbits of stars or gas clouds within a galaxy, which in turn requires knowing the distance of that galaxy; and (2) the orbital speeds of the stars or gas clouds, measured from the Doppler shifts of their spectral lines. That is enough information to use Kepler's third law and find the mass of the part of the galaxy contained within the star orbits with measured sizes and speeds (see Chapter 9). The rotation curve method works only for galaxies near enough to be well resolved. More distant galaxies appear so small that astronomers cannot measure the radial velocity at different points across the galaxy and must use other, less precise, methods to estimate masses.

The masses of galaxies cover a wide range. The smallest contain about 10^{-6} as much mass as the Milky Way, and the largest contain as much as 50 times more mass than the Milky Way.

Dark Matter in Galaxies

Given the size and luminosity of a galaxy, astronomers can make a rough guess as to the amount of matter it should contain. Astronomers know how much light stars

Figure 10.3

(a) In the artwork in the upper half of this diagram, the astronomer has placed the image of the galaxy over a narrow slit so that light from the galaxy can enter the spectrograph and produce a spectrum. A very short segment of the spectrum shows an emission line redshifted on the receding side of the rotating galaxy and blueshifted on the approaching side. Converting these Doppler shifts into velocities, the astronomer can plot the galaxy's rotation curve (right). (b) Real data are shown in the bottom half of this diagram. Galaxy NGC 2998 is shown over the spectrograph slit, and the segment of the spectrum includes three emission lines.

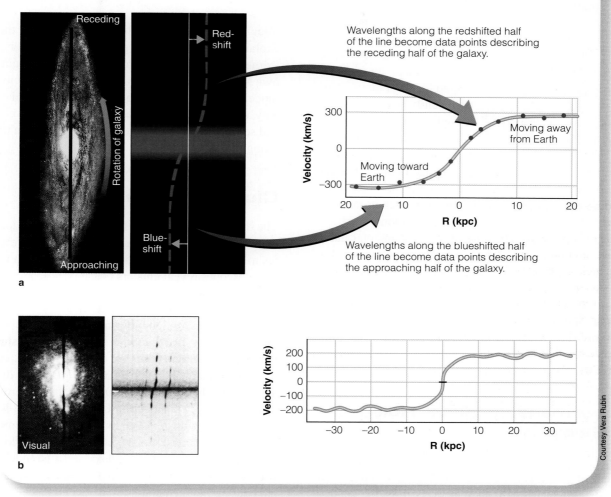

Wavelengths along the redshifted half of the line become data points describing the receding half of the galaxy.

Wavelengths along the blueshifted half of the line become data points describing the approaching half of the galaxy.

produce, and they know about how much matter there is between the stars, so it should be possible to estimate very roughly the mass of a galaxy from its luminosity. When astronomers measure the masses of galaxies, however, they often find that the measured masses are much larger than expected from the luminosities of the galaxies. You discovered this effect in the previous chapter when you studied the rotation curve of our own galaxy and concluded that it must contain large amounts of dark matter, especially in its outer regions (see Figure 9.9). Astronomer Vera Rubin found, in observations begun in the 1960s, that this also seems to be true of most nearby galaxies. Measured masses of galaxies amount to 10 to 100 times more mass than you can see.

Dark matter is difficult to detect, and it is even harder to explain. Some astronomers have suggested that dark matter consists of low-luminosity white dwarfs and brown dwarfs scattered through the halos of galaxies. Searches for white dwarfs and brown dwarfs in the halo of our galaxy have found a few but not enough to make up most of the dark matter. The dark matter can't be hidden in vast numbers of black holes and neutron stars, because astronomers don't see the X rays these objects would emit. The evidence indicates there is 10 to 100 times more dark matter than visible matter in galaxies, and if there were that many black holes they would produce X rays that would be easy to detect. Furthermore, recent images from the Chandra X-ray Observatory indicate

Rich galaxy cluster:
A cluster containing a thousand or more galaxies, usually mostly ellipticals, scattered over a volume only a few Mpc in diameter.

Poor galaxy cluster:
An irregularly shaped cluster that contains fewer than 1,000 galaxies, many of which are spiral, and no giant ellipticals.

Local Group: The small cluster of a few dozen galaxies that contains our Milky Way Galaxy.

that a collision between two galaxy clusters caused their gas and dark matter components to separate.

Because observations imply that the dark matter can't be composed of familiar objects or material, astronomers are forced to conclude that the dark matter is made up of unexpected forms of matter. Until recently neutrinos were thought to be massless, but studies now suggest they have a very small mass. Thus they can be part of the dark matter, but their masses are too low to make up all of the dark matter. There must be some other undiscovered form of matter in the universe that is detectable only by its gravitational field.

Dark matter remains one of the fundamental unresolved problems of modern astronomy. Observations of galaxies and clusters of galaxies reveal that 90 to 95 percent of the matter in the universe is dark matter. The universe you see—the kind of matter that you and the stars are made of—has been compared to the foam on an invisible ocean. You will return to this problem in Chapter 11 when you try to understand how dark matter affects the nature of the universe, its past, and its future.

Supermassive Black Holes in Galaxies

Doppler shift measurements show that the stars near the centers of many galaxies are orbiting very rapidly. To hold stars in such small, short-period orbits, the centers of those galaxies must contain masses of a million to a few billion solar masses, yet no object is visible. The evidence seems to require that the nuclei of many galaxies contain supermassive black holes. You saw in Chapter 9 that the Milky Way contains a supermassive black hole at its center. Evidently that is typical of galaxies. It is a common misconception that the orbits of stars throughout a galaxy are controlled by the central black hole. The masses of those black holes, large as they may seem, are negligible compared with a galaxy's mass. The 4.3-million-solar-mass black hole at the center of the Milky Way Galaxy contains only a thousandth of one percent of the total mass of the galaxy.

10.3 The Evolution of Galaxies

Your goal in this chapter has been to build a theory to explain the evolution of galaxies. In Chapter 9, you learned about one model that describes the origin of our own Milky Way Galaxy, plus recent modifications to that model. Presumably, other galaxies formed similarly. But why did some galaxies become spiral, some elliptical, and some irregular? Clues to that mystery lie in the clustering of galaxies.

Clusters of Galaxies

The distribution of galaxies is not entirely random. Galaxies tend to occur in clusters, ranging from a few galaxies to thousands. Deep photographs made with the largest telescopes reveal clusters of galaxies scattered out to the limits of visibility. This clustering of the galaxies can help you understand their evolution.

For this discussion, you can sort clusters of galaxies into two groups, rich and poor. **Rich galaxy clusters** contain over a thousand galaxies, mostly elliptical, crowded into a spherical volume about 3 Mpc (10^7 ly) in diameter. The Coma cluster (located 100 Mpc from Earth in the direction of the constellation Coma Berenices) is an example of a rich cluster, seen in Figure 10.4. It contains at least 1,000 galaxies, mostly E and S0 types. Close to its center are a giant elliptical galaxy and a large S0 galaxy. Rich clusters often contain one or more giant elliptical galaxies at their centers.

Poor galaxy clusters contain fewer than 1,000 galaxies and are irregularly shaped and less crowded toward the center. Our own **Local Group**, which contains the Milky Way, is a good example of a poor cluster (Figure 10.5). It contains a few dozen members scattered irregularly through a volume slightly more than 1 Mpc in diameter. Of the brighter galaxies, 14 are elliptical, 3 are spiral, and 4 are irregular.

Classifying galaxy clusters into rich and poor clusters reveals a fascinating and suggestive clue to the evolution of galaxies. In general, rich clusters tend to contain 80 to 90 percent E and S0 galaxies and few spirals. Poor clusters contain a larger percentage of spirals; and

Figure 10.4

(a) The Coma cluster of galaxies contains at least 1,000 galaxies and is especially rich in E and S0 galaxies. Two giant galaxies lie near its center. Only the central area of the cluster is shown in this image. If the cluster were visible in the sky, it would span eight times the diameter of the full moon. (b) In false colors, this X-ray image of the Coma cluster shows it filled and surrounded by hot gas. The box outlines the area of the cluster shown in panel (a).

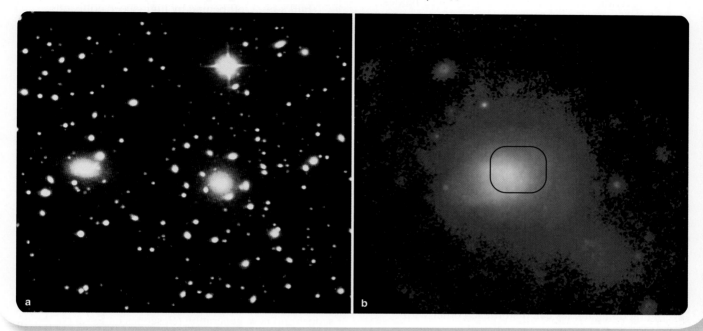

Figure 10.5

The Local Group. Our Milky Way Galaxy is located at the center of this diagram. The vertical lines giving distances from the plane of the Milky Way are solid above the plane and dashed below.

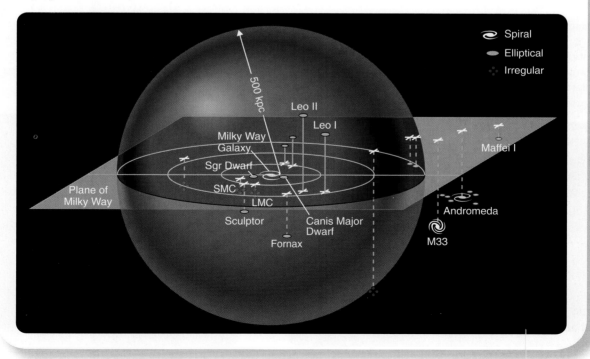

among isolated galaxies, those that are not in clusters, 80 to 90 percent are spirals. This suggests that a galaxy's environment is important in determining its structure and has led astronomers to suspect that one of the secrets to galaxy evolution lies in the collisions between galaxies.

Colliding Galaxies

Astronomers are finding more and more evidence to show that galaxies collide, interact, and merge. In fact, collisions among galaxies may dominate their evolution.

You should not be surprised that galaxies collide with each other. The average separation between galaxies is only about 20 times their diameter, so galaxies should bump into each other fairly often, astronomically speaking. In comparison, stars almost never collide because the typical separation between stars is about 10^7 times their diameter. A collision between two stars is about as likely as a collision between two gnats flitting about in a baseball stadium.

Study "Interacting Galaxies" on pages 208–209 and notice four important points:

1. Interacting galaxies can distort each other with tides, producing tidal tails and shells of stars. Interactions may help stimulate the formation of spiral arms. Large galaxies can even absorb smaller galaxies.

2. Galaxy interactions can trigger rapid star formation.

3. Evidence within galaxies in the form of star and gas motions and multiple nuclei reveals that many have suffered past interactions and mergers.

4. There are beautiful ring galaxies—understood to be bull's-eyes left behind by high-speed collisions.

Evidence of galaxy mergers is all around you. Our Milky Way Galaxy is a cannibal galaxy, snacking on the two Magellanic Clouds as they orbit around it. Its tides are also pulling apart two other small satellite galaxies, the Sagittarius and Canis Major Dwarf galaxies, producing great streamers of stars wrapped around the Milky Way. Almost certainly, our galaxy has dined on other small galaxies in the past.

The Origin and Evolution of Galaxies

The test of any scientific understanding is whether you can put all the evidence and theory together to tell the history of the objects studied. Can you describe the origin and evolution of the galaxies? Just a few decades ago, it would have been impossible, but the evidence from space telescopes and new-generation telescopes on Earth combined with advances in computer modeling and theory allow astronomers to outline the story of the galaxies.

Elliptical galaxies appear to be the product of galaxy mergers, which triggered star formation that used up the gas and dust. In fact, as shown in Figure 10.6,

Figure 10.6

Rapid star formation: (a) NGC 1569 is a starburst galaxy filled with clouds of young stars and supernovae. At least some starbursts are triggered by interactions between galaxies. (b) The inner parts of M64, known as the "Black Eye Galaxy," are filled with dust produced by rapid star formation. Radio Doppler shift observations show that the inner part of the galaxy rotates backward compared to the outer part of the galaxy, a product of a merger. Where the counter rotating parts of the galaxy collide, star formation is stimulated.

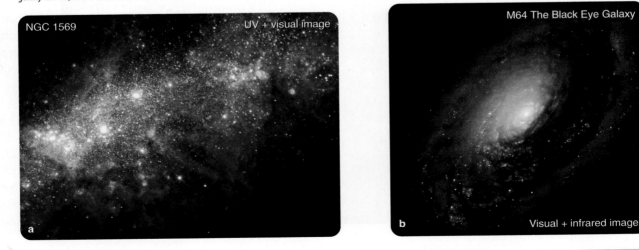

astronomers see star formation being stimulated to high levels in many galaxies. These **starburst galaxies** are very luminous in the infrared because a collision has triggered a burst of star formation that is heating the dust. The warm dust reradiates the energy in the infrared. The Antennae Galaxies (page 209) contain over 15 billion solar masses of hydrogen gas and will become a starburst galaxy as their ongoing merger continues to trigger rapid star formation.

A few collisions and mergers could leave a galaxy with no gas and dust from which to make new stars. Astronomers now suspect that most large ellipticals are formed by the merger of two or more galaxies. The dwarf ellipticals (too small to be formed by mergers) and irregulars may be fragments left over by the merger of larger galaxies.

In contrast, spirals seem never to have suffered major collisions. Their thin disks are delicate and would be destroyed by tidal forces in a collision with a massive galaxy. Also, they retain plenty of gas and dust and continue making stars. Our Milky Way Galaxy has evidently never merged with another large galaxy, although the Milky Way, as well as its large spiral neighbor the Andromeda Galaxy, both seem to have cannibalized smaller galaxies.

Barred spiral galaxies also may be the product of tidal interactions. Mathematical models show that bars are not stable and should eventually dissipate. It may take tidal interactions with other galaxies to regenerate the bars. Because well over half of all spiral galaxies have bars, you can suspect that these tidal interactions are common. Our Milky Way Galaxy is probably a barred spiral (see Figure 9.11), and interaction with its two Magellanic Cloud companions, or the more distant but very massive Andromeda Galaxy, could be the cause.

Other processes can alter galaxies. The S0 galaxies, which have disks and bulges like a spiral galaxy but no spiral arms, may have lost much of their gas and dust moving through the gas trapped in the dense clusters to which they belong. For example, X-ray observations show that the Coma cluster contains thin, hot gas between the galaxies (Figure 10.4). A galaxy moving through that gas would encounter a tremendous wind that could strip away its gas and dust.

Observations with the largest and most sophisticated telescopes take astronomers back to the age of galaxy formation. At great distances the look-back time is so large that they see the universe as it was soon after the galaxies began to form. There were more spirals then and fewer ellipticals. The observations show that galaxies were closer together then; about 33 percent of all distant galaxies are in close pairs, but only 7 percent of nearby (in other words, present-day) galaxies are in pairs. The observational evidence clearly supports the hypothesis that galaxies have evolved by merger.

The evolution of galaxies is not a simple process. A good theory helps you understand how nature works, and astronomers are just beginning to understand the exciting and complex story of the galaxies. Nevertheless, it is already clear that galaxy evolution has some resemblance to a pie-throwing contest and is just about as neat.

Starburst galaxy: A galaxy undergoing a rapid burst of star formation.

Active galaxy: A galaxy whose center emits large amounts of excess energy, often in the form of radio emission. Active galaxies have massive black holes in their centers into which matter is flowing.

Radio galaxy: A galaxy that is a strong source of radio signals.

Active galactic nucleus (AGN): The centers of active galaxies that are emitting large amounts of excess energy.

10.4 Active Galaxies and Quasars

Many galaxies have powerful energy sources in their nuclei that in some cases produce powerful jets and other outbursts. These are called **active galaxies**. By looking far away and back in time, astronomers have discovered that the origin of active galaxy energy sources and outbursts is closely related to the formation and history of galaxies.

The first type of active galaxy was discovered in the 1950s and named **radio galaxies** because these galaxies are sources of unusually strong radio waves. By the 1970s, astronomers had put space telescopes in orbit and discovered that radio galaxies are generally bright at many other wavelengths. The flood of energy pouring out of active galaxies originates almost entirely in their nuclei, which are referred to as **active galactic nuclei (AGN)**.

1 When two galaxies collide, they can pass through each other without stars colliding because the stars are so far apart relative to their sizes. Gas clouds and magnetic fields do collide, but the biggest effects may be tidal. Even when two galaxies just pass near each other, tides can cause dramatic effects, such as long streamers called **tidal tails**. In some cases, two galaxies can merge and form a single galaxy.

Small galaxy passing near a massive galaxy.

Tidal Distortion

Gravity of a second galaxy represented as a single massive object

1a When a galaxy swings past a massive object such as another galaxy, tides are severe. Stars near the massive object try to move in smaller, faster orbits while stars farther from the massive object follow larger, slower orbits. Such tides can distort a galaxy or even rip it apart.

Galaxy interactions can stimulate the formation of spiral arms

In this computer model, two uniform disk galaxies pass near each other.

The small galaxy passes behind the larger galaxy so they do not actually collide.

Tidal forces deform the galaxies and trigger the formation of spiral arms.

The upper arm of the large galaxy passes in front of the small galaxy.

A photo of the well-known Whirlpool Galaxy resembles the computer model.

Visual

Allen Beechel

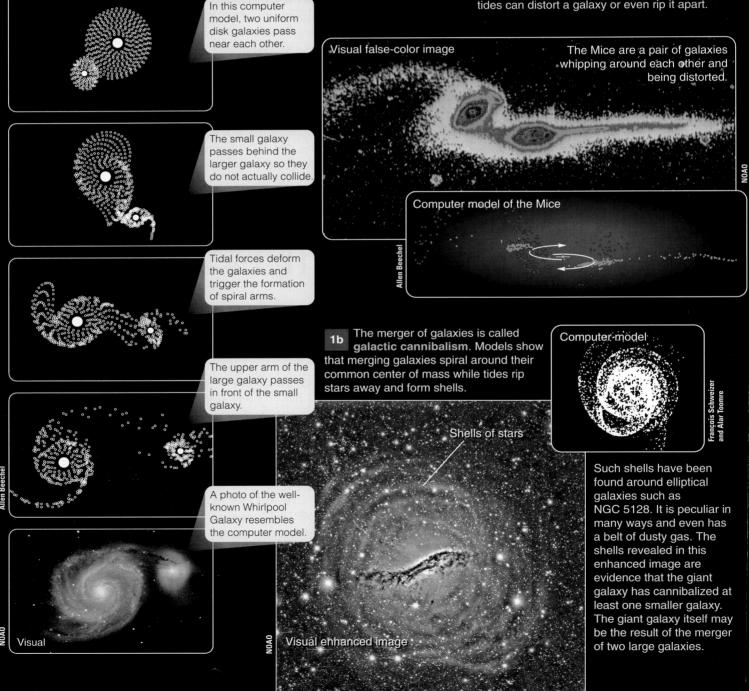

Visual false-color image

The Mice are a pair of galaxies whipping around each other and being distorted.

Allen Beechel

NOAO

Computer model of the Mice

1b The merger of galaxies is called **galactic cannibalism**. Models show that merging galaxies spiral around their common center of mass while tides rip stars away and form shells.

Computer model

François Schweizer and Alar Toomre

Shells of stars

Visual enhanced image

NOAO

Such shells have been found around elliptical galaxies such as NGC 5128. It is peculiar in many ways and even has a belt of dusty gas. The shells revealed in this enhanced image are evidence that the giant galaxy has cannibalized at least one smaller galaxy. The giant galaxy itself may be the result of the merger of two large galaxies.

2 The collision of two galaxies can trigger firestorms of star formation as gas clouds are compressed. Galaxies NGC 4038 and 4039 have been known for years as the Antennae because the long tails visible in Earth-based photos resemble the antennae of an insect. Hubble Space Telescope images reveal that the two galaxies are blazing with star formation. Roughly a thousand massive star clusters have been born.

Spectra show that the Antennae galaxies are 10 to 20 times richer in elements such as magnesium and silicon than the Milky Way. Such metals are produced by massive stars and spread by supernova explosions.

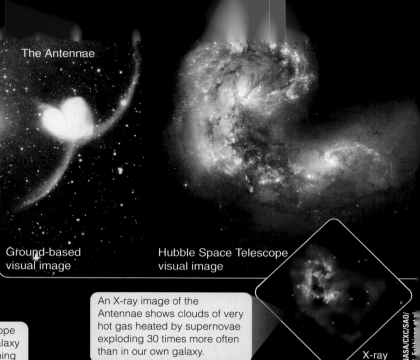

The Antennae

Brad Whitmore, STScI/NASA

Ground-based visual image

Hubble Space Telescope visual image

3 Evidence of past galaxy mergers shows up in the motions inside some galaxies. NGC 7251 is a highly distorted galaxy with tidal tails in this ground-based image.

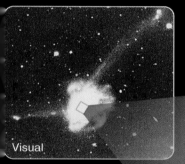

Visual

This rotation suggests that NGC 7251 is the remains of two oppositely rotating galaxies that merged about a billion years ago.

This Hubble Space Telescope image of the core of the galaxy reveals a small spiral spinning backward in the heart of the larger galaxy.

An X-ray image of the Antennae shows clouds of very hot gas heated by supernovae exploding 30 times more often than in our own galaxy.

NASA/CXC/SAO/
G. Fabbiano et al.

X-ray image

3a Radio evidence of past mergers: Doppler shifts reveal the rotation of the spiral galaxy M64. The upper part of the galaxy has a redshift and is moving away from Earth, and the bottom part of the galaxy has a blueshift and is approaching. A radio map of the core of the galaxy reveals that it is rotating backward. This suggests a merger long ago between two galaxies that rotate in opposite directions.

Rotation of galaxy M64

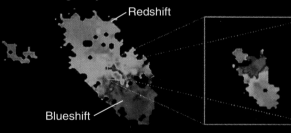

Redshift

Blueshift

Robin Braun, NRAO/AUI/NSF

Evidence of galactic cannibalism: Giant elliptical galaxies in rich clusters sometimes have multiple nuclei, thought to be the densest parts of smaller galaxies that have been absorbed and only partly digested.

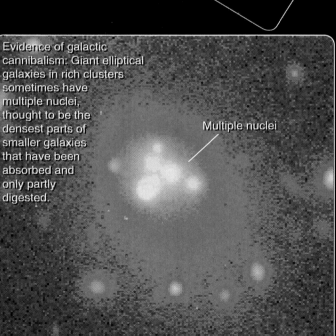

Multiple nuclei

Visual false-color image

4 The Cartwheel galaxy below was once a normal galaxy but is now a **ring galaxy**. One of its smaller companions has plunged through at high speed almost perpendicular at the Cartwheel's disk. That has triggered a wave of star formation, and the more massive stars have exploded leaving behind black holes and neutron stars. Some of those are in X-ray binaries, and that makes the outer ring bright in X-rays.

Purple = X-ray
Blue = UV
Green = Visible
Red = Infrared

Composite: NASA/JPL/
Caltech/P. Appleton et al.
x-ray: NASA/CXC/A. Wolter
& G. Trinchieri

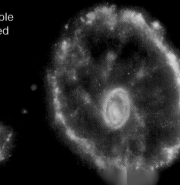

Seyfert Galaxies

In 1943, astronomer Carl K. Seyfert conducted a study of spiral galaxies in which he noted that about 2 percent of spirals have small, highly luminous nuclei in their bulges. Today, these **Seyfert galaxies** are recognized by the peculiar spectra of these luminous nuclei that contain broad emission lines of highly ionized atoms. Emission lines come from low-density gas; the presence of ionized atoms is evidence that the gas is very hot; broad spectral lines indicate large Doppler shifts produced by high gas velocities. The velocities of gas clouds at the centers of Seyfert galaxies are roughly 10,000 km/s, about 30 times greater than velocities at the center of normal galaxies. Something violent is happening in the cores of Seyfert galaxies.

Astronomers later discovered that the brilliant nuclei of Seyfert galaxies change brightness rapidly, in only a few hours or minutes, especially at X-ray wavelengths. As you saw in Chapter 8, an astronomical body cannot change its brightness significantly in a time shorter than the time it takes light to cross its diameter. If a Seyfert nucleus can change in a few minutes, then it cannot be larger in diameter than a few light-minutes. For comparison, the distance from Earth to the sun is 8 light minutes. Yet, despite their small size, the brightest Seyfert nuclei emit a hundred times more energy than the entire Milky Way Galaxy. Something in the centers of these galaxies not much bigger than Earth's orbit produces a galaxy's worth of energy.

Seyfert nuclei are three times more common in interacting pairs of galaxies than in isolated galaxies. Also, about 25 percent of Seyfert galaxies have peculiar shapes

---

| NAME | Statistical Evidence | No. 0 |

It Wouldn't Stand Up in Court: Statistical Evidence

Notice that some scientific evidence is statistical. For example, you might note that Seyfert galaxies are three times more likely to have a nearby companion than a normal galaxy is. This is statistical evidence because you can't be certain that any specific Seyfert galaxy will prove to have a companion. Yet the probability is higher than if it were a normal galaxy, and that leads you to suspect that interactions between galaxies are involved. Such statistical evidence can tell you something in general about the cause of Seyfert eruptions.

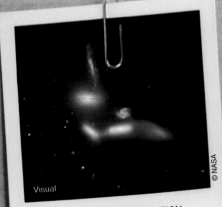

Visual
© NASA

GUILT BY ASSOCIATION IS STATISTICAL EVIDENCE AND CAN BE APPLIED TO GALAXIES.

Statistical evidence is common in science, but it is inadmissible as evidence of guilt in a criminal trial. No judge would allow you to argue that the defendant must be guilty of robbing a store in the middle of the night because the defendant is an astronomer and astronomers are statistically more likely to be awake at night. The American legal system is based on the concept of reasonable doubt, so statistics can't be applied to prove guilt. But you could use statistics to learn something about the general sleep habits of astronomers.

You can use statistical evidence in science if you do not demand that the statistics demonstrate anything conclusively about any single example. To continue the example of Seyfert galaxies, you can't demand that the statistics predict that any specific galaxy has a companion. Rather, you can use the statistical evidence to gain a general insight into the cause of active galactic nuclei. That is, you are trying to understand galaxies in general, not to convict any single galaxy.

Of course, if you surveyed only a few galaxies, your statistics might not be very good, and a critic might be justified in making the common complaint, "Oh, that's only statistics." For example, if you surveyed only four galaxies and found that three had companions, your statistics would not be very reliable. But if you surveyed 1,000 galaxies and found that 750 had companions, your statistics could be very good indeed, and your conclusions could be highly significant.

Scientists can use statistical evidence if it passes two tests. It cannot be used to draw conclusions about specific cases, and it must be based on large enough samples so the statistics are significant. With these restrictions, statistical evidence can provide deep insights into how nature works.

NGC 1410 and NGC 1409

bridge

IC 5283 and NGC 7469

tidal tails

suggesting tidal interactions with other galaxies, as in Figure 10.7. This statistical evidence hints that Seyfert galaxies may have been triggered into activity by collisions or interactions with companions. Some Seyferts are observed to be expelling matter in oppositely directed jets, a geometry you have seen on smaller scales when matter flows into neutron stars and black holes and forms an accretion disk plus jets.

The accumulated evidence leads modern astronomers to conclude that the core of a Seyfert galaxy contains a supermassive black hole, with a mass as high as a billion solar masses, plus a correspondingly large accretion disk. Gas in the centers of Seyfert galaxies is traveling so fast it would escape from a normal galaxy, and only large central masses could exert enough gravity to hold the gas inside the nuclei. Encounters with other galaxies could throw matter toward the black hole; and, as you learned in Chapter 8, lots of energy can be liberated by matter flowing through an accretion disk into a black hole. In Chapter 9 you learned that the Milky Way Galaxy contains a massive central black hole, but one that seems to be on a starvation diet and is therefore relatively inactive.

© NASA, ESA, Hubble Heritage Team, STScI/AURA-ESA/Hubble Collaboration/A. Evans, Univ. of Virginia, NRAO/Stony Brook

Figure 10.7
Some Seyfert galaxies interact with nearby companions and appear distorted with tidal tails and bridges.

NGC 1672

Seyfert galaxies have small, luminous nuclei

Double-Lobed Radio Sources

Beginning in the 1950s, radio astronomers found that some sources of radio energy in the sky consist of pairs of radio-bright regions. When optical telescopes studied the locations of these **double-lobed radio sources**, they revealed galaxies located between the two lobes. The geometry suggests that radio lobes are inflated by jets of excited gas emerging in two directions from the central galaxy. Statistical evidence indicates that jets and radio lobes, like Seyfert nuclei, are associated with interacting galaxies. The AGN jets seem to be related to matter falling into a central black hole via an accretion disk, although the details of this process are not understood.

The violence of these active galaxies is so great it can influence entire clusters of galaxies. The Perseus galaxy cluster contains thousands of galaxies and is one of the largest objects in the universe. One of its galaxies, NGC 1275, is among the largest galaxies known. It is pumping out jets of high-energy particles, heating the gas in the galaxy cluster, and inflating low-density bubbles that distort the huge gas cloud (Figure 10.8). The hot gas observed in galaxy clusters is heated to multimillion-degree temperatures as galaxy after galaxy goes through eruptive stages that can last for hundreds of millions of years. NGC 1275 is erupting now, and it is so powerful and has heated the surrounding gas to such high temperatures that the gas can no longer fall in and the galaxy has probably limited its own growth.

Quasars

The largest telescopes detect multitudes of faint points of light with peculiar emission spectra, objects called **quasars** (also known as *quasi-stellar objects*, or *QSOs*). Although astronomers now recognize quasars as extreme examples of AGN as well as some of the most distant visible objects in the universe, they were a mystery when they were first identified.

In the early 1960s, photographs of the location of some radio sources that resembled radio galaxies revealed only single starlike points of light. The first of these objects identified was 3C 48, and later the source 3C 273 was found. They were obviously not normal radio galaxies. Even the most distant photographable galaxies look fuzzy, but these objects looked like stars. Their spectra, however, were totally unlike stellar spectra, so the objects were called quasi-stellar objects.

For a few years, the spectra of quasars were a mystery. A few unidentifiable emission lines were superimposed on a continuous spectrum. In 1963, astronomer Maarten Schmidt calculated that if hydrogen Balmer lines were redshifted by $z = 0.158$ (meaning each line's change in wavelength divided by its lab wavelength equals 0.158), they would fit the observed lines in 3C 273's spectrum (Figure 10.9). Other quasar spectra quickly yielded to this type of analysis, revealing even larger redshifts.

To understand the significance of these large redshifts and the large velocities of recession they imply, recall the Hubble law from earlier in this chapter stating that galaxies have apparent velocities of recession proportional to their distances. The large redshifts of the quasars imply that they must be at great distances, some farther away than any known galaxy. Many quasars are evidently so far away that galaxies at those distances are very difficult to detect, yet the quasars are easily photographed. This leads to the conclusion that quasars must be ultraluminous, 10

Figure 10.8

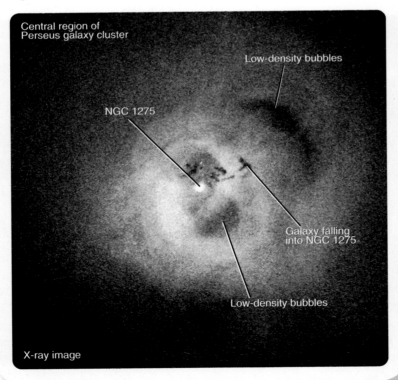

Central region of Perseus galaxy cluster

Low-density bubbles

NGC 1275

Galaxy falling into NGC 1275

Low-density bubbles

X-ray image

© NASA/CXC/IoA/A. Fabian et al.

Figure 10.9

This image of 3C 273 shows the bright quasar at the center surrounded by faint fuzz. Note the jet protruding to lower right. The spectrum of 3C 273 (top) contains three hydrogen Balmer lines redshifted by 15.8 percent. The drawing shows the unshifted positions of the lines.

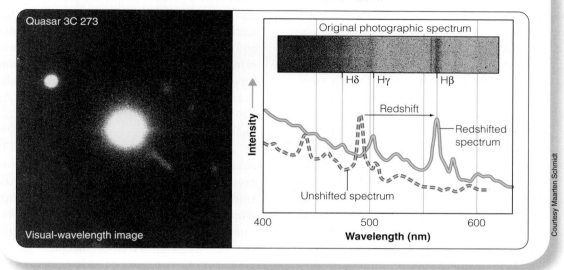

Quasar 3C 273

Visual-wavelength image

Original photographic spectrum

Hδ Hγ Hβ

Redshift

Redshifted spectrum

Unshifted spectrum

Intensity

400   500   600

Wavelength (nm)

Courtesy Maarten Schmidt

to 1,000 times the luminosity of a large galaxy.

Soon after quasars were discovered, astronomers detected fluctuations in their brightness over time scales of hours or minutes. The rapid fluctuations in quasars showed that they are small objects like AGN, only a few light-minutes or light-hours in diameter. Evidence, such as that seen in Figure 10.10, has accumulated that quasars are the most luminous AGN, located in very distant galaxies. For example, some quasars, like AGN, are at the centers of double radio lobes plus jets.

Perhaps you have a skeptical question about quasar distances at this point: "How can you be sure quasars really are that far away?" Astronomers faced with explaining how a small object could produce so much energy asked themselves the same question. In the early 1980s, astronomers were able to photograph faint nebulosity surrounding some quasars, which was called quasar fuzz. The spectra of quasar fuzz looked like the spectra of normal but very distant galaxies with the same redshift as the central quasar. In other cases, quasar light shines through the outskirts of a dis-

Figure 10.10

This radio image of quasar 3C 175 reveals that it is ejecting a jet and is flanked by radio lobes. Presumably you see only one jet because it is directed approximately toward Earth, and the other jet is invisible because it is directed away from Earth. The presence of jets and radio lobes indicates that quasars are the active cores of distant galaxies.

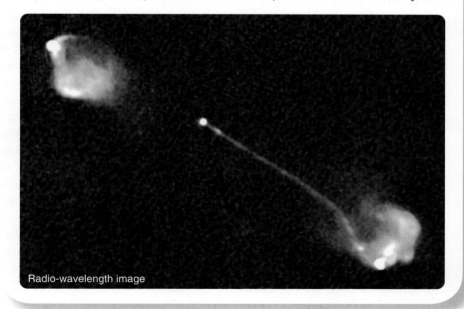

Radio-wavelength image

© NRAO/AUI/NSF

tant galaxy, and the quasar spectrum has extra absorption lines from gas at the redshift of the galaxy, smaller than the quasar's redshift. This means the quasar is farther away than the galaxy containing the gas that absorbed some of the quasar's light on its way to Earth. Both of these observations confirm that galaxy and quasar redshifts indicate distances in a mutually consistent way.

The Search for a Unified Model

Astronomers studying galaxies are now developing a **unified model** of AGN and quasars. A monster black hole is the centerpiece (Figure 10.11).

Even a supermassive black hole is quite small compared with a galaxy. A 10-million-solar-mass black hole would be only one-fifth the diameter of Earth's orbit. This means that matter in an accretion disk can get very close to the black hole, orbit very fast, and grow very hot. Theoretical calculations indicate that the disk immediately around the black hole is "puffed up," thick enough to hide the central black hole from some viewing angles. The hot inner disk seems to be the source of the jets often seen coming out of active galaxy cores, but the process by which jets are generated is not understood. Mathematical models indicate that the outer part of the disk is a fat, relatively cold torus (doughnut shape) of dusty gas.

According to the unified model depicted in Figure 10.11, what you see when you view the core of an AGN or QSO depends on how its accretion disk is tipped with respect to your line of sight.

1. If you view the accretion disk from the edge, you cannot see the central zone at all because the thick dusty torus blocks your view. Instead you see radiation emitted by gas lying above and below the central disk that is therefore relatively cool and moving relatively slowly, with small Doppler shifts. Thus, you see narrower spectral lines coming from what is called the narrow-line region.

2. If the accretion disk is tipped slightly, you may be able to see some of the intensely hot gas in the central cavity. This is called the broad-line region because the gas is hot and also orbiting at high velocities. The resulting high Doppler shifts spread out the spectral lines.

3. If you look directly into the central cavity around the black hole—down the dragon's throat, so to speak—you see the jet emerging perpendicular to the accretion disk and coming straight at you. Model calculations indicate this would result in a very luminous and highly variable source with few or no emission lines. This is the appearance of AGN known as **blazars**.

Unified model: An attempt to explain the different types of active galactic nuclei using a single model viewed from different directions.

Blazar: A type of active galaxy nucleus that is especially variable and has few or no spectral emission lines.

Astronomers are now using this unified model to sort out the different kinds of active galaxies and quasars so they can understand how they are related. For example, about 1 percent of quasars are strong radio sources, and the radio radiation may come from synchrotron radiation (see Chapter 8) produced in the high-energy gas and magnetic fields in the jets. Another example: Using infrared cameras to see through dust, astronomers observed the core of the double-lobed radio galaxy Cygnus A and found an object much like a quasar. Astronomers have begun to refer to such hidden objects as "buried quasars."

The unified model is far from complete. The detailed structure of accretion disks is poorly understood, as is the process by which the disks produce jets. Furthermore, the spiral Seyfert galaxies are clearly different from the giant elliptical galaxies that have double radio lobes. Unification does not explain all of the differences among active galaxies. Rather, it is a model that provides some clues to what is happening in AGN and quasars.

The Origin of Supermassive Black Holes

Naturally you are wondering where these supermassive black holes in the nuclei of galaxies came from, and that question is linked to a second question. What makes a supermassve black hole erupt? Answering those questions will help you understand how galaxies form.

Evidence is accumulating that most galaxies contain a supermassive black hole at their center. Even our own Milky Way Galaxy and the nearby Andromeda Galaxy contain central black holes. Only a few percent of galaxies, however, have obviously active galactic nuclei. That must mean that most of the supermassive black holes are dormant. Presumably they are not being fed large amounts of matter. A slow trickle of matter flowing into the supermassive black hole at the center of our galaxy could explain the relatively mild activity seen there. It would take a larger meal to trigger an eruption such as those seen in active galactic nuclei.

What could trigger a supermassive black hole to erupt? The answer is something that you studied back in Chapter 3—tides. You have seen in a previous section of this chapter how tides twist interacting galaxies and rip matter away into tidal tails. Active galaxies are often distorted: They have evidently been twisted by tidal forces as they interacted or merged with another galaxy. Mathematical models show that those interactions can also throw stars plus clouds of interstellar gas and dust inward toward the galaxies' centers. A sudden flood of

Figure 10.11

The features visible in the spectrum of an AGN depend on the angle at which it is viewed. The unified model, shown in cross section, suggests that matter flowing inward passes first through a large, opaque torus; then into a thinner, hotter disk; and finally into a small, hot cavity around the black hole. Telescopes viewing such a disk edge-on would see only narrow spectral lines from cooler gas, but a telescope looking into the central cavity would see broad spectral lines formed by the hot gas. This diagram is not to scale. The central cavity may be only 0.01 pc in radius, while the outer torus may be 1,000 pc in radius.

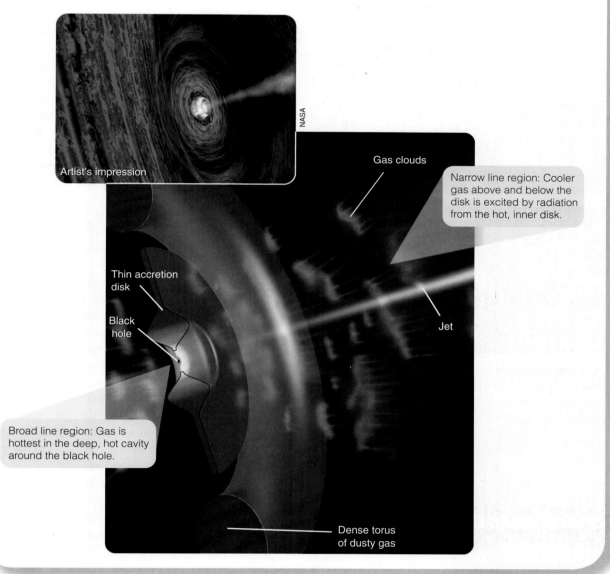

Artist's impression

NASA

Gas clouds

Narrow line region: Cooler gas above and below the disk is excited by radiation from the hot, inner disk.

Thin accretion disk

Black hole

Jet

Broad line region: Gas is hottest in the deep, hot cavity around the black hole.

Dense torus of dusty gas

matter flowing into a supermassive black hole would trigger it into eruption. Figure 10.12 shows how a passing star would be shredded and partially consumed by a supermassive hole. A steady diet of inflowing gas, dust, and an occasional star would keep an AGN powered by a supermassive black hole active.

A few dozen supermassive black holes have measured masses, and their masses are correlated with the masses of the host galaxies' central bulges. In each case, the mass of the black hole is about 0.5 percent the mass of the surrounding central bulge. Apparently, as a galaxy forms its central bulge, a certain fraction of the mass sinks to the center where it forms a supermassive black hole. All of that matter flowing together to form the black hole would release a tremendous amount of energy and trigger a violent eruption. Long ago, when galaxies were actively forming, the birth of the central bulges must have triggered many AGN. Later episodes of AGN activity could be triggered by interactions or mergers with other galaxies.

Figure 10.12

Artist's conception of the cause of an X-ray flare in the galaxy RX J1242-11 detected by X-ray telescopes in orbit around Earth. Equaling the energy of a supernova explosion, the flare was evidently caused when a star wandered too close to the 100-million-solar-mass black hole at the center of the galaxy. When tidal forces ripped the star apart, some of the mass fell into the black hole, and the rest was flung away. That sudden meal for the black hole was enough to trigger an outburst.

Star Falling into a Black Hole

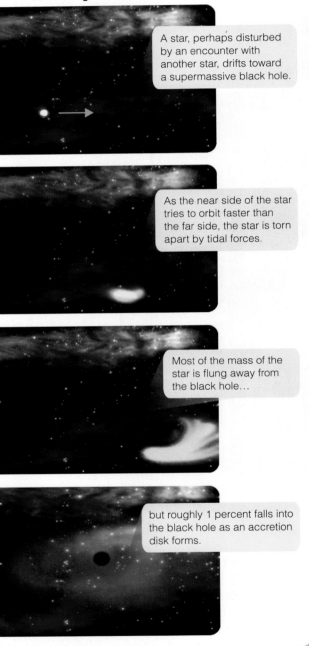

A star, perhaps disturbed by an encounter with another star, drifts toward a supermassive black hole.

As the near side of the star tries to orbit faster than the far side, the star is torn apart by tidal forces.

Most of the mass of the star is flung away from the black hole...

but roughly 1 percent falls into the black hole as an accretion disk forms.

© ESA

Quasars through Time

In the next chapter, you will see evidence that the universe began 13.7 billion years ago. Some quasars are over 10 billion light-years away, and because of their large look-back times they appear as they were when the universe was only 10 percent of its present age. The first clouds of gas that formed galaxies would have also made supermassive black holes at the centers of those galaxies' central bulges. The abundance of matter flooding into those early black holes could have triggered outbursts that are seen as quasars.

You should also note that galaxies were closer together when the universe was young and had not expanded very much. Because they were closer together, the forming galaxies collided more often, and you have seen how collisions between galaxies could throw matter into central supermassive black holes and trigger eruptions. Quasars are often located in host galaxies that are distorted as if they were interacting with other galaxies (Figure 10.13).

Quasars are most common with redshifts of a little over 2 and less common with redshifts above 2.7. The largest quasar redshifts are over 6, but such high-redshift quasars are quite rare. Evidently, if you looked at quasars with redshifts a bit over 2, you would be looking back to an age when galaxies were actively forming, colliding, and merging. In that era, quasars were about 1,000 times more common than they are now, although, even so, only a fraction of galaxies had quasars erupting in their cores at any one time. If you looked back to higher redshifts, you would see fewer quasars because you would be looking back to an age when the universe was so young it had not yet begun to form many galaxies and quasars.

Then where are all the dead quasars? There is no way to get rid of supermassive black holes. Astronomers have discovered that nearly all galaxies contain supermassive black holes, and those black holes may have suffered quasar eruptions when the universe was younger, galaxies were closer together, and infalling gas and dust were more plentiful. Quasar eruptions became less common as galaxies became more stable and as the abundance of gas and dust in the centers of galaxies was exhausted. Our own Milky Way Galaxy is a good example. It could have been a quasar long ago, but today its supermassive black hole is resting. Dormant black holes at the centers of galaxies today can be reawakened to become AGN by galaxy collisions.

Figure 10.13

The bright object at the center of each of these images is a quasar. Fainter objects near and around the quasars are galaxies distorted by collisions. Compare the ring-shaped galaxy in the first image below with the ring galaxy on page 209 and compare the tail in the third image with the Antennae Galaxies on page 209.

Visual-wavelength image

Visual

Visual

Visual

J. Bahcall, Institute for Advanced Study, Mike Disney, University of Wales, NASA

recap You are riding a small planet orbiting a humdrum star that is just one of at least 100 billion in the Milky Way Galaxy. You have just learned that there are at least 100 billion galaxies visible with existing telescopes, and that each of these galaxies contains roughly 100 billion stars. Humans fight wars over politics, religion, and economics; we do our work, play our games, and wash our laundry. It's all important stuff, but next time you are frantically rushing to a meeting, glance up at the sky.

When you look at galaxies, you are looking across voids deeper than human imagination. You can express such distances with numbers, but the distance is truly beyond human comprehension. Some people say astronomy makes them feel humble, but before you agree, consider that you can feel small without feeling humble. We humans live out our little lives on our little planet, but we are figuring out some of the biggest mysteries of the universe. We are exploring deep space and deep time and coming to understand what galaxies are and how they evolve. Most of all, we humans are beginning to understand what we are. That's something to be proud of.

Help yourself better understand how our galaxy fits into the universe as a whole by answering the following questions:

1. What types of galaxies exist?

2. How do astronomers measure the distances to galaxies, and how does that allow the sizes, luminosities, and masses of galaxies to be determined?

3. Why are there different kinds of galaxies, and how do galaxies evolve?

4. What is the energy source for active galaxies, what can trigger the activity, and what does that reveal about the history of galaxies?

Cosmology
in the 21st Century

L ook at your thumb. The matter in your thumb was present in the fiery beginning of the universe. **Cosmology**, the study of the universe as a whole, can tell you where your matter came from, and it can tell you where your matter is going. Cosmology is a mind-bendingly weird subject, and you can enjoy it for its strange ideas. It is fun to think about space stretching like a rubber sheet, invisible energy pushing the universe to expand faster and faster, and the origin of vast walls of galaxy clusters. Notice that this is better than speculation—it is all supported by evidence. Cosmology, however strange it may seem, is a serious and logical attempt to understand how the universe works, and it leads to wonderful insights into how you came to be a part of it.

11.1 Introduction to the Universe

Many people have an impression of the universe as a vast sphere filled with stars and galaxies. Your vision may be similar to the composite image in Figure 11.1 on page 220, but, as you begin exploring the universe, you need to become aware of your assumptions so they do not mislead you. The first step is to deal with an expectation so obvious that most people, for the sake of a quiet life, don't even think about it.

The Edge–Center Problem

In your daily life, you are accustomed to boundaries. Rooms have walls, athletic fields have boundary lines, countries have borders, oceans have shores. It is natural to think of the universe as having an edge, but that idea can't be right.

> **Cosmology:** The study of the nature, origin, and evolution of the universe.

If the universe had an edge, imagine going to the edge. What would you find there: A wall? A great empty space? Nothing? Even a child can ask: If there is an edge to space, what's beyond it? A true edge would have to be more than just an end of the

looking back

Since Chapter 1 you have been on an outward journey through the universe. You have studied the appearance of the night sky seen from Earth, the births and deaths of stars, and the interactions of galaxies. Now you have reached the limit of your journey in space and in time, and you can study the universe as a whole.

looking ahead

Once you have finished this chapter, you will have a modern insight into the nature of the universe, and it will be time to refocus on your local neighborhood in the universe—the subject of the rest of this book.

Big Bang. Conceptual computer artwork of the origin and evolution of the universe. The term Big Bang describes the initial expansion of all matter in the universe from an infinitely compact state 13.7 billion years ago. The initial conditions are not known, but less than a second after the beginning, temperatures were trillions of degrees Celsius and the primordial universe was much smaller than an atom. It has been expanding and cooling ever since.

distribution of matter, it would have to be an end of space itself; but then, what would happen if you tried to reach past or move past that edge?

An edge to the universe violates common sense, and modern observations (which you will study later in this chapter) indicate that the universe could be infinite and have no edge. Note that you find the centers of things—galaxies, globular clusters, oceans, pizzas—by referring to their edges. If the universe has no edge, then it cannot have a center.

It is a common misconception to imagine that the universe has a center, but, as you have just learned, that is impossible. As you study cosmology, you need to take care to avoid thinking that there is a center of the universe.

The Necessity of a Beginning

Of course you have noticed that the night sky is dark. That is an important observation because reasonable assumptions about the universe can lead to the conclusion that the night sky actually should glow blindingly bright! This conflict between observation and theory is called **Olbers's paradox** after Heinrich Olbers, an Austrian physician and astronomer who publicized the problem in 1826 (*problem* or *question* would be more accurate words than *paradox*). Olbers was not the first to pose the question, but it is named after him because modern cosmologists were not aware of the earlier discussions. You will be able to answer Olbers's question and understand why the night sky is dark by revising your assumptions about the universe.

The point Olbers made seems simple. Suppose you assume that the universe is infinite and filled with stars. (The clumping of stars into galaxies can be shown mathematically to make no difference.) If you look in any direction, your line of sight must eventually reach the surface of a star, as in Figure 11.2. Now think about trying to see out of a forest. When you are deep in a forest, every line of sight ends on a tree trunk, and you cannot see out of the forest. By analogy, every line of sight from Earth out into space should eventually end on the surface of a star, so the entire sky should be as bright

Figure 11.1

In this image of a typical spot on the sky, bright objects with spikes caused by diffraction in the telescope are nearby stars. All other objects are galaxies, ranging from the nearby face-on spiral at upper right to the most distant galaxies visible only in the infrared, shown as red in this composite image.

Visual + infrared image

© R. Williams, STScI HDF-South Team, NASA

as the surface of an average star—like suns crowded "shoulder to shoulder," covering the sky from horizon to horizon. It should not get dark at night.

Today, cosmologists believe they understand why the sky is dark. Olbers's paradox makes an incorrect prediction because it is based on a hidden assumption. The universe may be infinite in size but it is not eternal—that is, not infinitely old. That answer to Olbers's question was suggested by Edgar Allan Poe in 1848. He proposed that the night sky is dark because the universe is not infinitely old but instead began at some time in the past. The more distant stars are so far away that light from them has not yet reached Earth. That is, if you look far enough, the look-back time is greater than the age of the universe. The night sky is dark because the universe had a beginning.

This is a powerful idea because it clearly illustrates the difference between the universe and the observable universe. The universe is everything that exists, but the **observable universe** is the part that you can see. You will learn later that the universe is 13.7 billion years old. Because of that, the observable universe is limited by a light travel time of 13.7 billion years. Do not confuse the observable universe, which is finite, with the universe as a whole, which could be infinite.

Figure 11.2

Every direction you look in a forest eventually reaches a tree trunk, and you cannot see out of the forest. If the universe is infinite and filled with stars, then any line from Earth should eventually reach the surface of a star. This assumption predicts that the night sky should glow as brightly as the surface of the average star.

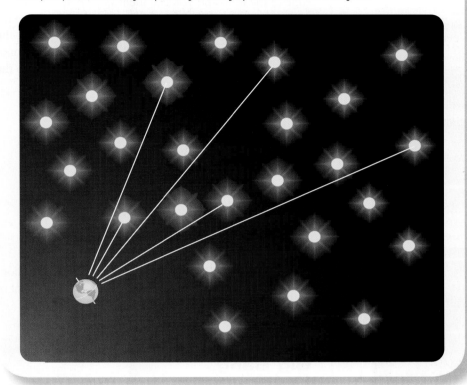

Cosmic Expansion

In 1929, Edwin P. Hubble published his discovery that the sizes of galaxy redshifts are proportional to galaxy distances (see Figure 10.2). Nearby galaxies have small redshifts, but more distant galaxies have larger redshifts. These redshifts imply that the galaxies are receding from each other. Figure 11.3 shows spectra of galaxies in clusters at various distances. The Virgo cluster is relatively nearby, and its redshift is small. The Hydra cluster is very distant, and its redshift is so large that the two dark lines formed by ionized calcium are shifted from near-ultraviolet wavelengths well into the visible part of the spectrum.

The expansion of the universe does not imply that Earth is at the center. To see why, think about raisin bread. As the

Figure 11.3

These galaxy spectra extend from the near-ultraviolet at left to the blue part of the visible spectrum at right. The two dark absorption lines of once-ionized calcium are prominent in the near-ultraviolet. The redshifts in galaxy spectra are expressed here as apparent velocities of recession. Note that the apparent velocity of recession is proportional to distance.

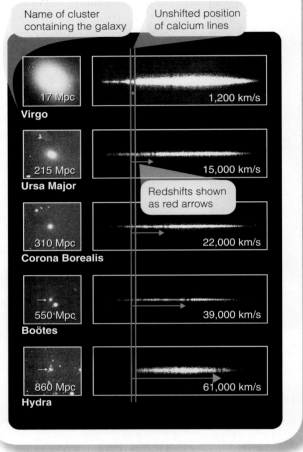

Name of cluster containing the galaxy

Unshifted position of calcium lines

17 Mpc **Virgo**	1,200 km/s
215 Mpc **Ursa Major**	15,000 km/s
310 Mpc **Corona Borealis**	22,000 km/s
550 Mpc **Boötes**	39,000 km/s
860 Mpc **Hydra**	61,000 km/s

Redshifts shown as red arrows

© Caltech

dough rises, it pushes the raisins away from each other uniformly at speeds that are proportional to their distances from each other. Two raisins that were originally close are pushed apart slowly, but two raisins that were far apart, having more dough between them, are pushed apart faster. If bacterial astronomers lived on a raisin in your raisin bread, they could observe the redshifts of the other raisins and derive a bacterial Hubble law. They would conclude that their universe was expanding uniformly. It does not matter which raisin the bacterial astronomers lived on, they would get the same Hubble law—no raisin has a special viewpoint. Similarly, astronomers in any galaxy will see the same law of expansion—no galaxy has a special viewpoint.

When you think about that loaf of bread, you see the edge of the loaf, and you can identify a center to the loaf of bread. The raisin bread analogy for the expanding universe no longer works when you consider the crust—the edge—of the bread. Remember that the universe cannot have an edge or a center, so there can be no center to the expansion.

Big bang: The high-density, high temperature state from which the expanding universe of galaxies began.

Hubble time: The age of the universe, equivalent to 1 divided by the Hubble constant. The Hubble time is the age of the universe if it has expanded since the big bang at a constant rate.

11.2 The Big Bang Theory

The expansion of the universe led astronomers to conclude that the universe must have begun with an event of astounding cosmic intensity.

Necessity of the Big Bang

Imagine that you have a video of the expanding universe, and you run it backward. You would see the galaxies moving toward each other. There is no center to the expansion of the universe, so you would not see galaxies approaching a single spot. Rather, you would see the space between galaxies disappearing, distances between all galaxies decreasing without the galaxies themselves moving, and eventually galaxies beginning to merge. If you ran your video far enough back, you would see the matter and energy of the universe compressed into a high-density, high-temperature state. You can conclude that the expanding universe began with expansion from that condition of extremely high density and temperature, which modern astronomers call the **big bang**.

How long ago did the universe begin? You can estimate the age of the universe with a simple calculation. If you must drive to a city 100 miles away and you can travel 50 miles per hour, you divide distance by rate of

© Ian McDonnell/iStockphoto.com

travel and learn the travel time—in this example, 2 hours. To find the age of the universe, you can divide the distance between galaxies by the speed with which they are separating and find out how much time was required for them to have reached their present separation. You will fine-tune your estimate later in this chapter, but for the moment you can conclude that basic observations of the recession of the galaxies require that the universe began with a big bang approximately 14 billion years ago. That estimated span is called the **Hubble time**.

Your instinct is to think of the big bang as an historical event, like the Gettysburg Address—something that happened long ago and can no longer be observed. But the look-back time makes it possible to observe the big bang directly. The look-back time to nearby galaxies is only a few million years, but the look-back time to more distant galaxies is a large fraction of the age of the universe (see Chapter 10). Suppose you were to look between and beyond the distant galaxies, back to the time of the big bang. You should be able to detect the hot gas that filled the universe long ago.

Again, do not think of an edge or a center when you think of the big bang. It is a very common misconception that the big bang was an explosion and that the galaxies are flying away from a center. Although your imagination tries to visualize the big bang as a localized event, you must keep firmly in mind that the big bang did not occur at a single place but filled the entire volume of the universe. You cannot point to any particular place and say, "The big bang occurred over there." At the time of the big bang, all of the galaxies, stars, and atoms in the observable universe were confined to a very small volume. That hot, dense state, the big bang, occurred everywhere, including right where you are now. The matter of which you are made was part of the big bang, so you are inside the remains of that event, and the universe continues to expand around you. In whatever direction astronomers look, at great distances they can see back to the age when the universe was filled with dense, hot gas (Figure 11.4).

No raisin has a special viewpoint.

Figure 11.4

This diagram shows schematically the expansion of a small part of the universe. Although the universe is now filled with galaxies, the look-back time distorts what you see. Nearby you can see galaxies, but at greater distances, the look-back time reveals the universe at earlier stages. At very great distances, the big bang is detectable as infrared and radio energy arriving from the hot gas that filled the universe soon after the big bang.

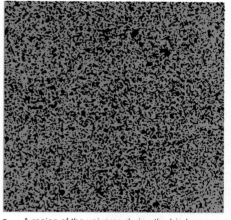

a   A region of the universe during the big bang

b   A region of the universe now

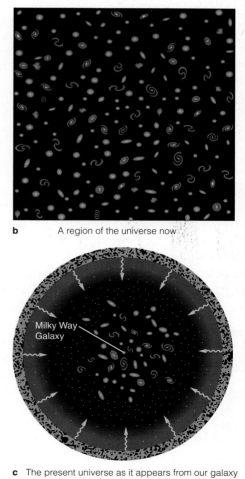

Milky Way Galaxy

c   The present universe as it appears from our galaxy

Evidence for the Big Bang— The Cosmic Microwave Background

The radiation that comes from great distance has a tremendous redshift. The most distant visible objects are faint galaxies and quasars, with redshifts of about 8, meaning the light from them arrives at Earth with wavelengths 9 times longer than when it started the journey. In contrast, the radiation from the hot gas of the big bang is calculated to have a redshift of about 1,100. That means the light emitted by the big bang gases arrives at Earth as far-infrared radiation and short-wavelength radio waves. You can't see it with your eyes, but it should be detectable with infrared and radio telescopes. Amazingly, the big bang can still be detected by the radiation it emitted.

In the mid-1960s two Bell Laboratories physicists, Arno Penzias and Robert Wilson, were measuring the radio brightness of the sky (Figure 11.5) when they discovered peculiar radio noise coming from all directions. Physicists George Gamow and Ralph Alpher had predicted in the 1940s that the big bang would have emitted blackbody radiation that should now be in the far-infrared and radio parts of the spectrum. Physicist Robert Dicke and his team at Princeton began building a receiver to detect that radiation in the early 1960s. When Penzias and Wilson learned about the preceding work, they realized the radio noise they had detected, now called the **cosmic microwave background (CMB)**, is actually radiation from the big bang. They received the 1978 Nobel Prize in physics for their discovery.

The detection of the background radiation was tremendously exciting, but astronomers wanted confirmation. Critical observations in the far-infrared checking whether the CMB really has a blackbody spectrum could

Figure 11.5

When the CMB was first detected in 1965, technology did not allow measurements at many wavelengths. Not until infrared detectors could be put in orbit was it conclusively shown that the background radiation, as predicted by theory, follows a blackbody curve very precisely.

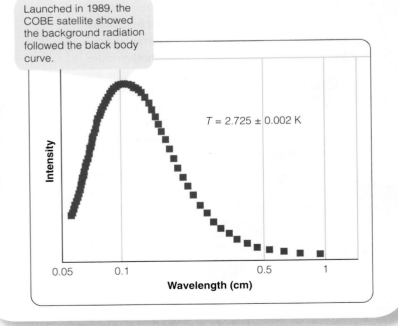

All-sky WMAP plot of tiny variations in the background radiation.

Launched in 1989, the COBE satellite showed the background radiation followed the black body curve.

$T = 2.725 \pm 0.002$ K

Intensity

Wavelength (cm)

In 1965, Arno Penzias (right) and Robert Wilson first detected the background radiation using the horn antenna behind them in this photograph.

Cosmic microwave background (CMB): Radiation from the hot matter of the universe soon after the big bang. The large redshift makes it appear to come from a blackbody with a temperature of 2.7 K.

not be made from the ground. It was not until January 1990 that satellite measurements confirmed that the CMB is blackbody radiation with an apparent temperature of 2.725 ± 0.002 K—in good agreement with theoretical predictions.

It may seem strange that the hot gas of the big bang appears to have a temperature 2.7 degrees above absolute zero, but recall the tremendous redshift. Observers on Earth see light that has a redshift of about 1,100—that is, the wavelengths of the photons are about 1,100 times longer than when they were emitted. The gas clouds that emitted the photons had a temperature of about 3,000 K, and they emitted blackbody radiation with a λ_{max} of about 1,000 nm (see Chapter 5 regarding Wien's law). The expansion of the universe has redshifted the wavelengths about 1,100 times longer, so λ_{max} is now about 1 million nm (1 mm). That is why the hot gas of the big bang seems to be 1,100 times cooler now, about 2.7 K.

> The Universe, as has been observed before, is an unsettlingly big place, a fact which for the sake of a quiet life most people tend to ignore.
>
> –Douglas Adams
> *The Restaurant at the End of the Universe*

Particles and Nucleosynthesis—The First Seconds and Minutes

Simple observations of the darkness of the night sky and the redshifts of the galaxies tell you that the universe must have had a beginning. Furthermore, you have seen that the CMB is clear evidence that conditions at the beginning were hot and dense. Theorists can combine these observations with knowledge from physics of how atoms and subatomic particles behave to work out the story of how the big bang occurred.

Cosmologists cannot begin their history of the big bang at time zero, because no one understands the behavior of matter and energy under such extreme conditions, but they can come amazingly close. If you could visit the universe when it was only one 10-millionth of a second old, you would find it filled with high-energy photons having a blackbody temperature well over 1 trillion (10^{12}) K and a density (using Einstein's equation $E = mc^2$ to calculate the mass equivalent to a certain amount of energy) greater than 5×10^{13} g/cm^3, nearly the density of an atomic nucleus.

If photons have enough energy, two photons can combine and convert their energy into a pair of particles—a particle of normal matter and a particle of **antimatter**. And when an antimatter particle meets its matching particle of normal matter, the two particles annihilate each other and convert their mass back into energy in the form of two gamma rays. In the early universe, photons had enough energy to produce proton–antiproton pairs or neutron–antineutron pairs. When these particles collided with their antiparticles, they converted their mass back into photons. Thus, the early universe was filled with a dynamic soup of energy flickering from photons into particles and back again.

Antimatter: Matter composed of antiparticles, which upon colliding with a matching particle of normal matter, annihilate and convert the mass of both particles into energy.

While all this went on, the expansion of the universe cooled the radiation. By the time the universe was 0.0001 second old, its blackbody temperature had fallen to 10^{12} K. At that point the average energy of the gamma-ray photons fell below the energy equivalent to the mass of a proton or a neutron, so the gamma rays could no longer produce such heavy particles, and the creation of protons and neutrons stopped. Those particles combined with their antiparticles and quickly converted most of the mass into photons.

You might guess from this that all of the protons and neutrons would have been annihilated with their antiparticles, and the universe should now consist of nothing but photons. However, for reasons that are poorly understood, a small excess of normal particles apparently existed. For every billion protons annihilated by antiprotons, one survived with no antiparticle to destroy it. Consequently, you live in a world of normal matter, and antimatter is very rare.

Although the gamma rays did not have enough energy to produce protons and neutrons after the universe fell below 10^{12} K, electron–positron pairs, having lower mass, could still be produced. (The positron is the antiparticle of the electron, mentioned previously in the context of nuclear fusion.) That continued until the universe was about 4 seconds old, at which time the expansion had cooled the gamma rays to the point where they could no longer create even electron–positron pairs. Most of the electrons and positrons combined to form photons, and only one in a billion elections survived. The protons, neutrons, and electrons of which our universe is made were produced during the first 4 seconds of its history.

This soup of hot gas and radiation continued to cool. By the time the universe was about 2 minutes old, protons

Figure 11.6

Photons scatter (bounce) off electrons (blue) easily but hardly at all from the much more massive protons (red). (a) When the universe was very dense and ionized, photons could not travel very far before they scattered off an electron. This means the gas was opaque. (b) As the universe expanded, the electrons were spread further apart, and the photons could travel farther; this made the gas more transparent. (c) After the era of recombination, most electrons were locked to protons to form neutral atoms, and the universe became almost completely transparent.

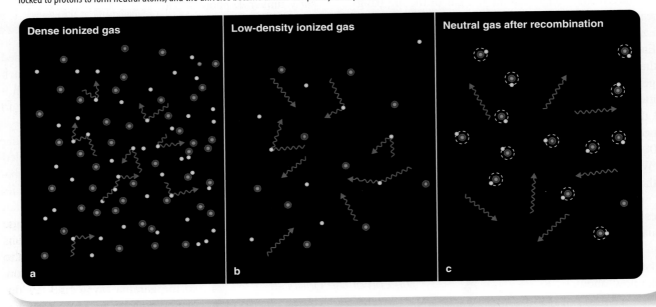

Dense ionized gas | **Low-density ionized gas** | **Neutral gas after recombination**

a   b   c

and neutrons could link to form deuterium, the nucleus of a heavy hydrogen atom, and not be broken apart again. By the end of the next minute, further reactions began converting deuterium into helium, but almost no heavier atoms could be built because there are no stable nuclei with atomic weights of 5 or 8 (in units of the hydrogen atom)—they fall apart as soon as they are created. Cosmic element building during the big bang had to proceed rapidly, step-by-step, like someone hopping up a flight of stairs. The lack of stable nuclei at atomic weights of 5 and 8 meant there were missing steps in the stairway, and the step-by-step reactions had great difficulty jumping over these gaps. Astronomers can calculate that the big bang produced a tiny amount of lithium (atomic weight 7) but no elements heavier than that.

By the time the universe was 3 minutes old, it had become so cool that most nuclear reactions had stopped, and by the time it was 30 minutes old the nuclear reactions had ended completely. At that time about 25 percent of the mass was in the form of helium nuclei, and the rest was in the form of hydrogen nuclei (protons). That is the abundance of hydrogen and helium observed today in the oldest stars. The cosmic abundance of hydrogen and helium were essentially fixed during the first minutes of the universe. The hydrogen nuclei in water molecules in your body have survived unchanged since

they formed during the first moments of the big bang. Heavier elements were built by nucleosynthesis inside later generations of stars (see Chapters 8 and 9).

Recombination and Reionization—The First Thousands and Millions of Years

At first, the universe was so hot that the gas was totally ionized, and the electrons were not attached to nuclei. Free electrons interact with photons so easily that a photon could not travel very far before it encountered an electron and was deflected (Figure 11.6a). The radiation and matter interacted continuously with each other and cooled together as the universe expanded.

As the young universe expanded, it went through three important changes. First, when the universe reached an age of roughly 50,000 years, the density of the energy present in the form of photons became less than the density of the gas. Before that time, matter could not clump together because the intense sea of photons smoothed the gas out. Once the density of the radiation fell below that of matter, the matter could begin to draw together under the influence of gravity and form the clouds that eventually became galaxy clusters and galaxies.

The expansion of the universe spread the particles of the ionized gas farther and farther apart. As the universe reached the age of about 400,000 years, the second important change began. As the density decreased and the falling temperature of the universe reached 3,000 K, protons were able to capture and hold free electrons to form neutral hydrogen, a process called **recombination**, although "combination" (for the first time) would be more accurate. As the free electrons were gobbled up into atoms, they could no longer deflect photons. The photons could travel easily through the gas, so the gas became transparent (Figure 11.6c), and the photons retained the blackbody temperature of 3,000 K that the gas and photons together had at the time of recombination. Those photons are what are observed today as the CMB, with a large redshift that makes their temperature now about 2.7 K.

Recombination left the gas of the big bang neutral, hot, dense, and transparent. At first the universe was filled with the glow of the hot gas, which would have been partly at visible wavelengths. As the universe expanded and cooled, the glow faded and the universe entered what cosmologists call the **dark age**, a period lasting hundreds of millions of years until the formation of the first stars. During the dark age, the universe expanded in darkness.

The dark age ended as the first stars began to form. The gas from which those first stars formed contained almost no metals and was consequently quite transparent. Mathematical models show that the first stars formed from this metal-poor gas would have been very massive, very luminous, and very short lived. That first burst of star formation produced enough ultraviolet light to begin ionizing the gas, and today's astronomers, looking back to the most distant visible quasars and galaxies, can see traces of that **reionization** of the universe (Figure 11.7). Reionization marks the end of the dark ages and the beginning of the age of stars and galaxies that continues today.

Look carefully at Figure 11.8 on the next page; it summarizes the story of the big bang, from the formation of helium in the first 3 minutes through energy–matter equality, recombination, and finally reionization of the gas. It may seem amazing that mere humans trapped on Earth can draw such a diagram, but remember that it is based on evidence and on the best understanding of how matter and energy interact.

Recombination: The stage, within 400,000 years of the big bang, when the gas became transparent to radiation.

Dark age: The period of time after the glow of the big bang faded into the infrared and before the birth of the first stars, during which the universe expanded in darkness.

Reionization: The stage in the early history of the universe when ultraviolet photons from the first stars ionized the gas filling space.

Figure 11.7

In this artist's conception of reionization, the first stars produce floods of ultraviolet photons that ionized the gas in expanding bubbles. Such a storm of star formation ended an age when the universe had expanded in darkness. Spectra of the most distant quasars reveal that those first galaxies were surrounded by neutral gas that had not yet been fully ionized. Thus the look-back time allows modern astronomers to observe the age of reionization.

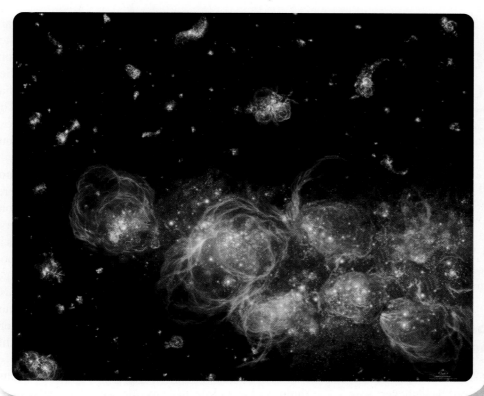

Figure 11.8

During the first few minutes of the big bang, some hydrogen was fused to produce helium, but the universe quickly became too cool for such fusion reactions to occur. The rate of cooling increased as matter began to dominate over radiation. Recombination freed the radiation from the influence of the gas, and reionization was caused by the birth of the first stars. Note how the exponential scale in time stretches early history and compresses recent history.

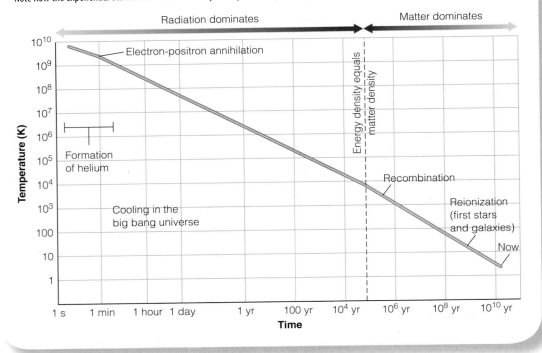

11.3 Space and Time, Matter and Energy

How can the big bang have happened everywhere? To solve the puzzle, you must put what seem to be reasonable expectations on hold and look carefully at how space and time behave on cosmic scales.

Looking at the Universe

The universe looks about the same whichever way you look. That is called **isotropy**. Of course, there are local differences. If you look toward a galaxy cluster you see more galaxies, but that is only a local variation. On the average, you see similar numbers of galaxies in every direction. Furthermore, the background radiation is also almost perfectly uniform across the sky. The universe is observed to be highly isotropic, almost exactly the same in all directions when viewed from our position.

The universe also seems to be homgenous; **homogeneity** is the property of being the same everywhere. Of course there are local variations. Some regions contain more galaxies and some less. Also, if the universe evolves, then at large look-back times you see galaxies at an earlier stage. If you account for these well-understood variations, then the universe seems to be, on average, the same everywhere. This is harder to check because you can't actually go to the locations of distant galaxies and check in detail that things are about the same there as here, but all astronomical observations indicate this is so.

Isotropy and homogeneity together lead to the **cosmological principle**, that any observer in any galaxy sees that the universe has the same general properties, after accounting for relatively minor local and evolutionary variations. The cosmological principle implies that there are no special places in the universe. What you see from the Milky Way Galaxy is typical of what all intelligent creatures see from their respective home galaxies. Furthermore, the cosmological principle is another way of saying that the universe has no center or edge. Such locations would be special places, and the cosmological principle means there are no special places.

The Cosmic Redshift

Distance is the separation between two points in space; time is the separation between two moments. Einstein's theories of special relativity and general relativity (published respectively in 1905 and 1916) describe how space and time are related and can be considered together as the fabric of the universe, called space-time. You can think of space-time as the canvas on which the universe is painted. Einstein's theories predict that the canvas of space-time can potentially expand (or contract), and, amazingly, that has been confirmed by observations.

The stretching of space-time explains one of the most important observations in cosmology—cosmological redshifts. Modern astronomers understand that, except for small local motions within clusters of galaxies, the galaxies are basically at rest and have kept approximately the same "address" in space since the big bang. The distances between them increase as space-time expands. Furthermore, as space-time expands, it stretches photons traveling through space to longer wavelengths, as you can see in Figure 11.9. Photons from distant galaxies spend more time traveling through space and are stretched more than photons from nearby galaxies. That is why redshift depends on distance. Note that objects—such as the Milky Way, Earth, and you—that are held together by gravity or electromagnetic forces do not expand as the universe expands.

Astronomers often express redshifts as if they were radial velocities, but the redshifts of the galaxies are not Doppler shifts. That is why this book is careful to refer to a galaxy's *apparent* velocity of recession. All a cosmological redshift tells you directly is how much the universe has expanded since the light began its journey to Earth. The formula to calculate the distance a photon has traveled, given its redshift, is complicated, and not all the parameters have been measured precisely. Nevertheless, the Hubble law does apply, and redshifts can be used to estimate the distances to galaxies.

Open universe: A model of the universe in which space-time is curved in such a way that the universe is infinite.

Closed universe: A model of the universe in which space-time is curved to meet itself and the universe is finite.

Flat universe: A model of the universe in which space-time is not curved.

Model Universes

Almost immediately after Einstein published his theory, theorists were able to solve the highly sophisticated mathematics to compute simplified descriptions of the behavior of space-time and matter. Those "model universes" dominated cosmology throughout the 20th century.

The equations allowed three general possibilities. Space-time might be curved in ways that are called an **open universe** or a **closed universe**, or space-time might have no overall curvature at all, a situation that is called a **flat universe**. Most people find the curved space-time models difficult to imagine. Fortunately, modern observations have shown that the flat universe model is almost certainly correct, so you don't have to wrap your brain around the curved models. Note that "flat" does not mean two dimensional. Rather, it means that the familiar rules of geometry you learned in elementary and middle

Figure 11.9

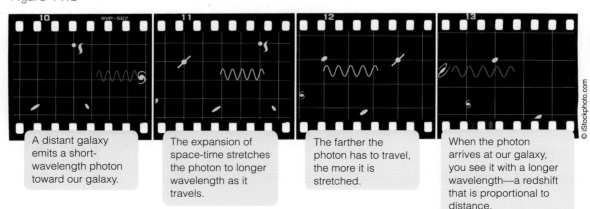

A distant galaxy emits a short-wavelength photon toward our galaxy.

The expansion of space-time stretches the photon to longer wavelength as it travels.

The farther the photon has to travel, the more it is stretched.

When the photon arrives at our galaxy, you see it with a longer wavelength—a redshift that is proportional to distance.

© iStockphoto.com

school, for instance that the circumference of a circle is 2π times its radius and that the interior angles of a triangle add up to 180 degrees, are true on the largest scales. Different rules of geometry would apply on large scales if the universe were curved.

A main criterion separating the three models is the average density of the universe, which according to general relativity, determines the overall curvature of space-time. If the average density of matter and energy in the universe equals what is called the **critical density**, calculated to be about 9×10^{-30} g/cm^3 (depending on the exact value of Hubble constant), space-time will be flat. If the average density is more than the critical density, the universe must be closed; and if it is less, the universe must be open.

The expansions versus time of the three different models are compared in Figure 11.10. The parameter R on the vertical axis is a measure of the extent to which the universe has expanded. You could think of it, essentially, as the average distance between galaxies. In the figure you can see that closed universes may expand and

then contract, and open universes may expand forever. Notice also that you can't find the actual age of the universe—the distance on the horizontal axis between the present and the beginning, when R was zero—from the expansion rate alone. You also need to know whether the universe is open, closed, or flat. The Hubble time, discussed earlier in this chapter as a rough estimate of the age of the universe, is actually the age the universe would be if it were totally open, which means it would have to contain almost no matter at all.

Dark Matter in Cosmology

Later in this chapter you will see how observational evidence indicates that the universe is probably flat. For now, you must solve a different problem. If the universe is flat, then its average density must equal the critical density, yet when astronomers added up the matter they could detect, they found only a few percent of the critical density. They wondered if the dark matter made up the rest.

In Chapters 9 and 10, you learned that our galaxy, other galaxies, and galaxy clusters have much stronger gravitational fields than expected based on the amount of visible matter. Even when you add in the nonluminous gas and dust that you expect to find, their gravitational fields are stronger than expected. Galaxies must contain dark matter—in fact, much more dark matter than normal matter (see Figure 9.9).

The protons and neutrons that make up normal matter, including Earth and you, belong to a family of subatomic particles called *baryons*. Modern evidence based on what can be determined about the products of nuclear reactions in the first few minutes after the big bang shows that the dark matter is not baryonic. If there were lots of baryons present during those early moments, they would have (1) collided with and destroyed deuterium nuclei and (2) collided with some of the helium to make lithium.

Figure 11.11 shows that the observed amount of deuterium sets a lower limit on the density of the universe, and the observed abundance of lithium-7 sets an upper limit.

Figure 11.10

Illustrations of some simple universe models that depend only on the effect of gravity. Under that assumption, open universe models expand without end, and the corresponding curves fall in the region shaded orange. Closed models expand and then contract again (red curve, one example). A flat universe (dotted line) marks the boundary between open and closed universe models. The relationship between estimated age and actual age of the universe depends on the geometry of space-time. The age of the universe for each model is shown on the graph as the horizontal distance from "Now" back to the time when the universe scale factor R equaled 0.

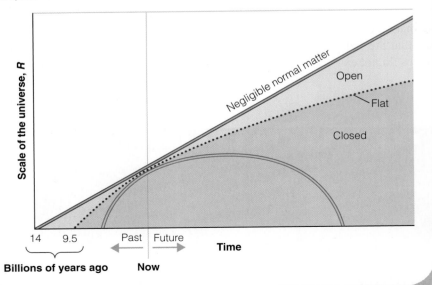

Those limits indicate that the baryons you and Earth and the stars are made of cannot add up to more than 4 percent of the critical density. Yet, observations show that galaxies and galaxy clusters contain as much as 30 percent of the critical density in the form of dark matter. Only a small amount of the matter in the universe can be baryonic, so the dark matter must be **nonbaryonic matter**.

The true nature of dark matter remains one of the mysteries of astronomy. The most successful models of galaxy formation require that the dark matter be made up of **cold dark matter**, meaning the particles move slowly and can clump into structures with sizes that explain the galaxies and galaxy clusters you see today.

Although the evidence is very strong that dark matter exists, it is not abundant enough by itself to make the universe flat. Dark matter appears to constitute no more than about 30 percent of the critical density. As you will see later in this chapter, there is more to the universe than meets the eye and more even than dark matter.

Figure 11.11

This diagram compares observation with theory. Theory predicts how much deuterium and lithium-7 you would observe for different densities of normal matter (red and blue curves). The observed density of deuterium falls in a narrow range shown at upper left and sets a lower limit on the possible density of normal matter. The observed density of lithium-7 sets an upper limit. This means the true density of normal matter must fall in a narrow range represented by the green column. Certainly, the density of normal matter is much less than the critical density.

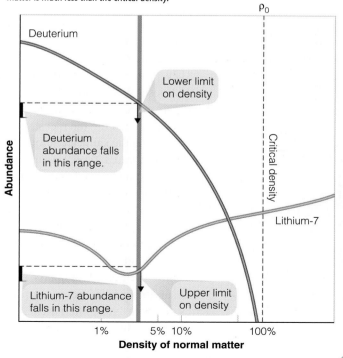

11.4 Modern Cosmology

If you are a little dizzy from the weirdness of expanding space-time and dark matter, make sure you are sitting down before you read further. As the 21st century began, astronomers made a discovery that startled all cosmologists: the expansion of the universe is actually accelerating. To get a running start on these new discoveries, you'll have to go back a couple of decades.

Inflation

By 1980, the big bang model was widely accepted, but it faced two problems that led to the development of an improved theory—a big bang model with an important addition.

One of the problems is called the **flatness problem**. The curvature of space-time seems to be near the transition between an open and a closed universe. That is, the universe seems approximately flat. It seems peculiar that the actual density of the universe is anywhere near the critical density that would make it flat. To be so near critical density now, the density of the universe during its first moments must have been very close, within 1 part in 10^{49}, of the critical density. *So, the flatness problem is: Why was the universe so close to exactly flat, with no space-time curvature, at the time of the big bang?*

The second problem with the original big bang theory is called the **horizon problem**. When astronomers correct for the motion of Earth, they find that the CMB is very isotropic, the same in all directions to a precision of better than 1 part in 1,000.

Nonbaryonic matter: Proposed dark matter made up of particles other than protons and neutrons (baryons).

Cold dark matter: Dark matter which is made of slow-moving particles.

Flatness problem: The peculiar circumstance that the early universe must have contained almost exactly the right amount of matter to make space-time flat.

Horizon problem: The circumstance that the primordial background radiation seems much more isotropic than can be explained by the standard big bang theory.

Inflationary big bang:
A version of the big bang theory, derived from grand unified theories of particle physics, that includes a rapid expansion when the universe was very young.

Yet, background radiation coming from two points in the sky separated by more than an angle of one degree is from two parts of the big bang far enough apart that they should not have been connected at any previous time. That is, when the CMB photons were released, the universe was not old enough for energy to have traveled at the speed of light from one of those regions to the other—the regions should always have been beyond each other's "horizon" and could not have exchanged heat to make their temperatures equal. *So, the horizon problem is: How did every part of the observable universe get to be so nearly the same temperature by the time of recombination?*

The key to these two problems and to other problems with the simple big bang model may lie with a modified model called the **inflationary big bang** that predicts there was a sudden extra expansion when the universe was very young, even more extreme than that predicted by the original big bang model. According to the inflationary universe model, the universe expanded and cooled until about 10^{-36} seconds after the big bang. Then, the universe became cool enough that the forces of nature (see Chapter 7), which at earlier extremely high temperatures would have behaved identically, began to differ from each other. Physicists calculate this would have released tremendous amounts of energy and suddenly inflated the universe by a factor of 10^{50} or larger. As a result, the part of the universe that is now visible from Earth, the entire observable universe, expanded rapidly from 35 orders of magnitude smaller than a proton to roughly a meter across and then continued a slower expansion to its present state.

That sudden inflation can solve both the flatness problem and the horizon problem. The inflation of the universe would have forced whatever curvature it had toward zero, just as inflating a balloon makes a small spot on its surface flatter. You now live in a universe that has almost perfectly flat space-time geometry because of that sudden inflation long ago. In addition, because the observable part of the universe was no larger in volume than an atom before inflation, it was small enough to have equalized its temperature by then. Now you live in a universe that has the same CMB temperature in all directions.

The inflationary theory predicts that the universe is almost perfectly flat. Observations, however, give evidence that the masses scientists know about (baryonic matter plus dark matter) add up only to about 30 percent of the amount needed to make space-time flat. Can there be more to the universe than baryonic matter and dark matter? What could be weirder than dark matter? Read on.

The Acceleration of the Universe

Both common sense and mathematical models suggest that, as the galaxies recede from each other, the expansion should be slowed by gravity trying to pull the galaxies toward each other. How much the expansion is slowed should depend on the amount of matter in the universe. If the density of matter is less than the critical density, the expansion should be slowed only slightly, and the universe should expand forever. If the density of matter in the universe is greater than the critical density, the expansion should be slowing down dramatically, and the universe should eventually begin contracting.

For decades, astronomers struggled to measure the distance to very distant galaxies directly, compare distances with redshifts, and thereby detect the slowing of the expansion. The rate of slowing would in turn reveal the true curvature of the universe. This was one of the key projects for the Hubble Space Telescope, and two teams of astronomers spent years in competition making the measurements using the same technique. They calibrated type Ia supernovae as standard candles by locating such supernovae occurring in nearby galaxies whose distances were known from Cepheid variables and other reliable distance indicators (see Chapter 10). Once the peak luminosity of type Ia supernovae had been determined, they could be used to find the distances of much more distant galaxies.

Both teams announced their results in 1998. They agreed that the expansion of the

© Ashok Rodrigues/iStockphoto.com

universe is not slowing down. Contrary to expectations, it is speeding up! In other words, the expansion of the universe is accelerating (Figure 11.12).

The announcement that the expansion of the universe is accelerating was totally unexpected, and astronomers immediately began testing it. The most likely problem was thought to be that the calibration of the supernovae by the original teams might be incorrect. However, that type of problem has been ruled out by measurements of more recently discovered supernovae at very great distances. Other possible problems have been checked and appear to have been eliminated. The universe really does seem to be expanding faster and faster.

Dark Energy and Acceleration

If the expansion of the universe is accelerating, then there must be a force of repulsion in the universe, and astronomers are struggling to understand what it could be. One possibility leads back to Albert Einstein.

When Einstein published his theory of general relativity in 1916, he noticed that his equations describing space-time implied that the universe should contract because of the gravitational attraction of galaxies for each other. In 1916 astronomers did not yet know that the universe was expanding, so Einstein thought he needed to balance the attractive force of gravity by adding

Figure 11.12

From the way supernovae fade over time, astronomers can identify those that are type Ia. Once calibrated, those supernovae can be compared with their redshifts, revealing that distant type Ia supernovae are about 25 percent fainter than expected. That must mean they are farther away than expected, given their redshifts. This is strong evidence that the expansion of the universe is accelerating.

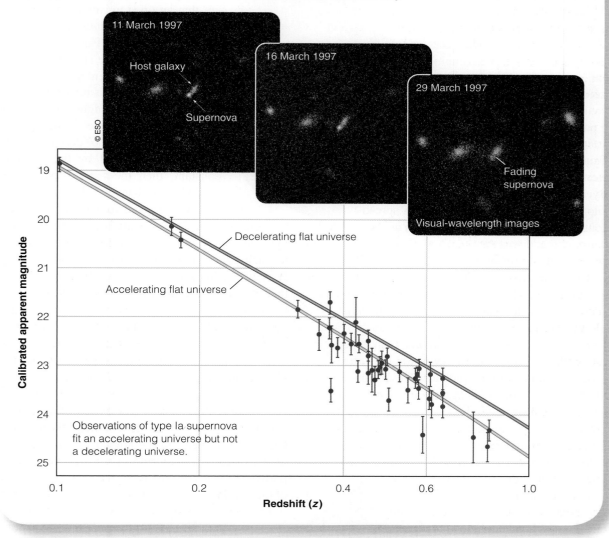

a constant to his equations called the **cosmological constant**, representing a force of repulsion that would make the universe hold still. Thirteen years later, in 1929, Edwin Hubble announced that the universe was expanding, and Einstein said that introducing the cosmological constant was his biggest blunder. Modern astronomers aren't so sure.

One explanation for the acceleration of the universe is that there is a cosmological constant after all, representing a real force that drives a continuing acceleration in the expansion of the universe. The cosmological constant, as its name implies, would be constant in strength over time. Another possibility is a type of energy that is not constant in strength over time. Astronomers have begun referring to this type of energy as **quintessence**. In either case, the observed acceleration is evidence that some form of energy, either a cosmological constant or quintessence, is spread throughout space. Astronomers refer to this as **dark energy**, energy that drives the acceleration of the universe but does not contribute to starlight or the CMB.

You will recall that acceleration and dark energy were first discovered when astronomers found that supernovae a few billion light-years away were slightly *fainter* than expected. Since then even more distant supernovae have been determined to be a bit *brighter* than expected, meaning they are not as far away as the redshifts of their galaxies would seemingly indicate. This means that sometime about 6 billion years ago the universe shifted gears from deceleration to acceleration. The careful calibration of type Ia supernovae allow astronomers to observe this change from deceleration to acceleration. This discovery has the important consequence of increasing previous estimates of the age of the universe by several billion years.

Dark energy can also help you understand the lack of curvature of space-time. The theory of inflation makes the specific prediction that the universe is flat. Dark energy seems to fit with that prediction. As mentioned earlier in this chapter, energy and matter are equivalent, so dark energy is equivalent to mass spread through space. Baryonic matter plus dark matter makes up about a third of the critical density, and dark energy appears to make up two-thirds. That is, when you include dark energy, the total mass-plus-energy density of the universe equals the critical density, making the universe flat.

The Fate of the Universe

For many years, cosmologists enjoyed saying: "Geometry is destiny." Thinking about models of open, closed, and flat universes, they concluded that the density of a model universe determines its geometry, and its geometry determines its fate. In other words, they were sure that an open universe must expand forever, and a closed universe must eventually begin contracting. That is true, however, only if the universe is ruled by gravity. If the power of dark energy dominates gravity, then geometry is not destiny, and even a closed universe might expand forever.

The ultimate fate of the universe depends on the nature of dark energy. If dark energy is described by the cosmological constant, then the force driving acceleration does not change with time, and our flat universe will expand forever. The galaxies will get farther and farther apart, use up their gas and dust making stars, and the stars will ultimately all die, until each galaxy is isolated, burnt out, dark, and alone. If, however, dark energy is described by quintessence, then its strength could increase with time, and the universe expansion may accelerate faster and faster as space pulls the galaxies away from each other and eventually pulls the galaxies apart, then pulls the stars apart, and finally rips individual atoms apart. This has been called the **big rip**. Don't worry. Even if a big rip is in the future, nothing will be happening for at least 30 billion years.

There will probably be no big rip. Important observations made by the Chandra X-Ray Observatory have been used to measure the amount of hot gas and dark matter in 25 galaxy clusters. These observations are important for two reasons. First, the redshifts and distance of these galaxies confirm that the universe expansion initially slowed down, but then shifted gears about 6 billion years ago and is now accelerating. The results are also important because they are almost good enough to rule out quintessence. If dark energy is described by the cosmological constant and not by quintessence, then there will be no big rip (Figure 11.13). Notice that this method does not depend on type Ia supernovae at all, so it is independent confirmation that acceleration is real. When a theory is confirmed by observations of many different types, scientists have much more confidence that it is a true description of nature.

The Origin of Structure

On the largest scales, the universe is isotropic, meaning it looks the same in all directions. On smaller scales, there are irregularities. Galaxies are grouped in clusters ranging from a few galaxies to thousands, and those clusters appear to be grouped into **superclusters**. The Local Supercluster, in which we live, is a roughly disk-shaped swarm of galaxy clusters 50 to 75 Mpc in diameter. By measuring the redshifts and positions of thousands of galaxies in great slices across the sky, astronomers have been able to create maps revealing that the superclusters are not scattered at random. They are distributed in long, narrow filaments and thin walls that outline great voids nearly empty of galaxies, as demonstrated in Figure 11.14. These aggregations and gaps are referred to as **large-scale structure**.

This large-scale structure is a problem because the cosmic microwave background radiation is very uniform, and that means the gas of the big bang must have been extremely uniform at the time of recombination. Yet the look-back time to the farthest known galaxy clusters, galaxies, and quasars is about 95 percent of the way back to the big bang. How did the uniform gas at the time of recombination coagulate so quickly to form galaxy clusters, galaxies, and supermassive black holes in the centers of galaxies so early in the history of the universe? The answer appears to lie in the characteristics of dark matter.

Baryonic matter is so rare in the universe that it did not have enough gravity to pull itself together quickly after the big bang. As you have read earlier, astronomers propose that dark matter is non-baryonic and therefore immune to the smoothing effect of the intense radiation that prevented normal matter from contracting. Dark matter was able to collapse into clouds and then pull in the normal matter to begin the formation of galaxies, clusters, and superclusters. Mathematical models, such

> **Supercluster:** Cluster of galaxy clusters.
>
> **Large-scale structure:** The distribution of clusters and superclusters of galaxies in filaments and walls enclosing voids.

Figure 11.13

X-ray observations of hot gas in galaxy clusters confirm that in its early history the universe was decelerating because gravity was stronger than the dark energy. As expansion weakened the influence of gravity, dark energy began to produce acceleration. The evidence is not conclusive, but it most directly supports the cosmological constant and weighs against quintessence, which means the universe may not face a big rip. This diagram is only schematic, and the two curves are drawn separated for clarity; at the present time the two curves have not diverged from each other.

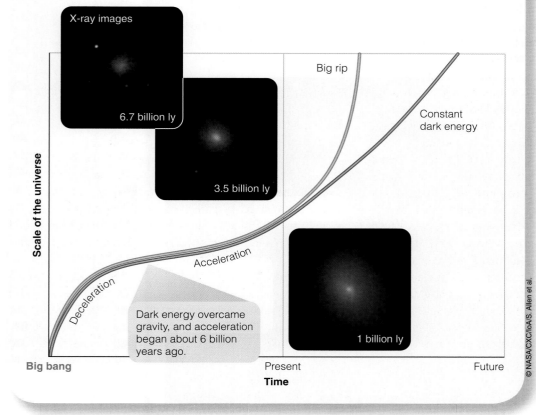

X-ray images

6.7 billion ly

3.5 billion ly

1 billion ly

Big rip

Constant dark energy

Acceleration

Deceleration

Dark energy overcame gravity, and acceleration began about 6 billion years ago.

Scale of the universe

Big bang   Present   Future

Time

© NASA/CXC/IoA/S. Allen et al.

as Figure 11.15, have attempted to describe this process, and cold dark matter does seem capable of jump-starting the formation of structure.

But what started the clumping of the dark matter? Theorists say that space is filled with tiny, random quantum mechanical fluctuations smaller than the smallest atomic particles. At the moment of inflation, those tiny fluctuations would have been stretched to become very large but very subtle variations in gravitational fields that could have later led to the formation of clusters, filaments, and walls. The structure you see in Figure 11.14 may be the ghostly traces of quantum fluctuations in the infant universe.

CMB Irregularities and the Curvature of Space-Time

Observations of tiny irregularities in the background radiation can also reveal details about inflation and acceleration. In fact, the inflationary theory of the universe makes very specific predictions about the sizes of the irregularities an observer on Earth should see in the CMB. The Wilkinson Microwave Anisotropy Probe (WMAP), a space infrared telescope, has made extensive observations of the type required to test those predictions.

The background radiation is very isotropic—it looks almost exactly the same in all directions. However, when the average intensity is subtracted from each spot on the sky, small irregularities are evident. That is, some spots on the sky look a tiny bit hotter and brighter than other spots (Figure 11.16).

Cosmologists can analyze those irregularities in the intensity of the CMB using sophisticated mathematics to measure how often spots of different sizes occur. The analysis confirms that spots about 1 degree in diameter are the most common, but spots of other sizes occur as well, and it is possible to plot a graph such as Figure 11.17 to show how common irregularities of various sizes are.

Figure 11.14

Nearly 70,000 galaxies are plotted in this double slice of the universe. The nearest galaxies are shown in red and the more distant in green and blue. The Sloan Great Wall is almost 1.4 billion light-years long and is the largest known structure in the universe.

Approx. 3 billion ly

Sloan Great Wall

Earth

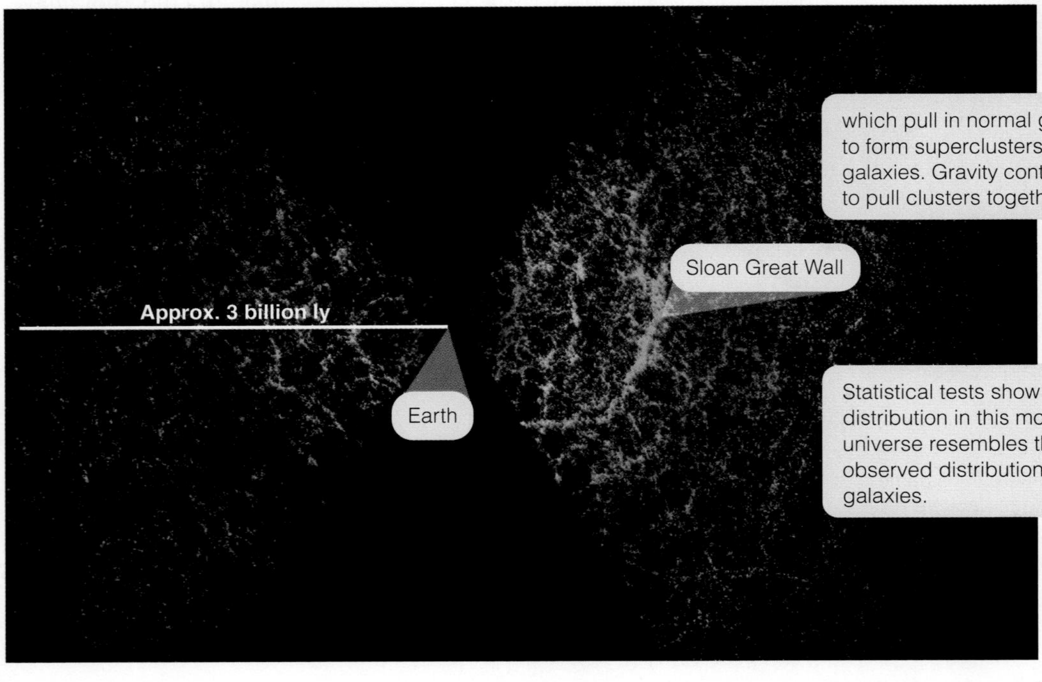

© Sloan Digital Sky Survey

Figure 11.15

Mathematical Model of the Growth of Structure in the Universe.

Soon after the big bang, radiation and hot gas are almost uniformly spread through the universe.

Cold dark matter, immune to the influence of light, can contract to form clouds…

which pull in normal gas to form superclusters of galaxies. Gravity continues to pull clusters together.

Statistical tests show the distribution in this model universe resembles the observed distribution of galaxies.

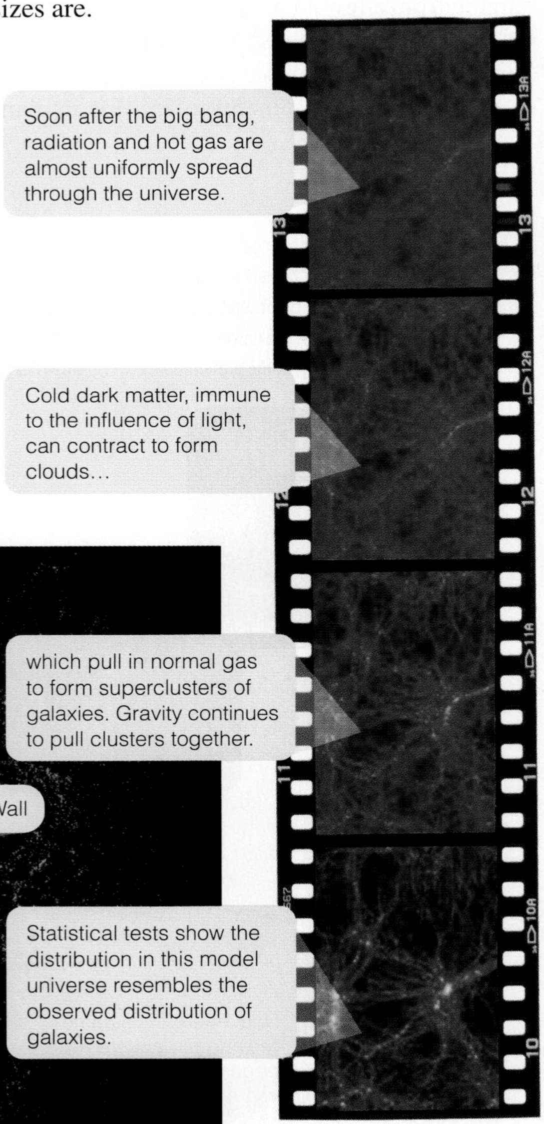

Adapted from a model by Kauffmann, Colberg, Diaferio, & White: Max-Planck Institut für Astronomie / © iStockphoto.com

Figure 11.16

You can see the difference yourself. Compare the observations of the irregularities in the background radiation at right with the three simulations at left. The grids represent the curvature of space-time for closed, flat, and open models, respectively. The observed size of the CMB irregularities fits best with cosmological models having flat geometry. Detailed mathematical analysis confirms your visual impression: The universe is flat.

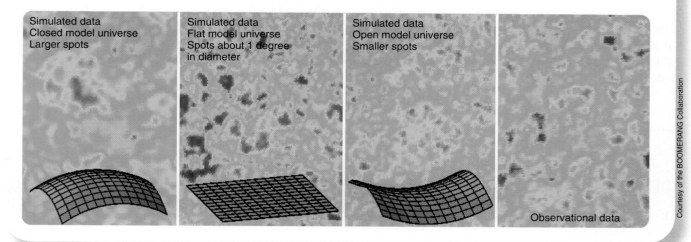

Simulated data
Closed model universe
Larger spots

Simulated data
Flat model universe
Spots about 1 degree
in diameter

Simulated data
Open model universe
Smaller spots

Observational data

Courtesy of the BOOMERANG Collaboration

Figure 11.17

This graph shows how commonly irregularities of different sizes occur in the cosmic microwave background radiation. Irregularities of about 1 degree in diameter are most common. Models of the universe that are open or closed are ruled out. The data fit a flat model of the universe very well. Crosses on data points show the uncertainty in the measurements.

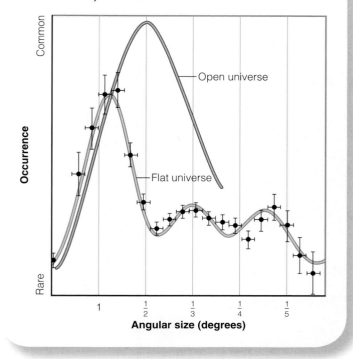

Theory predicts that most of the irregularities in the hot gas of the big bang should be about one degree in diameter if the universe is flat. If the universe were open, the most common irregularities would be smaller. The size of the irregularities in the cosmic background radiation show that the observations fit the flat universe model well, as you can see in Figures 11.16 and 11.17. You learned earlier in this chapter that a flat universe is most easily explained by inflation. Not only is the theory of inflation supported, an exciting result itself, but these data show that the universe is flat, which indirectly confirms the existence of dark energy and the acceleration of the universe.

The results from the WMAP observations make a complicated curve in Figure 11.17, and details of the wiggles tell cosmologists a great deal about the universe. The curve shows that the universe is flat, accelerating, and will expand forever. The age of the universe derived from the data is 13.7 billion years. Furthermore, the smaller peaks in the curve reveal that the universe contains 4 percent baryonic (normal) matter, 23 percent dark matter, and 73 percent dark energy. The Hubble constant is confirmed to be 71 km/s/Mpc. The inflationary theory is confirmed, and the data support the cosmological constant version of dark energy, although quintessence is not quite ruled out. The dark matter needs to be "cold" (meaning, in this context, slowly-moving) in order to clump together rapidly enough after the big bang to make the first galaxies, quasars, and galaxy clusters.

Please reread the preceding paragraph. It is mind-blastingly amazing. WMAP and other studies of the cosmic microwave background radiation and the distribution of galaxies have revolutionized cosmology. At last, astronomers have accurate observations against which to test theories. The basic constants are known to a precision of a few percent.

On reviewing these results, one cosmologist announced that "Cosmology is solved!"—but that might be premature. Scientists don't understand dark matter or the dark energy that drives the acceleration, so over 95 percent of the universe is not understood. Hearing this, another astronomer suggested that a better phrase is "Cosmology in crisis!" Certainly there are further mysteries to be explored, but cosmologists are growing more confident that they can describe the origin and evolution of the universe.

RECAP

You have traced the origin of the universe, the creation of the chemical elements, the birth of galaxies, and the births and deaths of stars. You now have a perspective that few humans share, and you have tried to tackle some of the biggest and most difficult ideas in science.

As you review the history and structure of the universe, it is wise to recognize the mysteries that remain; but note that they are mysteries that may be solved and not mysteries that are unknowable. Only a century ago, humanity didn't know there were other galaxies, or that the universe was expanding, or that stars generate energy by nuclear fusion. Human curiosity has solved many of the mysteries of cosmology and will solve more during your lifetime.

Test your understanding of the material you've experienced by answering these questions:

1. Does the universe have an edge and a center?

2. How do you know that the universe began with a big bang?

3. How does the universe expand?

4. How has the universe evolved, and what will be its fate?

"It's easy to read, it outlines important topics, and it's relevant. Thanks for the good stuff on the website, I think it will **really help with tests**.

– Thomas Scholtes, Student at University of Maryland, College Park

REVIEW

HE DID

ASTRO puts a multitude of study aids at your fingertips. After reading the chapters, check out these resources for further help:

• **Chapter Review Cards**, found in the back of your book, include section summaries, Key Term definitions, and selected review questions for each chapter.

• **Online printable flash cards** give you three additional ways to check your comprehension of key astronomy concepts.

Other great ways to help you study include **animations of key exhibits, additional Web links,** and **interactive quizzes**.

You can find it all at **4ltrpress.cengage.com/astro**.

Chapter 12

The Origin
of the Solar System

The solar system is your home in the universe. Because humans are an intelligent species, we have the ability and the responsibility to wonder what we are. Our kind have inhabited this solar system for several million years, but only within the last hundred years have we begun to understand what a solar system is.

12.1 The Great Chain of Origins

You are linked through a great chain of origins that leads backward through time to the first instant when the universe began 13.7 billion years ago. The gradual discovery of the links in that chain is one of the most exciting adventures of the human intellect. In earlier chapters, you studied some of that story: the origin of the universe in the big bang, the formation of galaxies, the origin of stars, and the production of the chemical elements. Here you will explore further and consider the origin of planets.

The History of the Atoms in Your Body

By the time the universe was a few minutes old, the protons, neutrons, and electrons in your body had come into existence (see Chapter 11). You are made of very old matter.

Although those particles formed quickly, they were not linked together to form the atoms that are common today. Most of the matter was hydrogen, and about 25 percent of the mass was helium. Very few of the heavier atoms were made in the big bang. Although your body does not contain helium, it does contain many of those ancient hydrogen atoms unchanged since the universe began.

Within a few hundred million years after the big bang, matter began to collect to form galaxies containing billions of stars. You have learned how nuclear reactions inside stars combine low-mass atoms such as hydrogen to make heavier atoms. Generations of stars

©Tom Falch/Stone+/Getty Images / © James Steidl/iStockphoto.com

looking back

You have become an expert on the universe. You have studied the appearance, origin, structure, and evolution of stars, galaxies, and the universe itself. But so far, your studies have left out one important class of objects—planets. Now it is time for you to correct that omission; after all, you live on a planet.

looking ahead

One reason you should learn about the origin of the solar system is that you live here, but there is another reason. In the next two chapters you will explore each of the planets in more detail. By studying the origin of the solar system before studying individual planets, you will have a framework for understanding these fascinating worlds.

astro

cooked the original particles, fusing them into atoms such as carbon, nitrogen, and oxygen (see Chapters 8 and 9). Those are common atoms in your body. Even the calcium atoms in your bones were assembled inside stars.

Most of the iron in your body was produced by carbon fusion in type Ia supernovae and by the decay of radioactive atoms in the expanding matter ejected by type II supernovae. Some atoms heavier than iron (for example, iodine), critical in the functioning of your thyroid gland, were created by rapid nuclear reactions that can occur only during supernova explosions (see Chapter 8). Elements uncommon enough to be expensive, such as gold, silver, and platinum in the jewelry that humans wear, also were produced during the violent deaths of rare massive stars.

Our galaxy contains at least 100 billion stars, including the sun. It formed from a cloud of gas and dust about 5 billion years ago, and the atoms in your body were part of that cloud. How the sun took shape, how the cloud gave birth to the planets, how the atoms in your body found their way onto Earth and into you is the story of this chapter. As you explore the origin of the solar system, keep in mind the great chain of origins that created the atoms. As the geologist Preston Cloud remarked, "Stars have died that we might live."

The Origin of the Solar System

Astronomers have a theory for the origin of our solar system that is consistent both with observations of the solar system and with observations of star formation, and now they are refining the details.

The **solar nebula theory** supposes that planets form in rotating disks of gas and dust around young stars. You have seen clear evidence that disks of gas and dust are common around young stars. This idea is so comprehensive and explains so many observations that it can be considered to have 'graduated' from being just a hypothesis to being properly called a theory. Bipolar flows from protostars (see Chapter 7) were the first evidence of such disks, but modern techniques, like those used by the Hubble telescope in Figure 12.1, can image the disks directly. Our own planetary system formed in such a disk-shaped cloud around the sun. When the sun became luminous enough, the remaining gas and dust were blown away into space, leaving the planets orbiting the sun.

According to the solar nebula theory, Earth and the other planets of our solar system formed billions of years ago as the sun condensed from the interstellar medium. This theory predicts that most stars should have planets because planet formation is a natural part of star formation, and therefore planets should be very common in the universe, probably more common than stars.

Figure 12.1

Dust disks around young stars are evident in these Hubble Space Telescope infrared images. The stars are so young that material is still falling inward and being illuminated by the light from the star. Dark bands across the nebulae (arrows) are dense clouds of gas and dust from which planets may form.

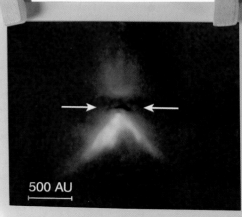

500 AU

DG Tau B

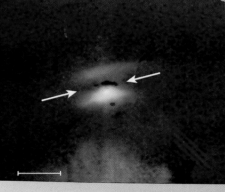

Haro 6-5B

The Solar Nebula Theory

A rotating cloud of gas contracts and flattens...

to form a thin disk of gas and dust around the forming sun at the center.

Planets grow from gas and dust in the disk and remain in orbit when the disk clears.

12.2 A Survey of the Planets

To explore consequences of the solar nebula theory, astronomers search the present solar system for evidence of its past. In the next two sections you will survey the solar system and compile a list of its most significant characteristics, potential clues to how it formed.

You should begin with the most general view of the solar system. It is, in fact, almost entirely empty space (see Figure 1.7). Imagine that you reduce the solar system until Earth is the size of a small grain of table salt, about 0.3 mm (0.01 in.) in diameter. The sun is the size of a small plum 4 m (13 ft) from Earth. Jupiter is an apple seed 20 m (66 ft) from the sun. Neptune, at the edge of the solar system, is a large grain of sand located 120 m (400 ft) from the central plum. You can see that planets are tiny specks of matter scattered around the sun—the last significant remains of the solar nebula.

Revolution and Rotation

The planets revolve around the sun in orbits that lie close to a common plane. (Recall from Chapter 2 that the words *revolve* and *rotate* refer to different motions: A planet revolves around the sun but rotates on its axis.) The orbit of Mercury, the closest planet to the sun, is tipped 7.0° to Earth's orbit. The rest of the planets' orbital planes are inclined by no more than 3.4°. As you can see, the solar system is basically "flat" and disk shaped.

The rotation of the sun and planets on their axes also seems related to the same overall direction of motion. The sun rotates with its equator inclined only 7.2° to Earth's orbit, and most of the other planets' equators are tipped less than 30°. The rotations of Venus and Uranus are peculiar, however. Venus rotates backward compared with the other planets, and Uranus rotates on its "side" (with its equator almost perpendicular to its orbit). You will

IRAS 04248+2612

IRAS 04302+2247

243

explore these planets in detail in Chapters 13 and 14, but later in this chapter you will learn a hypothesis about how they may have acquired their peculiar rotations.

Apparently, the preferred direction of motion in the solar system—counterclockwise as seen from the north—is related to the rotation of the disk of material that became the planets. All the planets revolve around the sun in that direction; and, with the exception of Venus and Uranus, they rotate on their axes in that same direction. Furthermore, nearly all of the moons in the solar system, including Earth's moon, orbit around their planets in that direction. With only a few exceptions, most of which astronomers think they understand, revolution and rotation in the solar system both follow a common theme.

Two Kinds of Planets

Perhaps the most striking clue to the origin of the solar system comes from the obvious division of the planets into two categories, the small Earthlike worlds and the giant Jupiter-like worlds. The difference is so dramatic that you are led to say, "Aha, this must mean something!" Study "Terrestrial and Jovian Planets" on pages 246–247 and notice three important points:

1. The two kinds of planets are distinguished by their location. The four inner (Terrestrial) planets are quite different from the outer four (Jovian) planets.

2. Almost every solid surface in the solar system is covered with craters.

3. The two groups of planets are also distinguished by individual properties such as rings, clouds, and moons. Any theory of the origin of the planets needs to explain these properties.

The division of the planets into two families is a clue to how our solar system formed. The present properties of individual planets, however, don't tell everything you need to know about their origins. The planets have all evolved since they formed. For further clues you can look at smaller objects that have remained largely unchanged since the birth of the solar system.

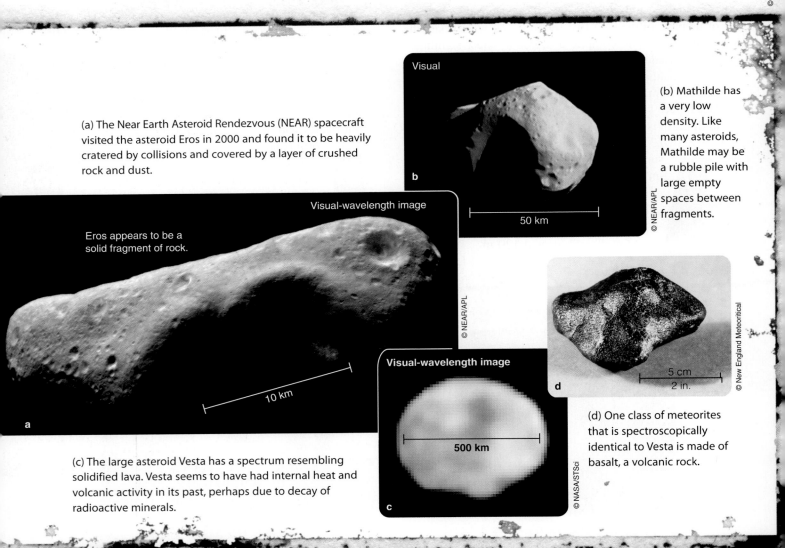

(a) The Near Earth Asteroid Rendezvous (NEAR) spacecraft visited the asteroid Eros in 2000 and found it to be heavily cratered by collisions and covered by a layer of crushed rock and dust.

Visual

(b) Mathilde has a very low density. Like many asteroids, Mathilde may be a rubble pile with large empty spaces between fragments.

50 km

Visual-wavelength image

Eros appears to be a solid fragment of rock.

10 km

Visual-wavelength image

500 km

5 cm
2 in.

(d) One class of meteorites that is spectroscopically identical to Vesta is made of basalt, a volcanic rock.

(c) The large asteroid Vesta has a spectrum resembling solidified lava. Vesta seems to have had internal heat and volcanic activity in its past, perhaps due to decay of radioactive minerals.

12.3 Space Debris: Planet Building Blocks

The sun and planets are not the only remains of the solar nebula. The solar system is littered with three kinds of space debris: asteroids, comets, and meteoroids. Although these objects represent a tiny fraction of the mass of the system, they are a rich source of information about the origin of the planets.

Asteroids

The **asteroids**, sometimes called minor planets, are small rocky worlds, most but not all of which orbit the sun in a belt between the orbits of Mars and Jupiter. More than 100,000 asteroids have orbits that are charted, of which at least 2,000 follow orbits that bring them into the inner solar system where they can occasionally collide with a planet. You will learn in the next section that Earth has been struck many times in its history. Other asteroids share Jupiter's orbit, while some have been found beyond the orbit of Saturn.

About 200 asteroids are more than 100 km (60 mi) in diameter, and tens of thousands are estimated to be more than 10 km (6 mi) in diameter. There are probably a million or more that are larger than 1 km (0.6 mi) and billions that are smaller. Because even the largest are only a few hundred kilometers in diameter, Earth-based telescopes can detect no details on their surfaces, and even the Hubble Space Telescope can image only the largest features.

Photographs returned by robotic spacecraft and space telescopes show that asteroids are generally irregular in shape and battered by impact cratering. In fact, some asteroids appear to be rubble piles of broken fragments, and a few asteroids are known to be double objects or to have small moons in orbit around them. These are understood by astronomers to be evidence of multiple collisions among the asteroids. A few larger asteroids show signs of volcanic activity on their surfaces that may have happened when those asteroids were young.

Astronomers recognize the asteroids as debris left over by a planet that failed to form at a distance of about 3 AU from the sun. A good theory should explain why a planet didn't form there, but instead left behind a belt of construction material.

> **Asteroid:** Small, rocky world. Most asteroids orbit between Mars and Jupiter in the asteroid belt.
>
> **Comet:** One of the small, icy bodies that orbit the sun and produce tails of gas and dust when they approach the sun.

Comets

In contrast to the rocky asteroids, the brightest **comets** are impressively beautiful objects, as you can see in Figure 12.2. Most comets are faint, however, and

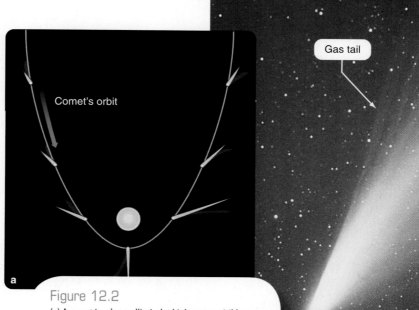

Visual-wavelength image

Gas tail

Dust tail

Comet's orbit

a

b

Figure 12.2

(a) A comet in a long, elliptical orbit becomes visible when the sun's heat vaporizes its ices and pushes the gas and dust away in separate tails. (b) A comet may remain visible in the evening or morning sky for weeks as it moves through the inner solar system. Comet West was in the sky during March 1976.

Terrestrial and Jovian Planets

1 The distinction between the Terrestrial planets and the Jovian planets is dramatic. The inner four planets, Mercury, Venus, Earth, and Mars, are **Terrestrial planets**, meaning they are small, dense, rocky worlds with little or no atmosphere. The outer four planets, Jupiter, Saturn, Uranus, and Neptune, are **Jovian planets**, meaning they are large, low-density worlds with thick atmospheres and liquid or ice interiors.

Planetary orbits to scale. The Terrestrial planets lie quite close to the sun, whereas the Jovian planets are spread far from the sun.

Mercury
Venus
Earth
Mars
Asteroids
Jupiter
Saturn
Uranus
Neptune

The planets and the sun to scale. Saturn's rings would just reach from Earth to the moon.

Mercury
Venus
Earth
Moon
Mars
Sun
Jupiter
Saturn
Uranus
Neptune

1a Of the Terrestrial planets, Earth is most massive, but the Jovian planets are much more massive. Jupiter is over 300 Earth masses, and Saturn is nearly 100 Earth masses. Uranus and Neptune are 15 and 17 Earth masses.

Mercury is only 40 percent larger than Earth's moon, and its weak gravity cannot retain a permanent atmosphere. Like the moon, it is covered with craters from meteorite impacts.

Mercury

NASA

© UC Regents/Lick Observatory

Earth's moon

*key terms

in orange are defined on your Chapter Review Card.

2 Craters are common on all of the surfaces in the solar system that are strong enough to retain them. Earth has about 150 impact craters, but many more have been erased by erosion. Besides the planets, the asteroids and nearly all of the moons in the solar system are scarred by craters. Ranging from microscopic to hundreds of kilometers in diameter, these craters have been produced over the ages by meteorite impacts. When astronomers see a rocky or icy surface that contains few craters, they know that the surface is young.

Mercury is so close to the sun it is difficult to study from Earth. The Mariner 10 and MESSENGER spacecraft flew past Mercury in 1974 and 2008–9, respectively, and were able to take detailed close-up photos of the planet's surface.

Moon

The surface of Venus is not visible through its cloudy atmosphere, but radar maps reveal a dry world of craters and volcanoes.

Mercury

These five worlds are shown in proper relative size.

Earth

3 The Terrestrial planets have densities like that of rock or metal. The Jovian planets all have low densities, and Saturn's density is only 70 percent the density of water. It would float in a big-enough bathtub.

The atmospheres of the Jovian planets are turbulent, and some are marked by great storms such as the Great Red Spot on Jupiter, but the atmospheres are not deep. If Jupiter were shrunk to a few centimeters in diameter, its atmosphere would be no deeper than the fuzz on a badly worn tennis ball.

Mars

Mars has a thin atmosphere and little water. Craters and volcanoes are common on its desert surface.

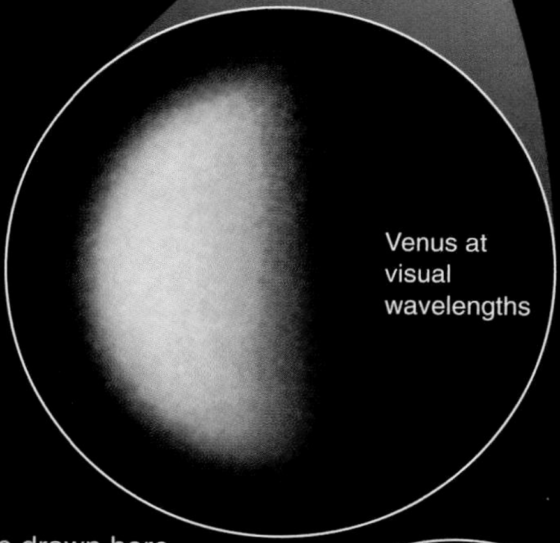

Venus (radar image)

These Jovian worlds are shown in proper relative size.

3a The interiors of the Jovian planets contain small cores of heavy elements such as metals, surrounded by a liquid. Jupiter and Saturn contain hydrogen forced into a liquid state by the high pressure. Less-massive Uranus and Neptune contain heavy-element cores surrounded by partially solid water mixed with heavy material such as rocks and

Venus at visual wavelengths

Jupiter

Great Red Spot

The Terrestrial planets are drawn here to the same scale as the Jovian planets.

The Jovian planets have extensive systems of satellites. For example, Jupiter is orbited by four large moons, discovered by Galileo in 1610, and dozens of smaller moons discovered up to the present day.

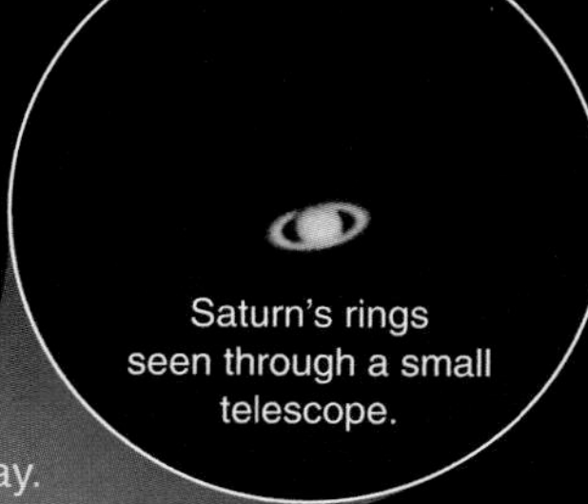

Saturn's rings seen through a small telescope.

Neptune

Uranus

3b All four Jovian planets have ring systems. Saturn's rings are made of ice particles. The rings of Jupiter, Uranus, and Neptune are made of dark rocky particles. Terrestrial planets have no rings.

NASA

Saturn

difficult to locate even at their brightest. A comet may take months to sweep through the inner solar system, during which time it appears as a glowing head with an extended tail of gas and dust.

The beautiful tail of a comet can be longer than an AU, but it is produced by an icy nucleus (sometimes described as a dirty snowball) only a few tens of kilometers in diameter. The nucleus remains frozen and inactive while it is far from the sun. As the nucleus moves along its elliptical orbit into the inner solar system, the sun's heat begins to vaporize the ices, releasing gas and dust. The pressure of sunlight and the solar wind pushes the gas and dust away, forming a long tail. The gas and dust respond differently to the forces acting on them so they sometimes separate into two separate sub-tails (Figure 12.2b). The motion of the nucleus along its or-

bit, the pressure of sunlight, and the outward flow of the solar wind cause the tails of comets to point always approximately away from the sun (Figure 12.2a).

Comet nuclei contain ices of water and other **volatile** (easily vaporized) compounds such as carbon dioxide, carbon monoxide, methane, ammonia, and so on. These ices are the kinds of compounds that should have condensed from the outer solar nebula, and that makes astronomers think that comets are ancient samples of the gases and dust from which the outer planets formed.

Five spacecraft flew past the nucleus of Comet Halley when it passed through the inner solar system in 1985 and 1986. Since then spacecraft have visited the nuclei of several other comets. Images show that comet nuclei are irregular in shape and very dark, with jets of gas and dust spewing from active regions on the nuclei (Figure 12.3). In general, these nuclei are darker than a lump of coal, which suggests they have composition similar to certain

Figure 12.3

Visual-wavelength images made by spacecraft and by the Hubble Space Telescope show how the nucleus of a comet produces jets of gases from regions where sunlight vaporizes ices.

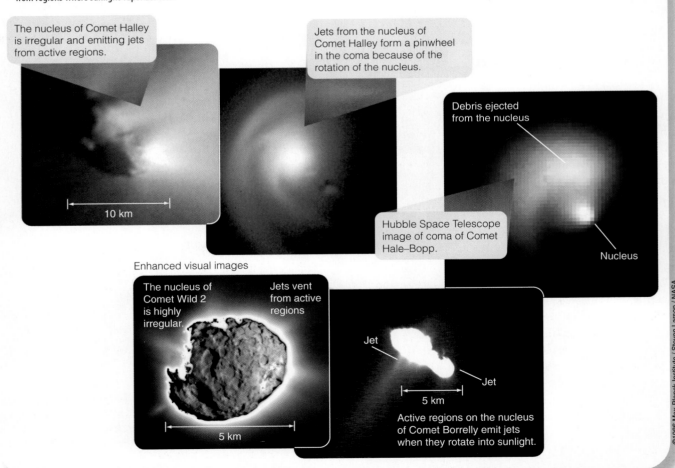

The nucleus of Comet Halley is irregular and emitting jets from active regions.

10 km

Jets from the nucleus of Comet Halley form a pinwheel in the coma because of the rotation of the nucleus.

Debris ejected from the nucleus

Hubble Space Telescope image of coma of Comet Hale–Bopp.

Nucleus

Enhanced visual images

The nucleus of Comet Wild 2 is highly irregular.

Jets vent from active regions

5 km

Jet

Jet

5 km

Active regions on the nucleus of Comet Borrelly emit jets when they rotate into sunlight.

Figure 12.4

Iron meteorites are very heavy for their size and have a dark, irregular surface.

A stony-iron meteorite cut and polished reveals a mixture of iron and rock.

Stony meteorites tend to have a fusion crust caused by melting in Earth's atmosphere.

a

b

c

Chondrules are small, glassy spheres found in chondrites.

d

dark, water- and carbon-rich meteorites that are described in the next section.

Since 1992, astronomers have discovered roughly a thousand small, dark, icy bodies orbiting in the outer fringes of the solar system beyond Neptune. This collection of objects is called the **Kuiper belt** after the Dutch-American astronomer Gerard Kuiper (pronounced KIE-per), who predicted their existence in the 1950s. There are probably 100 million bodies larger than 1 km in the Kuiper belt, and any successful theory should explain how they came to be where they are. Astronomers have evidence that some comets, those that move in the plane of the solar system with relatively short orbital periods, come from the Kuiper belt.

Meteoroids, Meteors, Meteorites

Unlike the stately comets, **meteors** flash across the sky in momentary streaks of light. They are commonly called "shooting stars." Of course, they are not stars but small bits of rock and metal falling into Earth's atmosphere and bursting into incandescent vapor about 80 km (50 mi) above the ground because of friction with the air. This vapor condenses to form dust, which settles slowly to the ground, adding about 40,000 tons per year to Earth's mass.

Technically, the word *meteor* refers to the streak of light in the sky. In space, before its fiery plunge, the object is called a **meteoroid**. Most meteoroids are the size of dust, grains of sand, or tiny pebbles. Almost all

the meteors you see in the sky are produced by meteoroids that weigh less than 1 gram. Only rarely is a meteoroid massive and strong enough to survive its plunge, reach Earth's surface, and become what is called a **meteorite**.

Meteorites can be divided into three broad categories. *Iron* meteorites are solid chunks of iron and nickel. *Stony* meteorites are silicate masses that resemble Earth rocks. *Stony-iron* meteorites are mixtures of iron and stone. These types of meteorites are illustrated in Figure 12.4.

One type of stony meteorite called **carbonaceous chondrites** has a chemical composition that resembles a cooled lump of the sun with the hydrogen and helium removed. These meteorites generally contain abundant volatile compounds including significant amounts of carbon and water, and may have similar composition to comet nuclei. Heating would have modified and driven off these fragile compounds, so carbonaceous chondrites must not have been heated

Kuiper belt: The collection of icy planetesimals orbiting in a region from just beyond Neptune out to 50 AU or more.

Meteor: A small bit of matter heated by friction to incandescent vapor as it falls into Earth's atmosphere.

Meteoroid: A meteor in space before it enters Earth's atmosphere.

Meteorite: A meteor that survives its passage through the atmosphere and strikes the ground.

Carbonaceous chondrite: Stony meteorite that contains small glassy spheres called chondrules and volatiles. These chondrites may be the least-altered remains of the solar nebula still present in the solar system.

Meteor shower: A display of meteors that appear to come from one point in the sky, understood to be cometary debris.

Half-life: The time required for half of the radioactive atoms in a sample to decay.

since they formed. Astronomers conclude that carbonaceous chondrites, unlike the planets, have not changed since they formed, and thus give direct information about the early solar system.

You can find evidence of the origin of meteors through one of the most pleasant observations in astronomy by watching a **meteor shower**, a display of meteors that are clearly related by a common origin. For example, the Perseid meteor shower occurs each year in August, and during the height of the shower you might see as many as 40 meteors per hour. The Perseid shower is so named because all its meteors appear to come from a point in the constellation Perseus. Meteor showers are seen when Earth passes near the orbit of a comet (Figure 12.5). The meteors in meteor showers must be produced by dust and debris released from the icy head of the comet. In contrast, the orbits of some meteorites have been calculated to lead back into the asteroid belt.

An important reason to mention meteorites here is for one specific clue they can give you concerning the solar nebula: Meteorites can reveal the age of the solar system.

12.4 The Story of Planet Formation

The challenge for modern planetary astronomers is to compare the characteristics of the solar system with the solar nebula theory and tell the story of how the planets formed.

The Age of the Solar System

According to the solar nebula theory, the planets should be about the same age as the sun. The most accurate way to find the age of a rocky body is to bring a sample into the laboratory and determine its age by analyzing the radioactive elements it contains.

When a rock solidifies, the process of cooling causes it to incorporate known proportions of the chemical elements. A few of these elements are radioactive and decay into other elements, called daughter elements or isotopes. The **half-life** of a radioactive element is the time it takes for half of the radioactive atoms to decay into the daughter elements. For example, potassium-40 decays into daughter isotopes calcium-40 and argon-40 with a

Figure 12.5

Meteors in a meteor shower enter Earth's atmosphere along parallel paths, but, much like lines on a road, perspective makes them appear to diverge from a radiant point in the sky.

Meteors in a shower are debris left behind as a comet's icy nucleus vaporizes. The rocky and metallic bits of matter spread along the comet's orbit. If Earth passes through such material, you can see a meteor shower.

half-life of 1.3 billion years. Another example, uranium-238, decays with a half-life of 4.5 billion years to lead-206 and other isotopes. As time passes, the abundance of a radioactive element in a rock gradually decreases, and the abundances of the daughter elements gradually increase (Figure 12.6). You can estimate the original abundances of the elements in the rock from rules of chemistry and observations of rock properties in general. Thus, measuring the present abundances of the parent and daughter elements allows you to find the age of the rock. This works best if you have several radioactive element "clocks" that can be used as independent checks on each other.

Of course, to find a radioactive age, you need a sample in the laboratory, and the only celestial bodies from which scientists have samples are Earth, the moon, Mars, and meteorites. The oldest Earth rocks so far discovered and dated are tiny zircon crystals from Australia that are 4.4 billion years old. The surface of Earth is active, and the crust is continually destroyed and reformed from material welling up from beneath the crust (see Chapter 13). Consequently, the age of these oldest rocks tells you only that Earth is *at least* 4.4 billion years old.

Unlike Earth's surface, the moon's surface is not being recycled by constant geologic activity, so you can guess that more of it might have survived unaltered since early in the history of the solar system. The oldest moon rocks brought back by the Apollo astronauts are 4.48 billion years old. That means the moon must be at

Figure 12.6

The radioactive atoms in a mineral sample (red) decay into daughter atoms (blue). Half the radioactive atoms are left after one half-life, a fourth after two half-lives, an eighth after three half-lives, and so on.

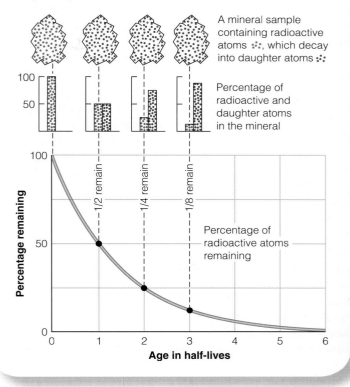

least 4.48 billion years old. Although no one has yet been to Mars, over a dozen meteorites found on Earth have been identified by their chemical composition as having come from Mars. The oldest has an age of approximately 4.5 billion years. Mars must be at least that old.

The most important source for determining the age of the solar system is meteorites. As you learned earlier, carbonaceous chondrite meteorites have compositions indicating that they have not been heated much or otherwise altered since they formed. They have a range of ages with a consistent and precise upper limit of 4.56 billion years. This figure is widely accepted as the age of the solar system and is often rounded to 4.6 billion years. That is in agreement with the age of the sun, which is estimated to be 5 billion years plus or minus 1.5 billion years using mathematical models of the sun's interior (see Chapter 7) that are completely independent of meteorite radioactive ages.

Apparently, all the bodies of the solar system formed at about the same time some 4.6 billion years ago. You can add this as the final item to your list of characteristic properties of the solar system that need to be explained by a good theory of its formation (Table 12.1).

Table 12.1 Characteristic Properties of the Solar System

1. Disk shape of the solar system
   Orbits in nearly the same plane
   Common direction of revolution and rotation

2. Two planetary types
   Terrestrial—inner planets; small, high density
   Jovian—outer planets; large, low density

3. Planetary rings and large satellite systems around the Jovian planets, not around the Terrestrial planets

4. Space debris—asteroids, comets, and meteors
   Composition: two types, rocky versus icy
   Orbits: two types, inner versus outer solar system

5. Common age of about 4.6 billion years for Earth, the moon, Mars, meteorites, and the sun

Chemical Composition of the Solar Nebula

Everything astronomers know about the solar system and star formation suggests that the solar nebula was a fragment of an interstellar gas cloud. Such a cloud would have been mostly hydrogen with some helium and minor traces of the heavier elements.

This is precisely what you see in the composition of the sun (see Table 5.1). Analysis of the solar spectrum shows that the sun is mostly hydrogen, with a quarter of its mass being helium and only about 2 percent being heavier elements. Of course, nuclear reactions have fused some hydrogen into helium, but that happens in the sun's core and has not affected its surface composition. That means the composition of the sun's atmosphere revealed in its spectrum is essentially the composition of the solar nebula gases from which it formed.

You can see that same solar nebula composition reflected in the chemical compositions of the planets. The composition of Jupiter and the other Jovian planets resembles the composition of the sun. Furthermore, if you allowed low-density gases to escape from a blob of sun-stuff, the remaining heavier elements would resemble the composition of Earth and the other Terrestrial planets, as well as meteorites.

Uncompressed density: The density a planet would have if its gravity did not compress it.

Condensation of Solids

The key to understanding the process that converted the nebular gas into solid matter is the observed variation in density among solar system objects. You have already noted that the four inner planets are high-density, Terrestrial bodies, whereas the outer, Jupiter-like planets are low-density, giant planets. This division is due to the different ways gases condensed into solids in the inner and outer regions of the solar nebula.

Even among the four Terrestrial planets, you will find a pattern of slight differences in density. The **uncompressed densities**—the densities the planets would have if their gravity did not compress them, or to put it another way, the densities of their original construction material—can be calculated from the actual densities and masses of each planet. In general, the closer a planet is to the sun, the higher is its uncompressed density.

This density variation is understood to have originated when the solar system first formed solid grains. The kind of matter that condensed in a particular region would depend

Reconstructing the Past from Evidence and Hypothesis

Scientists often solve problems in which they must reconstruct the past. Some of these reconstructions are obvious, such as an archaeologist excavating the ruins of a burial tomb, but others are less obvious. In each case, success requires the interplay of hypotheses and evidence to re-create a past that no longer exists.

Astronomers reconstruct the past when they use evidence gathered from meteorites to study the origin of the solar system, but a biologist studying a centipede is also reconstructing the past. Although a biologist's focus might at first seem to be anatomy, the biologist must reconstruct the ancient environment that gave rise to the centipedes.

The astronomer's problem is not just to understand what the planets are like but to understand how they got that way. That means planetary astronomers must look at the evidence they can see today and reconstruct a history of the solar system, a past that is quite different from the present. If you had a time machine, it would be a fantastic adventure to go back and watch the planets form. Time machines are impossible, but scientists can use the grand interplay of evidence and theory, the distinguishing characteristic of science, to journey back billions of years and reconstruct a past that no longer exists.

on the temperature of the gas there. In the inner regions, the temperature seems to have been 1,500 K or so. The only materials that can form grains at this temperature are compounds with high melting points, such as metal oxides and pure metals, which are very dense, corresponding to the composition of Mercury. Farther out in the nebula it was cooler, and silicates (rocky material) could condense. These are less dense than metal oxides and metals, corresponding more to the compositions of Venus, Earth, and Mars. Somewhere further from the sun there was a boundary called the **ice line** beyond which the water vapor could freeze to form ice. Not much farther out, compounds such as methane and ammonia could condense to form other ices. Water vapor, methane, and ammonia were abundant in the solar nebula, so beyond the ice line, the nebula was filled with a blizzard of ice particles, and those ices have low densities like the Jovian planets.

The sequence in which different materials condensed from the gas at different distances from the sun is called the **condensation sequence** (Table 12.2). This is really a hypothesis that says the planets, forming at different distances from the sun, would have accumulated from different kinds of materials. The original chemical composition of the solar nebula should have been roughly the same throughout the nebula. The important factor was temperature: The inner nebula was hot, and only metals and rock could condense there, whereas the cold outer nebula could form lots of ices in addition to metals and rock. The ice line seems to have been between Mars and Jupiter, and it separates the formation of the dense Terrestrial planets from that of the lower-density Jovian planets.

Formation of Planetesimals

In the development of a planet, three groups of processes operate to collect solid bits of matter—rock, metal, or ices—into larger bodies called **planetesimals** that eventually build the planets. The study of planet building is the study of these three groups of processes: condensation, accretion, and gravitational collapse, which will be described in detail in this section.

According to the solar nebula theory, planetary development in the solar nebula began with the growth of dust grains. These specks of matter, whatever their composition, grew from microscopic size first by condensation, then by accretion.

A particle grows by **condensation** when it adds matter one atom or molecule at a time from a surrounding gas. Snowflakes, for example, grow by condensation in Earth's atmosphere. In the solar nebula, dust grains were

Table 12.2 **The Condensation Sequence**

Temperature (K)	Condensate	Object (Estimated Temperature of Formation; K)
1,500	Metal oxides	Mercury (1,400)
1,300	Metallic iron and nickel	
1,200	Silicates	
1,000	Feldspars	Venus (900)
680	Troilite (FeS)	Earth (600) Mars (450)
175	H_2O ice	Jovian (175)
150	Ammonia–water ice	
120	Methane–water ice	
65	Argon–neon ice	Pluto (65)

continuously bombarded by atoms of gas, some of which stuck to the grains.

The second process is **accretion**, the sticking together of solid particles. You may have seen accretion in action if you have walked through a snowstorm with big, fluffy flakes. If you caught one of those "flakes" on your mitten and looked closely, you saw that it was actually made up of many tiny, individual flakes that had collided as they fell and accreted to form larger particles. Model calculations indicate that in the solar nebula the dust grains were, on the average, no more than a few centimeters apart, so they collided frequently and accreted into larger particles.

There is no clear distinction between a very large grain and a very small planetesimal, but you can

Ice line: Boundary beyond which vapor could freeze to form ice.

Condensation sequence: The sequence in which different materials condense from the solar nebula depending on distance from the sun.

Planetesimal: One of the small bodies that formed from the solar nebula and eventually grew into protoplanets.

Condensation: The growth of a particle by addition of material from surrounding gas, atom by atom.

Accretion: The sticking together of solid particles to produce a larger particle.

consider an object a planetesimal when its diameter approaches a kilometer or so, like the size of a typical small asteroid or comet.

Objects in the solar nebula larger than a centimeter were subject to new processes that tended to concentrate them. For example, collisions with the surrounding gas and with each other would have caused growing planetesimals to settle into a thin disk estimated to have been only about 0.01 AU thick in the central plane of the rotating nebula. This concentration of material would have made further planetary growth more rapid. Computer models show that the rotating disk of particles should have been gravitationally unstable and would have been disturbed by spiral density waves much like those found in spiral galaxies. This would have further concentrated the planetesimals and helped them coalesce into objects up to about 100 km (60 mi) in diameter.

Through these processes the nebula became filled with trillions of solid particles ranging in size from pebbles to small planets. As the largest began to exceed 100 km in diameter, new processes began to alter them, and a new stage in planet building began, the formation of protoplanets.

Growth of Protoplanets

The coalescing of planetesimals eventually formed **protoplanets**, massive objects destined to become planets. As these larger bodies grew, new processes began making them grow faster and altered their physical structure.

If planetesimals collided at orbital velocities, it is unlikely that they would have stuck together. The average orbital velocity in the solar system is about 10 km/s (22,000 mph). Head-on collisions at this velocity would have vaporized the material. However, the planetesimals were all moving in the same direction in the nebular plane and didn't collide head-on. Instead, they merely "rubbed shoulders," so to speak, at low relative velocities. Such gentle collisions would have been more likely to combine them than to shatter them.

The largest planetesimals would grow the fastest because they had the strongest gravitational field and could more easily attract ad-

Protoplanet: Massive object, destined to become a planet, resulting from the coalescence of planetesimals in the solar nebula.

Gravitational collapse: The process by which a forming body such as a planet gravitationally captures gas rapidly from the surrounding nebula.

Differentiation: The separation of planetary material inside a planet according to density.

Outgassing: The release of gases from a planet's interior.

ditional material. Computer models indicate that these planetesimals grew quickly to protoplanetary dimensions, sweeping up more and more material.

Protoplanets had to begin growing by accumulating solid material because they did not have enough gravity to capture and hold large amounts of gas. In the warm solar nebula, the atoms and molecules of gas were traveling at velocities much larger than the escape velocities of modest-size protoplanets. Because of that, in their early development, the protoplanets could only grow by attracting solid bits of rock, metal, and ice. Once a protoplanet approached a mass of 15 Earth masses or so, it could begin to grow by **gravitational collapse**, the rapid accumulation from the nebula of large amounts of infalling gas.

In its simplest form, the theory of Terrestrial protoplanet growth supposes that all the planetesimals in each orbital zone had about the same chemical composition. The planetesimals accumulated gradually to form a planet-size ball of material that was of homogeneous composition throughout. As the planet formed, heat accumulated in its interior both from the impacts of infalling planetesimals and from decay of short-lived radioactive elements. This heat eventually melted the planet and allowed it to differentiate. **Differentiation** is the separation of material according to density. When the planet melted, the heavy metals such as iron and nickel settled to the core, while the lighter silicates floated to the surface to form a low-density crust. The story of planet formation from accretion of planetesimals followed by differentiation is shown in Figure 12.7.

This process depends on the presence of short-lived radioactive elements whose rapid decay would have released enough heat to melt the interior of planets. Astronomers know such radioactive elements were present because the oldest meteorites contain daughter isotope magnesium-26 that is produced by the decay of aluminum-26 with a half-life of only 0.74 million years. The aluminum-26 and similar short-lived radioactive elements are gone now, but they must have been produced in a supernova explosion that occurred no more than a few million years before the formation of the solar nebula. In fact, many astronomers suspect that this supernova explosion compressed nearby gas and triggered the formation of stars, one of which became the sun. Thus our solar system may exist because of a supernova explosion that occurred about 4.6 billion years ago.

If planets formed and were later melted by radioactive decay, gases released from the planet's interior would have formed an atmosphere. The creation of a planetary atmosphere from a planet's interior is called **outgassing**. Models of the formation of the Earth indicate that the

Figure 12.7

This simple model of accretion assumes planets formed from planetesimals that were of uniform composition, containing both metals and rocky material, and the planets later differentiated, meaning that they melted and separated into layers by density and composition.

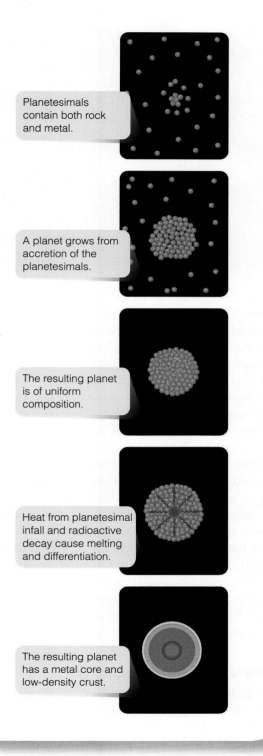

Planetesimals contain both rock and metal.

A planet grows from accretion of the planetesimals.

The resulting planet is of uniform composition.

Heat from planetesimal infall and radioactive decay cause melting and differentiation.

The resulting planet has a metal core and low-density crust.

local planetesimals would not have included much water, so some astronomers now think that Earth's water and some of its present atmosphere accumulated late in the formation of the planets as Earth swept up volatile-rich planetesimals forming in the cooling solar nebula. Such icy planetesimals may have formed in the outer parts of the solar nebula and been scattered by encounters with the Jovian planets, creating a solar system–wide bombardment of comets.

According to the solar nebula theory, the Jovian planets began growing by the same processes that built the Terrestrial planets, but the outer solar nebula not only contained solid bits of metals and silicates, it also included abundant ices. The Jovian planets grew rapidly and quickly became massive enough to grow by gravitational collapse, drawing in large amounts of gas from the solar nebula. Ices could not condense as solids at the locations of the Terrestrial planets, so those planets developed slowly and never became massive enough to grow by gravitational collapse.

The Jovian planets must have grown to their present size in less than about 10 million years, which is when astronomers calculate that the sun become hot and luminous enough to blow away the gas remaining in the solar nebula. The Terrestrial planets grew from solids and not from the gas, so they could have continued to grow by accretion from the solid debris left behind when the gas was blown away. Model calculations indicate the process of planet formation was almost completely finished by the time the solar system was about 30 million years old.

Heavy bombardment: The intense cratering that occurred sometime during the first 0.5 billion years in the history of the solar system.

Continuing Bombardment of the Planets

Astronomers have good reason to believe that comets and asteroids can hit planets. Meteorites hit Earth every day, and occasionally a large one can form a crater (Figure 12.8). Earth is marked by about 150 known meteorite craters. In a sense, this bombardment represents the slow continuation of the accretion of the planets. Earth's moon, Mercury, Venus, Mars, and most of the moons in the solar system are covered with craters. A few of these craters have been formed recently by the steady rain of meteorites that falls on all the planets in the solar system, but most of the craters appear to have been formed more than 4 billion years ago in what is called the **heavy bombardment**, when much of the remaining solid debris in the solar nebula were swept up by the planets. In the next

Figure 12.8   Barringer Meteorite Crater, Flagstaff, AZ

raised rim
from impact

© M. A. Seeds

approx. 1 mi.

© USGS

chapter you will learn that the heavy bombardment may have had more than one episode.

Sixty-five million years ago, at the end of the Cretaceous period, over 75 percent of the species on Earth, including the dinosaurs, went extinct. Scientists have found a thin layer of clay all over the world that was laid down at that time, and it is rich in the element iridium—common in meteorites but rare in Earth's crust. This suggests that a large impact altered Earth's climate and caused the worldwide extinction. Mathematical models indicate that a major impact would eject huge amounts of pulverized rock high above the atmosphere. As this material fell back, Earth's atmosphere would be turned into a glowing oven of red-hot meteorites streaming through the air, and the heat would trigger massive forest fires around the world. Soot from such fires has been found in the final Cretaceous clay layers. Once the firestorms cooled, the remaining dust in the atmosphere would block sunlight and produce deep darkness for a year or more, killing off most plant life. Other effects, such as acid rain and enormous tsunamis (tidal waves), are also predicted by the models.

Geologists have located a crater at least 180 km (110 mi) in diameter centered near the village of Chicxulub (pronounced cheek-shoe-lube) in the northern Yucatán region of Mexico (Figure 12.9). Although the crater is completely covered by sediments, mineral samples show that it contains shocked quartz typical of impact sites and that it is the right age. The impact of an object 10 to 14 km in diameter formed the crater about 65 million years ago, just when the dinosaurs and many other species died out. Most Earth scientists now consider this to be the scar of the impact that ended the Cretaceous period.

Earthlings watched in awe during six days in the summer of 1994 as 20 or more fragments from the head of comet Shoemaker–Levy 9 slammed into Jupiter and produced impacts equaling millions of megatons of TNT. Each impact created a fireball of hot gases and left behind dark smudges that remained visible for months afterward (Figure 12.10). Another such impact produced a single scar observed in 2009. Major impacts on Earth occur less often because Earth is smaller, but they are inevitable.

We are sitting ducks. All of human civilization is spread out over Earth's surface and exposed to anything that falls out of the sky. Meteorites, asteroids, and comets bombard Earth, producing impacts that vary from dust settling on rooftops to blasts capable of destroying all life. In this case, the scientific evidence is conclusive and highly unwelcome. Statistically you are quite safe. The chance that a major impact will occur during your lifetime is so small that it is hard to estimate. But the consequences of such an impact are so severe that humanity should be preparing. One way to prepare is to find those **NEOs** (Near Earth Objects) that could hit this planet, map their orbits in detail, and identify any that are dangerous.

The Jovian Problem

The solar nebula theory has been very successful in explaining the formation of the solar system. But there are some problems, and the Jovian planets are the troublemakers.

NEO: A small solar system body (asteroid or comet) with an orbit near enough to Earth that it poses some threat of eventual collision.

The gas and dust disks around newborn stars don't last long. In Chapter 7 you saw images of dusty gas disks around the young stars in the Orion nebula (page 141). Those disks are being evaporated by the intense ultraviolet radiation from hot stars within the nebula. Astronomers estimate that most stars form in clusters containing some massive stars, so this evaporation process must happen to most disks. Even if a disk did not evaporate quickly, model calculations predict that the gravitational influence of the crowded stars in a cluster should quickly strip away the outer parts of the disk. This is a troublesome observation because it seems to mean that planet-forming disks around young stars are unlikely to last longer than a few million years, and many must evaporate within 100,000 years or so. That's not long enough to grow a Jovian planet by the processes in the solar nebula theory. Yet, as you will learn in a later section of this chapter, Jovian planets are common in the universe.

A modification to the solar nebula theory has come from mathematical models of the solar nebula. The results show that the rotating gas and dust of the solar nebula could have become unstable and formed outer planets by direct gravitational collapse. That is, massive planets may have been able to form directly from the nebula gas without first accreting a dense core from planetesimals. Jupiters and Saturns can form in these models within a few hundred years. If the Jovian planets formed in this way, they could have formed quickly, long before the solar nebula disappeared.

This new insight into the formation of the outer planets may help explain the formation of Uranus and Neptune. They are so far from the sun that accretion could not have built them rapidly. It is hard to understand how they could have reached their present mass in a region where the material should have been sparse and orbital speeds are slow. Theoretical calculations show that Uranus and Neptune might instead have formed closer to the sun, in the region of Jupiter and Saturn, and then moved outward by gravitational interactions with the other planets or with planetesimals in the Kuiper belt. In any case, the formation of Uranus and Neptune is part of the Jovian problem.

The original solar nebula theory proposes that the planets formed by accreting a solid core and then, if they became massive enough, by gravitational collapse of nebula gas. The new theories suggest that some of the outer planets could have skipped the core accretion phase.

Explaining the Characteristics of the Solar System

Now you have learned enough to put all the pieces of the puzzle together and explain the distinguishing characteristics of the solar system in Table 12.1 on page 251.

The disk shape of the solar system is inherited from the solar nebula. The sun and planets revolve and rotate in the same direction because they formed from the same rotating gas cloud. The orbits of the planets lie in the same plane because the rotating solar nebula collapsed into a disk, and the planets formed in that disk.

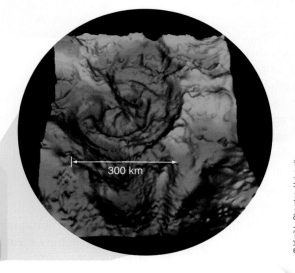

© Virgil L. Sharpton, University of Alaska, Fairbanks

300 km

United States

Mexico

Chicxulub crater

Yucatán

Figure 12.9

The giant impact scar buried in Earth's crust near the village of Chicxulub in the Yucatán region of Mexico was formed about 65 million years ago by the impact of a large comet or asteroid. The gravity map above shows the extent of the crater hidden below limestone deposited long after the impact.

Figure 12.10

The solar nebula theory calls on continuing evolutionary processes to gradually build the planets. Scientists call this type of explanation an **evolutionary theory**. In contrast, a **catastrophic theory** invokes special, sudden, even violent, events. For example, as you learned earlier, Uranus rotates on its side and Venus rotates backward. Both these pecularities could have been caused by off-center impacts of massive planetesimals on each planet as they were forming—an explanation of the catastrophic type. On the other hand, computer models suggest that the sun can produce tides in the thick atmosphere of Venus that eventually could have resulted in the reversal of that planet's rotation—an explanation of the evolutionary type.

The second item in Table 12.1, the division of the planets into Terrestrial and Jovian worlds, can be understood through the condensation sequence. The Terrestrial planets formed in the inner part of the solar nebula, where the temperature was high and only compounds such as metals and silicates could condense to form solid particles. That produced the small, dense Terrestrial planets. In contrast, the Jovian planets formed in the

Evolutionary theory: An explanation of a phenomenon involving slow, steady processes of the sort seen happening in the present day.

Catastrophic theory: An explanation of a phenomenon involving special, sudden, perhaps violent, events.

outer solar nebula, where the lower temperature allowed the gas to form large amounts of ices, perhaps three times more ices than silicates. That allowed the Jovian planets to grow rapidly and become massive, low-density worlds. Also, Jupiter and Saturn are so massive they eventually were able to grow even larger by drawing in gas directly from the solar nebula. The Terrestrial planets could not do this because they never became massive enough.

The **heat of formation** (the energy released by infalling matter) was tremendous for these massive planets. Jupiter must have grown hot enough to glow with a luminosity of about 1 percent that of the present sun. However, because it never got hot enough to start nuclear fusion as a star would, it never generated its own energy. Jupiter is still hot inside. In fact, both Jupiter and Saturn radiate more heat than they absorb from the sun, so they are evidently still cooling.

A glance at the solar system suggests that you should expect to find a planet between Mars and Jupiter at the present location of the asteroid belt. Mathematical models show Jupiter grew into such a massive planet that it was able to gravitationally disturb the motion of nearby planetesimals. The bodies that should have formed a planet between Mars and Jupiter were broken up, thrown into the sun, or ejected from the solar system due to the gravitational influence of massive Jupiter. The asteroids seen today are the last remains of those rocky planetesimals.

The comets, in contrast, are evidently the last of the icy planetesimals. Some may have formed in the outer solar nebula beyond Neptune, but many probably formed among the Jovian planets, where ices could condense easily. Mathematical models show that the massive Jovian planets could have ejected some of these icy planetesimals into the far outer solar system, to a region called the **Oort cloud**, from which comets come with very long periods and orbits highly inclined to the plane of the solar system.

The icy Kuiper belt objects, including Pluto, appear to be ancient planetesimals that formed in the outer solar system but were never incorporated into a planet. They orbit slowly far from the light and warmth of the sun and, except for occasional collisions, have not changed much since the solar system was young.

© Christian Carroll/iStockphoto.com

In Table 12.1, point 3 notes that all four Jovian worlds have ring systems, and you can understand this by considering the large mass of these worlds and their remote location in the solar system. A large mass makes it easier for a planet to hold onto orbiting ring particles; and, being farther from the sun, the ring particles are not as easily swept away by the pressure of sunlight and the solar wind. It is hardly surprising, then, that the Terrestrial planets, low-mass worlds located near the sun, have no planetary rings.

The last entry in Table 12.1 is the common ages of solar system bodies, and the solar nebula theory has no difficulty explaining that characteristic. The theory predicts that the planets formed at the same time as the sun and should all have roughly the same age.

Heat of formation: In planetology, the heat released by infalling matter during the formation of a planetary body.

Oort cloud: The hypothetical source of comets, a swarm of icy bodies understood to lie in a spherical shell extending to 100,000 AU from the sun.

12.5 Planets Orbiting Other Stars

Are there planets orbiting other stars? Yes, that is now certain. Are there planets like Earth? The evidence so far makes that seem likely.

Planet-Forming Disks around Other Stars

You have already learned about dense disks of gas and dust around stars that are forming. For example, at least 50 percent of the stars in the Orion nebula are encircled by dense disks of gas and dust with more than enough mass to make planetary systems like ours. The Orion star-forming region is only a few million years old, so it does not seem likely that planets have finished forming there yet. Nevertheless, the important point for astronomers is that disks of gas and dust that could become planetary systems are a common feature of star formation.

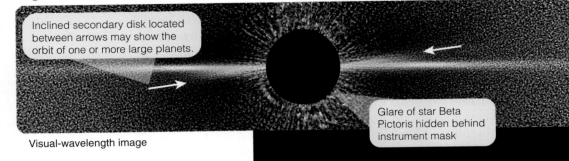

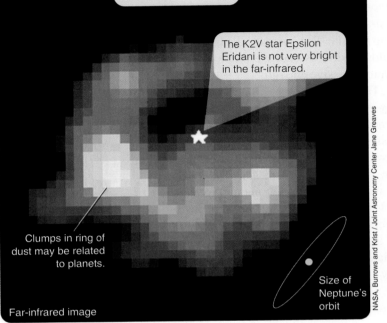

Figure 12.11

Inclined secondary disk located between arrows may show the orbit of one or more large planets.

Visual-wavelength image

Glare of star Beta Pictoris hidden behind instrument mask

The K2V star Epsilon Eridani is not very bright in the far-infrared.

Clumps in ring of dust may be related to planets.

Far-infrared image

Size of Neptune's orbit

NASA, Burrows and Krist / Joint Astronomy Center Jane Greaves

Debris Disks

In addition to these planet-forming disks around young stars, infrared astronomers have found cold, low-density dust disks around older stars such as Vega and Beta Pictoris. Although much younger than the sun, these stars are on the main sequence and have completed their formation, so they are clearly in a later stage than the newborn stars in Orion. These low-density disks generally have even lower-density inner zones where planets may have formed, as you can see in Figure 12.11. Such tenuous dust disks are sometimes called **debris disks** because they are understood to be dusty debris released in collisions among small bodies such as comets, asteroids, and Kuiper belt objects. Our own solar system contains such dust, and astronomers have evidence the sun has an extensive debris disk of cold dust extending far beyond the orbits of the planets.

Few planets orbiting stars with debris disks have been detected so far, but the presence of dust particles with lifetimes shorter than the ages of the stars assures you that remnant planetesimals like asteroids and comets are present as the sources of the dust. If planetesimals are there, then you can expect that there are also planets orbiting those stars. Many of the debris disks have details of structure and shape that are probably caused by the gravity of planets orbiting within or at the edges of the debris (Figure 12.11).

Notice the difference between the two kinds of disks that astronomers have found. The low-density dust disks such as the one around Beta Pictoris are produced by dust from collisions among remnant planetesimals such as comets, asteroids, and Kuiper belt objects. Such disks are evidence that planetary systems have already formed. In contrast, the dense disks of gas and dust such as those seen round the stars in Orion are sites where planets could be forming right now.

Debris disk: A disk of dust found by infrared observations around some stars. The dust is debris from collisions among asteroids, comets, and Kuiper belt objects.

Extrasolar planet: A planet orbiting a star other than the sun.

Extrasolar Planets

A planet orbiting another star is called an **extrasolar planet**. Such a planet would be quite faint and difficult to detect so close to the glare of its star. But there are ways to find these planets. To see how, all you have to do is imagine walking a dog.

You will remember that Earth and its moon orbit around their common center of mass, and two stars in a binary system orbit around their center of mass. When a planet orbits a star, the star moves very slightly as it orbits the center of mass of the planet–star system. Think of someone walking a poorly trained dog on a leash; the dog runs around pulling on the leash, and even if it were an invisible dog, you could plot its path by watching how its owner was jerked back and forth. Astronomers can detect a planet orbiting another star by watching how the star moves as the planet tugs on it.

The first planet orbiting a sunlike star detected this way was discovered in 1995. It orbits the star 51 Pegasi.

As the planet circles the star, the star wobbles slightly, and this very small motion of the star is detectable as Doppler shifts in the star's spectrum, as shown in Figure 12.12a. This is the same technique used to study spectroscopic binary stars (see Chapter 6). From the motion of the star and estimates of the star's mass, astronomers can deduce that the planet has half the mass of Jupiter and orbits only 0.05 AU from the star. Half the mass of Jupiter amounts to 160 Earth masses, so this is a large planet, larger than Saturn. Note also that it orbits very close to its star.

Astronomers were not surprised by the announcement that a planet orbited 51 Pegasi; for years they had assumed that many stars had planets. Nevertheless, they greeted the discovery with typical skepticism. That skepticism led to careful tests of the data and further observations that confirmed the discovery. In fact, over 400 planets have been discovered in this way, including at least three planets orbiting the star Upsilon Andromedae (Fgure 12.12b) and five orbiting 55 Cancri—true planetary systems. More than forty such multiple-planet systems have been found.

Another way to search for planets is to look for changes in the brightness of the star when the orbiting planet crosses in front of or behind the star. The decrease in light is very small, but it is detectable, and astronomers have used this technique to detect a few planets as they crossed in front of their stars. This is the same technique used to study eclipsing binary stars (see Chapter 6). From the amount of light lost, astronomers can tell that all of these planets are roughly the size of Jupiter. The Spitzer infrared space telescope confirmed the existence of two extrasolar planets by observing them pass behind their parent stars. These planets are hot and emit significant infrared radiation. As they orbit their respective stars, astronomers detect variation in the amount of infrared emission from the systems that indicates the temperatures and sizes of the planets. Measurements of these and other planets that pass in front of and behind

Figure 12.12

Just as someone walking a lively dog is tugged around, the star 51 Pegasi is tugged around by the planet that orbits it every 4.2 days. The wobble is detectable in precision observations of its Doppler shift. Someone walking three dogs is pulled about in a more complicated pattern, and you can see something similar in the Doppler shifts of Upsilon Andromedae, which is orbited by at least three planets.

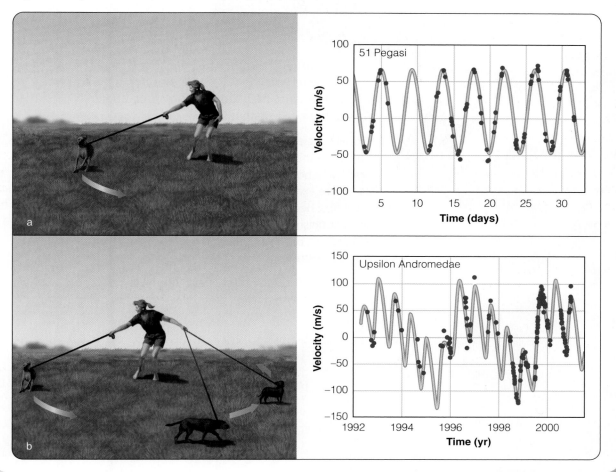

Scientists: Courteous Skeptics

"Scientists are just a bunch of skeptics who don't believe in anything." That is a common misconception among people who don't understand the methods and goals of science. Yes, scientists are skeptical about new ideas and discoveries, but they do hold strong beliefs about how nature works. Scientists are skeptical not because they want to disprove everything but because they are searching for the truth and want to be sure that a new description of nature is true before it is accepted.

Skeptical scientists question every aspect of a new discovery. They may wonder if another scientist's instruments were properly calibrated or if the scientist's mathematical models are correct. Other scientists will want to repeat the work themselves using their own instruments to see if they can obtain the same results. Observations are tested, discoveries are confirmed, and only an idea that survives many of these tests begins to be accepted as a scientific truth.

Scientists are prepared for this kind of treatment at the hands of other scientists. In fact, they expect it. Among scientists it is not bad manners to say, "Really, how do you know that?" or "Why do you think that?" or "Show me the evidence!" And it is not just new or surprising claims that are subject to such scrutiny. Astronomers had long expected to discover planets orbiting other stars. But when a planet was finally discovered circling 51 Pegasi, astronomers were skeptical—not because they thought the observations were flawed but because that is how science works.

The goal of science is to tell stories about nature. Some people use the phrase "telling a story" to describe someone who is telling a fib. But the stories that scientists tell are exactly the opposite; perhaps you could call them antifibs, because they are as true as scientists can make them. Skepticism eliminates stories with logical errors, flawed observations, or misunderstood evidence and eventually leaves only the stories that best describe nature.

Skepticism is not a refusal to hold beliefs. Rather, it is a way for scientists to find those natural principles worthy of belief.

NO WAY!

their stars reveal that they have Jupiter-like diameters as well as masses, so astronomers conclude they have Jovian densities and compositions.

The planets discovered so far tend to be massive and have short periods because lower-mass planets or longer-period planets are harder to detect. Low-mass planets don't tug on their stars very much, and present-day spectrographs can't detect the very small velocity changes that these gentle tugs produce. Planets with longer periods are harder to detect because Earth's astronomers have not been making high-precision observations for a long enough time yet. Jupiter takes 11 years to circle the sun once, so it will take years for astronomers to see the longer-period wobbles produced by planets lying farther from their stars. You should not be surprised that the first extrasolar planets discovered are massive and have short periods. This is called a selection effect.

The new planets may seem odd for another reason. In our own solar system, the large planets formed farther from the sun where the solar nebula was colder and ices could condense. How could big planets form so near their stars? Model calculations predict that planets that form in an especially dense disk of matter could spiral inward as they sweep up gas or planetesimals. That means it is possible for a few planets to become the massive, short-period planets that are detected most easily.

Actually getting an image of a planet orbiting another star is about as easy as photographing a bug crawling on the bulb of a searchlight miles away. Planets are small and dim and get lost in the glare of the stars they orbit. Nevertheless, as you can see in Figure 12.13, during 2008 astronomers managed to image planets around two A-type stars using the Gem-

ini and Keck telescopes on the ground and the Hubble Space Telescope.

The discovery of extrasolar planets gives astronomers added confidence in the solar nebula theory. The theory predicts that planets are common, and astronomers are finding them orbiting many stars.

© C. Marois et al./National Research Council of Canada/ AURA/Keck Observatories / © P. Kalas et al./STScI/NASA

recap

The matter you are made of came from the big bang, and it has been cooked into a wide range of atoms inside stars. Now you can see how those atoms came to be part of Earth. Your atoms were in the cloud of gas that formed the sun 4.6 billion years ago, and nearly all of that matter contracted to form the sun, but a small amount left behind in a disk formed planets. In the process, your atoms became part of Earth.

Are there other planet-walkers like you in the Universe? Now you know that planets are common, and you can reasonably suppose that there are more planets in the universe than there are stars, suggesting there may indeed be planet-walkers living on other worlds.

Based on your exploration of our solar system in space and time, you should be able to answer these five essential questions:

1. What is the theory explaining the origin of the solar system?

2. What are the observed properties of the solar system that must be explained by the theory of its origin?

3. How do planets form?

4. What are asteroids and comets, what are their connections to meteors and meteorites, and what clues to they give about the origin of the solar system?

5. What do astronomers know about other planetary systems?

Figure 12.13

HR 8799 Planetary System
(Sept. 2008)

70 AU

40 AU

25 AU

a

b

120 AU

Fomalhaut b Planet

2006
2004

Fomalhaut System

Hubble Space Telescope • ACS/HRC

NASA, ESA, and P. Kalas (University of California, Berkeley)

Comparative
Planetology of the Terrestrial Planets

If you had been the first person to step onto the surface of the moon, what would you have said? Neil Armstrong responded to the historic significance of the first human step on the surface of another world. Buzz Aldrin was second, and he responded to the moon itself. It *is* desolate, and it *is* magnificent. But it is not unusual. Many places in the universe probably look like Earth's moon, and people may someday walk on such worlds and compare them with the moon. The comparison of one planet with another is called **comparative planetology**, and it is one of the best ways to analyze the worlds in our solar system. You will learn much more by comparing planets than you could by studying them individually.

13.1 A Travel Guide to the Terrestrial Planets

In this chapter you will visit five worlds, and this preliminary section will be your guide to important features and comparisons.

Five Worlds

You are about to visit Earth, Earth's moon, Mercury, Venus, and Mars. It may surprise you that the moon is on your itinerary. It is, after all, just a natural satellite orbiting Earth and isn't one of the planets. But the moon is a fascinating world of its own, a planetlike object two-thirds the size of Mercury, and it makes a striking comparison with the other worlds on your list.

Figure 13.1 on page 266 compares the five worlds you are about to study. The first feature you should notice is diameter. The moon is small, and Mercury is not much bigger. Earth and Venus are large and similar in size to each other, and Mars is a medium-sized world. You will discover that size is a critical factor in determining a world's personality: Small worlds tend to be internally cold and geologically

Comparative planetology: Understanding planets by searching for and analyzing contrasts and similarities among them.

looking back

In the preceding chapter, you learned how our solar system formed as a by-product of the formation of the sun. You also saw how distance from the sun determines the general character of each planet. Now you are ready to visit the planets and get to know them as individuals.

looking ahead

Once you are familiar with the family of the Terrestrial planets, you will be ready to meet a more peculiar group of characters. The next chapter will introduce you to the worlds of the outer solar system.

That's one small step for [a] man… one giant leap for mankind.
NEIL ARMSTRONG

Beautiful, beautiful. Magnificent desolation…
EDWIN E. ALDRIN, JR.

astro

Mantle: The layer of dense rock and metal oxides that lies between the molten core and Earth's surface or a similar layer in another planet.

Core, Mantle, and Crust

Notice that these Terrestrial worlds are made up of rock and metal. They are all differentiated, with rocky, low-density crusts, high-density metal cores, and **mantles** composed of dense rock between the cores and crusts.

As you learned in Chapter 12, when the planets formed, their surfaces were subjected to heavy bombardment by leftover planetesimals and fragments; the cratering rate then was as much as 10,000 times what it is at present. You will see lots of craters on the Terrestrial

dead, but larger worlds can be geologically active. (Note that the term *geology*, which originally referred only to Earth, can be used to refer to other worlds as well.)

worlds, especially on Mercury and the moon. Notice that heavily cratered surfaces are old. For example, if a lava flow covered up some cratered landscape to make a new surface after the end of the heavy bombardment, few craters could be formed afterward on that surface because most of the debris in the solar system was gone. When you see a smooth plain on a planet, you can assume that surface is younger than the cratered areas.

One important way you can study a planet is by following the energy. The heat in the interior of a planet may be left over from the formation of the planet, or it may be heat generated by radioactive decay, but it must flow outward toward the cooler surface where it is radiated into space. In flowing outward, the heat can cause convection currents in the mantle, magnetic fields, plate motions, quakes, faults, volcanism, mountain building, and more. Heat flowing upward through the cooler crust makes a large world like Earth geologically active. In

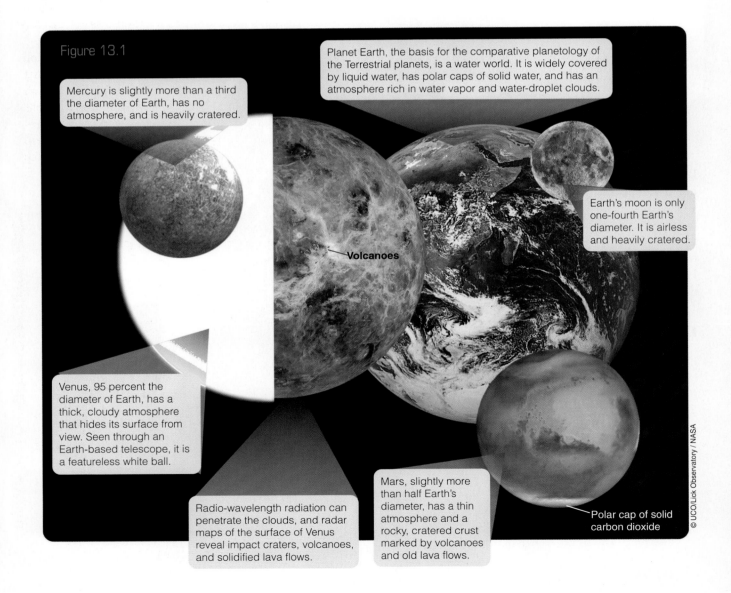

Figure 13.1

Mercury is slightly more than a third the diameter of Earth, has no atmosphere, and is heavily cratered.

Planet Earth, the basis for the comparative planetology of the Terrestrial planets, is a water world. It is widely covered by liquid water, has polar caps of solid water, and has an atmosphere rich in water vapor and water-droplet clouds.

Volcanoes

Earth's moon is only one-fourth Earth's diameter. It is airless and heavily cratered.

Venus, 95 percent the diameter of Earth, has a thick, cloudy atmosphere that hides its surface from view. Seen through an Earth-based telescope, it is a featureless white ball.

Radio-wavelength radiation can penetrate the clouds, and radar maps of the surface of Venus reveal impact craters, volcanoes, and solidified lava flows.

Mars, slightly more than half Earth's diameter, has a thin atmosphere and a rocky, cratered crust marked by volcanoes and old lava flows.

Polar cap of solid carbon dioxide

© UCO/Lick Observatory / NASA

contrast, the moon and Mercury, both small worlds, cooled fast so they have little heat flowing outward now and are relatively inactive.

Atmospheres

When you look at airless Mercury and the moon in Figure 13.1, you can see their craters and plains and mountains, but the surface of Venus is completely hidden by a cloudy atmosphere much thicker than Earth's. Mars, the medium-sized Terrestrial planet, has a relatively thin atmosphere.

You might ponder two questions. First, why do some worlds have atmospheres while others do not? You will discover that both size and temperature are important. The second question is more complex. Where did these atmospheres come from? To answer that question, you will have to study the geological history of these worlds.

13.2 Earth: The Active Planet

Earth is the basis for your comparative study of the Terrestrial planets, so you should pretend to visit it as if you didn't live here. It is an active planet with a molten interior and heat flowing outward that powers volcanism, earthquakes, and an active crust. Almost 75 percent of Earth's surface is covered by liquid water, and the atmosphere contains a significant amount of oxygen.

Earth's Interior

From what you know of the formation of Earth (see Chapter 12), you would expect it to have differentiated, but in science, evidence rules. What does the evidence reveal about Earth's interior?

Earth's mass divided by its volume tells you its average density, about 5.5 g/cm³, but the density of Earth's rocky crust is only about half that. Clearly, a large part of Earth's interior must be made of material denser than rock.

Each time an earthquake occurs, seismic waves travel through the interior and register on seismographs all over the world. Analysis

of these waves shows that Earth's interior is divided into a metallic core, a dense rocky mantle, and a thin, low-density crust.

The core has a density of 14 g/cm³, denser than lead; mathematical models indicate it is composed of iron and nickel at a temperature of roughly 6,000 K. The core of Earth is as hot as the gaseous surface of the sun, but high pressure keeps the metal a solid near the middle of the core and a liquid in its outer parts. Two kinds of seismic waves show that the outer core is liquid. The **P waves** travel like sound waves, and they can penetrate a liquid; but **S waves** travel as a side-to-side vibration that can travel along the surface of a liquid but not through it. Scientists can deduce the size of the liquid core by observing where S waves get through and where they don't, as you can see in Figure 13.2.

Earth's magnetism gives you further clues about the core. The presence of a magnetic field is evidence that part of Earth's core must be a liquid metal. Convection currents stir the molten liquid, and because it is a very good conductor of electricity and is rotating as Earth rotates, it generates a magnetic field through the dynamo effect—a different version of the process that creates the

P wave: A type of seismic wave involving compression and decompression of the material through which it passes.

S wave: A type of seismic wave involving lateral motion of the material through which it passes.

Figure 13.2

P and *S* waves give you clues to Earth's interior. No direct *S* waves reach the far side, which shows that part of Earth's core is liquid. The size of the *S* wave "shadow" tells you the size of the liquid outer core.

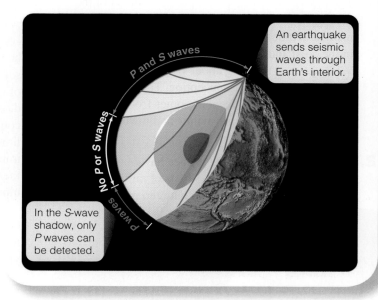

An earthquake sends seismic waves through Earth's interior.

In the *S*-wave shadow, only *P* waves can be detected.

sun's magnetic field (see Chapter 5).

Earth's mantle is a deep layer of dense rock that lies between the molten core and the solid crust. Models indicate the mantle material has the properties of a solid but is capable of flowing slowly, like asphalt used in paving roads, which shatters if struck with a sledgehammer, but bends slightly under the weight of a truck. Just below Earth's crust, where the pressure is less than at greater depths, the mantle flows most easily.

The Earth's rocky crust is made up of low-density rocks. The crust is thickest under the continents, up to 60 km (35 mi) thick, and thinnest under the oceans, where it is only about 10 km (6 mi) thick.

Earth's Active Crust

Earth's crust is composed of lower-density rock that floats on the mantle. The image of rock floating may seem odd, but recall that the rock of the mantle is denser than crust rocks. Also, just below the crust, the mantle rock tends to be more fluid, so sections of low-density crust do indeed float on the mantle like great lily pads floating on a pond.

The motion of the crust and the erosive action of water make Earth's crust highly active and changeable. Read "The Active Earth" on pages 270–271 and notice three important points:

1. The motion of crustal plates produces much of the geological activity on Earth. Earthquakes, volcanism, and mountain building are linked to motions in the crust and the location of plate boundaries. While you are thinking about volcanoes, you can correct a common misconception. The molten rock that emerges from volcanoes comes from pockets of melted rock in the upper mantle and lower crust and not from the molten core.

2. The continents on Earth's surface have moved and changed over periods of hundreds of millions of years. A hundred million years is only 0.1 billion years, $\frac{1}{45}$ of the age of Earth, so sections of Earth's crust are in geologically rapid motion.

3. Notice that most of the geological features you know—mountain ranges, the Grand Canyon, and even the outline of the continents—are recent products of Earth's active surface.

The Earth's surface is constantly renewed, and small crystals called zircons are the oldest known Earth material. They are from western Australia and are 4.4 billion years old. Most of the crust is much younger than that: The mountains and valleys you see around you are probably no more than a few tens or hundreds of millions of years old.

Earth's Atmosphere

When you think about Earth's atmosphere, you should consider three questions: How did it form? How has it evolved? How are humans changing it? Answering these questions will help you understand other planets as well as our own.

Our planet's first atmosphere, its **primary atmosphere**, was once thought to contain gases from the solar nebula, such as hydrogen and methane. Modern studies, however, indicate that the planets formed hot, so gases such as carbon dioxide, nitrogen, and water vapor would have been outgassed from (been cooked out of) the rock and metal (see Chapter 12). In addition, the final stages of planet building may have seen Earth and the other planets accreting planetesimals rich in volatile materials such as water, ammonia, and carbon dioxide. Thus the primary atmosphere must have been rich in carbon dioxide, nitrogen, and water vapor. The atmosphere you breathe today is a **secondary atmosphere** produced later in Earth's history.

Soon after Earth formed, it began to cool; once it cooled enough, oceans began to form, and carbon dioxide began to dissolve in the water. Carbon dioxide is highly soluble in water—which explains the easy manufacture of carbonated beverages. As the oceans removed carbon dioxide from the atmosphere, it reacted with dissolved compounds in the ocean water to form silicon dioxide, limestone, and other mineral sediments. Thus, the oceans transferred the carbon dioxide from the atmosphere to the seafloor and left air richer in other gases, especially nitrogen.

That removal of carbon dioxide is critical to Earth because an atmosphere rich in carbon dioxide can trap heat by a process called the **greenhouse effect**. When visible-wavelength sunlight shines through the glass roof of a greenhouse, it heats the interior. Infrared radiation from the warm interior can't get out through the glass,

© iStockphoto.com

Figure 13.3

The greenhouse effect: (a) Visual-wavelength sunlight can enter a greenhouse and heat its contents, but the longer-wavelength infrared radiation cannot get out. (b) The same process can heat a planet's surface if its atmosphere contains greenhouse gases such as CO_2. (c) The concentration of CO_2 in Earth's atmosphere as measured in Antarctic ice cores remained roughly constant for thousands of years until the beginning of the Industrial Revolution around the year 1800. Since then it has increased by more than 30 percent. (Note that the bottom of the vertical scale is not at zero.)

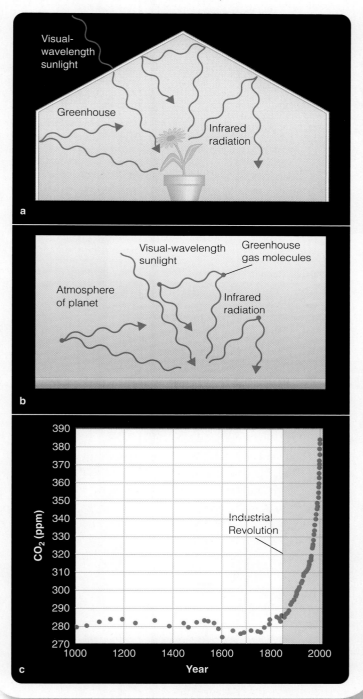

Graph adapted from a figure by Etheridge, Steele, Langenfelds, Francey, Barnola, and Morgan.

heat is trapped in the greenhouse, and the temperature climbs until the glass itself grows warm enough to radiate heat away as fast as sunlight enters, as seen in Figure 13.3a. (Of course, a real greenhouse also retains its heat because the walls prevent the warm air from mixing with the cooler air outside.) This is also called the "parked car effect," for obvious reasons.

Like the glass roof of a greenhouse, a planet's atmosphere can allow sunlight to enter and warm the surface. Note the similarity between the flow of light into the greenhouse in Figure 13.3a and into the atmosphere in Figure 13.3b. Carbon dioxide and other greenhouse gases such as water vapor and methane are opaque to infrared radiation, so an atmosphere containing enough of these greenhouse gases can trap heat and raise the temperature of a planet's surface.

It is a common misconception that the greenhouse effect is always bad. Without the greenhouse effect, Earth would be colder by at least 30°C (54°F), with a planetwide average temperature far below freezing. The problem is that human civilization is adding greenhouse gases to those that were already in the atmosphere. For 4 billion years, Earth's oceans and plant life have been absorbing carbon dioxide and burying it in the form of carbonates such as limestone and in carbon-rich deposits of coal, oil, and natural gas. In the last century or so, human civilization has begun digging up those fuels, burning them for energy, and releasing the carbon back into the atmosphere as carbon dioxide (Figure 13.3c). This process is steadily increasing the carbon dioxide concentration in the atmosphere and warming Earth's climate in what is called **global warming**. Evidence from proportions of carbon isotopes and oxygen in the atmosphere shows that most of the added CO_2 is the result of burning fossil fuels.

Global warming is a critical issue not just because it affects agriculture. It changes climate patterns that will warm some areas and cool other areas. It addition, the warming is melting permanently frozen ices in the polar caps, causing sea levels to rise. A rise of just a few feet will flood major land areas. When you visit Venus, you will see a planet dominated by a runaway greenhouse effect.

Greenhouse effect: The process by which a carbon dioxide atmosphere traps heat and raises the temperature of a planetary surface.

Global warming: The gradual increase in the surface temperature of Earth caused by human modifications to Earth's atmosphere.

1 Our world is an astonishingly active planet. Not only is it rich in water and therefore subject to rapid erosion, but its crust is divided into moving sections called plates. Where plates spread apart, lava wells up to form new crust; where plates push against each other, they crumple the crust to form mountains. Where one plate slides over another, you see volcanism. This process is called **plate tectonics**, referring to the Greek word for "builder." (An architect is literally an arch builder.)

A typical view of planet Earth

Mountains are common on Earth, but they erode away rapidly because of the abundant water.

William K. Hartmann

Janet Seeds

A **rift valley** forms where continental plates begin to pull apart. The Red Sea has formed where Africa has begun to pull away from the Arabian peninsula.

Midocean rise

Red Sea

Midocean rise

National Geophysical Data Center

1a Evidence of plate tectonics was first found in ocean floors, where plates spread apart and magma rises to form **midocean rises** made of rock called **basalt**, a rock typical of solidified lava. Radioactive dating shows that the basalt is younger near the midocean rise. Also, the ocean floor carries less sediment near the midocean rise. As Earth's magnetic field reverses back and forth, it is recorded in the magnetic fields frozen into the basalt. This produces a magnetic pattern in the basalt that shows that the seafloor is spreading away from the midocean rise.

1b A **subduction zone** is a deep trench where one plate slides under another. Melting releases low-density magma that rises to form volcanoes such as those along the northwest coast of North America, including Mt. St. Helens.

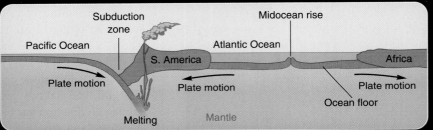

Subduction zone

Midocean rise

Pacific Ocean

Atlantic Ocean

S. America

Africa

Plate motion

Plate motion

Plate motion

Ocean floor

Melting

Mantle

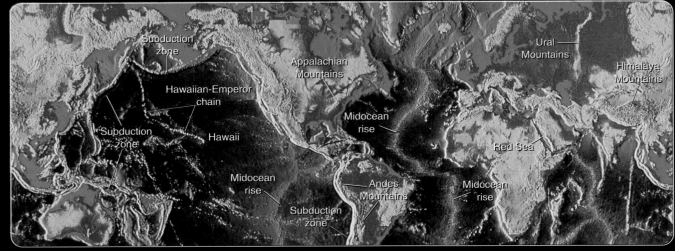

Subduction zone

Ural Mountains

Appalachian Mountains

Himalaya Mountains

Hawaiian-Emperor chain

Midocean rise

Subduction zone

Hawaii

Red Sea

Midocean rise

Andes Mountains

Midocean rise

Subduction zone

1c Hot spots caused by rising magma in the mantle can poke through a plate and cause volcanism such as that in Hawaii. As the Pacific plate has moved northwestward, the hot spot has punched through to form a chain of volcanic islands, now mostly worn below sea level. **Folded mountain ranges** can form where plates push against each other. For example, the Ural Mountains lie between Europe and Asia, and the Himalaya Mountains are formed by India pushing north into Asia. The Appalachian Mountains are the remains of a mountain range pushed up when North America collided with Africa.

1d The floor of the Pacific Ocean is sliding into subduction zones in many places around its perimeter. This pushes up mountains such as the Andes and triggers earthquakes and active volcanism all around the Pacific in what is called the Ring of Fire. In places such as southern California, the plates slide past each other, causing frequent earthquakes.

Hawaii

Yellow lines on this globe mark plate boundaries. Red dots mark earthquakes since 1980. Earthquakes within the plate, such as those at Hawaii, are related to volcanism over hot spots in the mantle.

Continental Drift

Not long ago, Earth's continents came together to form one continent.

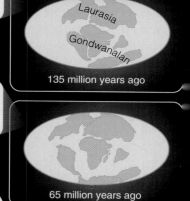

Pangaea

200 million years ago

Pangaea broke into a northern and a southern continent.

Laurasia

Gondwanalan

135 million years ago

Notice India moving north toward Asia.

65 million years ago

The continents are still drifting on the highly plastic upper mantle.

Today

2 The floor of the Atlantic Ocean is not being subducted. It is locked to the continents and is pushing North and South America away from Europe and Africa at about 3 cm per year, a motion called *continental drift*. Radio astronomers can measure this motion by timing pulsars from European and from American radio telescopes. Roughly 200 million years ago, North and South America were joined to Europe and Africa. Evidence of that lies in similar fossils and similar rocks and minerals found in the matching parts of the continents. Notice how North and South America fit against Europe and Africa like a puzzle.

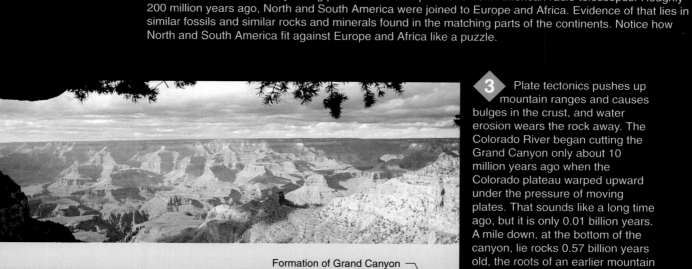

Formation of Grand Canyon

Formation of Earth

Age of dinosaurs

Heavy bombardment

Breakup of Pangaea

?

Oldest fossil life

First animals emerge on land

4.6   4   3   2   1   Now

Billions of years ago

3 Plate tectonics pushes up mountain ranges and causes bulges in the crust, and water erosion wears the rock away. The Colorado River began cutting the Grand Canyon only about 10 million years ago when the Colorado plateau warped upward under the pressure of moving plates. That sounds like a long time ago, but it is only 0.01 billion years. A mile down, at the bottom of the canyon, lie rocks 0.57 billion years old, the roots of an earlier mountain range that stood as high as the Himalayas. It was pushed up, worn away to nothing, and covered with sediment long ago. Many of the geological features we know on Earth have been produced by very recent events.

Oxygen in Earth's Atmosphere

When Earth was young, its atmosphere had no free oxygen. Oxygen is very reactive and quickly forms oxides in the soil, so activity of plant life is needed to keep a steady supply of oxygen in the atmosphere. Photosynthesis makes energy for the plant by absorbing carbon dioxide and releasing free oxygen. About 2 to 2.5 billion years ago plants began to produce oxygen faster than chemical reactions could remove it from the atmosphere. Atmospheric oxygen then increased rapidly.

Because there is oxygen (O_2) in the atmosphere now, there is also a layer of ozone (O_3) at altitudes of 15 to 30 km. Many people hold the common misconception that ozone is bad because they hear it mentioned as part of smog. Breathing ozone is bad for you, but you need the ozone layer in the upper atmosphere to protect you from harmful UV photons. Certain compounds called chlorofluorocarbons (CFCs) used in refrigeration and industry can destroy ozone when they leak into the atmosphere. Since the late 1970s, the ozone concentration has been falling, and the intensity of harmful UV radiation at Earth's surface has been increasing year by year. Note that ozone depletion is an additional Earth environmental issue that is separate from global warming. While this poses an immediate problem for public health on Earth, it is also of interest astronomically. When you visit Mars, you will see the effects of an atmosphere without ozone.

A Short Geological History of Earth

As Earth formed in the inner solar nebula, it passed through three stages that also describe the histories of the other Terrestrial planets to varying extents. When you try to tell the story of each planet in our solar system, you pull together all the known facts plus hypotheses and try to make them into a logical history of how the planet got to be the way it is. Of course, your stories will be incomplete because scientists don't yet understand all the factors affecting the history of the planets.

The first stage of planetary evolution is *differentiation*, the separation of each planet's material into layers according to density. Some of that differentiation may have occurred very early as the heat released by infalling matter melted the growing Earth. Some of the differen-

Maria (mare): One of the lunar lowlands filled by successive flows of dark lava, from the Latin for "sea."

tiation, however, may have occurred later as radioactive decay released more heat and further melted Earth, allowing the denser metals to sink to the core.

The second stage, *cratering and giant basin formation*, could not begin until a solid surface formed. The heavy bombardment in the early solar system cratered Earth just as it did the other Terrestrial planets. Some of the largest craters, called basins, were likely big enough to break through to the upper mantle where rocks are partly molten. That molten rock would have welled up through cracks to fill some of the basins and craters. On planets with liquid water, the basins may also have been the first oceans. As the debris of planet formation cleared away, the rate of impacts and crater formation fell to its present low rate.

The third stage, *slow surface evolution*, has continued for at least the past 3.5 billion years. Earth's surface is constantly changing as sections of crust slide over and past each other, push up mountains, and shift continents. In addition, moving air and water erode the surface and wear away geological features. Almost all traces of the earlier stages of differentiation and cratering have been destroyed by the active crust and erosion. Life apparently started on Earth around the beginning of the slow surface evolution stage, and the secondary atmosphere began to replace the primary atmosphere. That may be unique to Earth and may not have happened on the other Terrestrial planets.

Terrestrial planets pass through these stages, but differences in masses, temperature, and composition emphasize some stages over others and produce surprisingly different worlds.

13.3 The Moon

You can't go for a stroll on the moon without a spacesuit. There is no air, and the temperature difference from sunshine to shade is extreme. Check your Celestial Profile Card for the moon's data.

Lunar Geology

You could visit two kinds of terrain on the moon. The dark gray areas visible from Earth are the smooth lunar lowland which astronomers named **maria** (plural of **mare**, pronounced mah-ray), drawing on the Latin word

The Present Is the Key to the Past

You can learn about the history of Earth by looking at the present condition of Earth's surface. The position and composition of various rock layers in the Grand Canyon, for example, tell you that the western United States was once at the floor of an ocean. This principle of geology was first formulated in the late 1700s, and it continues to be relevant today as you try to understand other worlds such as Venus and Mars.

In the late 18th century, naturalists first recognized that the present gave them clues to the history of Earth. This was astonishing because most people assumed that Earth had no history. That is, they assumed either that Earth had been created in its present state, or that Earth was eternal. In either case, people commonly assumed that the hills and mountains they saw around them had always existed more or less as they were. The 18th-century naturalists began to see evidence that the hills and mountains were not eternal but were the result of past processes and were slowly changing. That gave birth to the idea that Earth had a history and stimulated the invention of modern geology as a way of understanding Earth.

What Copernicus, Kepler, and Newton did for the heavens in the 1500s and 1600s, the first geologists did for Earth beginning in the late 1700s. Of course, the invention of geology as the study of Earth led directly to the modern attempts to understand the geology of other worlds.

Geologists and astronomers share a common goal: They are attempting to reconstruct the past. Whether you study Earth, Venus, or Mars, you are looking at the present evidence and trying to reconstruct the past history of the planet by drawing on observations and logic to test each step in the story. How did Venus get to be covered with lava? How did Mars lose its atmosphere? The final goal of planetary astronomy is to draw together all of the available evidence (the present) to tell the story (the past) of how the planet got to be the way it is. Those first geologists of the late 1700s would be fascinated by the stories planetary astronomers tell today.

for *seas*. You could also visit the lighter-colored mountainous lunar highlands. The moon looks quite bright in the night sky seen from Earth. In fact, the **albedo** of the near side of the moon is only 0.06, meaning it reflects only 6 percent of the light that hits it. The moon looks bright only in contrast to the night sky. In reality, it is a dark gray world.

Wherever you go on the moon, you will find craters. The highlands are marked heavily by craters, but the smooth lowlands contain fewer craters. The craters on the moon were formed by impacts, as evidenced by their distinguishing characteristics such as shape and the material **(ejecta)** they have spread across the moon's surface. Craters range in size from giant basins hundreds of kilometers across to microscopic pits found in moon rocks. Most of the craters on the moon are old; they were formed long ago when the solar system was young. This is clear from comparison of ages of moon rocks collected from the heavily cratered highlands and the somewhat younger, less heavily cratered maria.

Twelve Apollo astronauts visited both the moon's maria and highlands in six expeditions between 1969 and 1972 (Figure 13.4). Most of the rocks they found, both in the highlands and the maria, were typical of hardened lava. The maria are actually ancient basalt lava flows. The highlands, in contrast, are composed of low-density rock, for example, **anorthosite**, a light-colored rock that contributes to the highlands' bright contrast with the dark maria and that would be among the first material to solidify and float to the top of molten rock. Many rocks found on the moon are **breccias**, made up of

Albedo: The ratio of the amount of light reflected from an object to the amount of light received by the object. Albedo equals 0 for perfectly black and 1 for perfectly white.

Ejecta: Pulverized rock scattered by meteorite impacts on a planetary surface.

Anorthosite: Aluminium- and calcium-rich silicate rock found in the lunar highlands.

Breccia: Rock composed of fragments of older rocks bonded together.

Figure 13.4

Rocks from the moon show that the moon's surface formed in a molten state, that it was heavily fractured by cratering when it was young, and that it is now affected mainly by micrometeorite erosion.

Rocks exposed on the lunar surface become pitted by micrometeorites.

Vesicular basalt contains bubbles frozen into the rock when it was molten.

A breccia is formed by rock fragments bonded together by heat and pressure.

© NASA

fragments of broken rock cemented together under pressure. The breccias show how extensively the moon's surface has been shattered by meteorites, as does the surface layer of powdery dust kicked up by the astronaut's boots.

The Origin of Earth's Moon

Over the last two centuries, astronomers developed three different hypotheses for the origin of Earth's moon. The *fission hypothesis* proposed that the moon broke from a rapidly spinning proto-Earth. The *condensation hypothesis* suggested that Earth and its moon condensed from the same cloud of matter in the solar nebula. The *capture hypothesis*

Large-impact hypothesis: Hypothesis that the moon formed from debris ejected during a collision between Earth and a large planetesimal.

suggested that the moon formed elsewhere in the solar nebula and was later captured by Earth. Each of these previous ideas had problems and failed to survive comparison with all the evidence.

In the 1970s, a new hypothesis was proposed that combined some aspects of the three previous hypotheses. The **large-impact hypothesis** proposes that the moon formed when a planetesimal estimated to have been at least as large as Mars smashed into the proto-Earth. Model calculations indicate that this collision would have ejected a disk of debris into orbit around Earth that would have quickly formed the moon.

This would explain a number of phenomena. If the collision occurred off-center, it would have spun the Earth–moon system rapidly and would thus explain the present high angular momentum. If the proto-Earth and impactor had already differentiated, the ejected

The Large-Impact Hypothesis

A protoplanet nearly the size of Earth differentiates to form an iron core.

Another body that has also formed an iron core strikes the larger body and merges, trapping most of the iron inside.

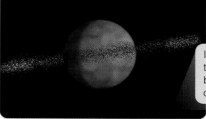

Iron-poor rock from the mantles of the two bodies forms a ring of debris.

Volatiles are lost to space as the particles in the ring begin to accrete into larger bodies.

Eventually the moon forms from the iron-poor and volatile-poor matter in the disk.

material would have been mostly iron-poor mantle and crust, which would explain the moon's low density and iron-poor composition. Furthermore, the material would have lost much of its volatile components while it was in space, so the moon also would have formed lacking those materials. Such an impact would have melted the

proto-Earth, and the material falling together to form the moon would have been heated hot enough to melt. This fits the evidence that the highland anorthosite in the moon's oldest rocks formed by differentiation of large quantities of molten material. The large-impact hypothesis passes tests of comparison with the known evidence and is now considered likely to be correct.

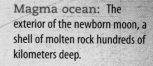

Magma ocean: The exterior of the newborn moon, a shell of molten rock hundreds of kilometers deep.

Multiringed basin: Large impact feature (crater) containing two or more concentric rims formed by fracturing of the planetary crust.

Late heavy bombardment: The sudden temporary increase in the cratering rate in our solar system that occurred about 4 billion years ago.

The History of Earth's Moon

The history of Earth's moon since its formation is dominated by a single fact—the moon is small, only one-quarter the diameter of Earth. The escape velocity is low, so it has been unable to hold any atmosphere, and it cooled rapidly as its internal heat flowed outward into space.

The Apollo moon rocks are the source of information about the timing of events during the moon's history. For example, they show that the moon must have formed in a mostly molten state. Planetary geologists now refer to the exterior of the newborn moon as a **magma ocean**. Denser materials sank toward the center, and low-density minerals floated to the top to form a low-density crust. In this way the moon partly differentiated. The radioactive ages of the moon rocks show that the surface solidified about 4.4 billion years ago.

The second stage, cratering and basin formation, began as soon as the crust solidified, and the older highlands show that cratering was intense for approximately 0.5 billion years—during the heavy bombardment at the end of the solar system's period of planet building. The moon's crust was shattered, and the largest impacts formed giant **multiringed basins** hundreds of kilometers in diameter (Figure 13.5). There is some evidence indicating that roughly 4 billion years ago there was a sudden temporary increase in the cratering rate. Astronomers refer to this as the **late heavy bombardment** and hypothesize that it could have been caused by the final accretion and migration of Uranus and Neptune (see Chapter 12) scattering remnant planetesimals across the solar system to collide with the moon and planets. About 3.8 billion years ago the cratering rate fell rapidly to the current low rate.

The tremendous impacts that formed the lunar basins cracked the crust as deep as 10 kilometers and led to flooding by lava. Though Earth's moon cooled rapidly after its formation, radioactive decay heated the subsurface material, and part of it melted, producing lava that followed the cracks up into the giant basins (Figure 13.5a). Studies of rocks brought back from the moon by Apollo astronauts show that the basins were flooded by successive lava flows of dark basalts from roughly 4 to 2 billion years ago, forming the maria. The moon's crust is thinner on the side toward Earth, perhaps due to tidal effects. Consequently, while lava flooded the basins on the Earth-facing side, it was unable to rise through the thicker crust to flood the lowlands on the far side.

The third stage, slow surface evolution, was limited both because the moon cooled rapidly and because

Figure 13.5

Much of the near side of the moon is marked by great, generally circular lava plains called maria. The crust on the far side is thicker, and there was much less flooding. Even the huge Aitken Basin near the lunar south pole contains little lava flooding. In these maps color marks elevation, with red the highest regions and purple the lowest.

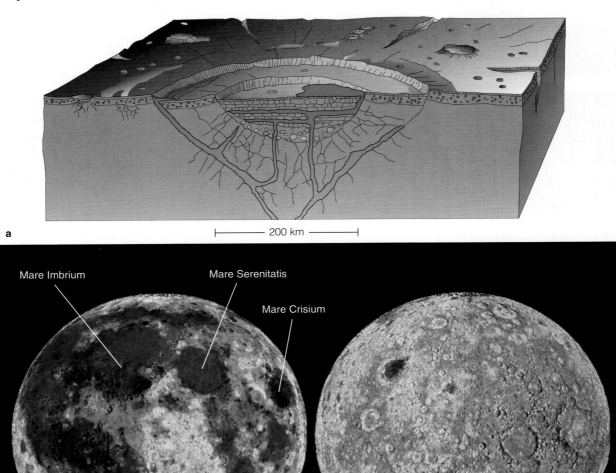

a   ⊢—— 200 km ——⊣

Mare Imbrium   Mare Serenitatis

Mare Crisium

Aitken Basin

Near side of moon   **Far side of moon**

b

it lacks water. Flooding on Earth included water, but the moon has never had an atmosphere and thus has never had liquid water. With no air and no water, erosion is limited to the constant bombardment of **micrometeorites** and rare larger impacts. Indeed, a few meteorites found on Earth have been identified as moon rocks ejected from the moon by impacts within the last few million years. As the moon lost its internal heat, volcanism died down, and the moon became geologically dead. Its crust never divided into moving plates—there are no folded mountain ranges—and the moon is now a "one-plate" planet.

13.4 Mercury

Like Earth's moon, Mercury is small and nearly airless. Your Celestial Profile Card has Mercury's data. Mercury cooled too quickly to develop plate tectonics, so you will find it also is a cratered, dead world.

Spacecraft at Mercury

Mercury orbits close to the sun and is difficult to observe from Earth, so little was known about it until 1974–1975 when the Mariner 10 spacecraft flew past Mercury three times and revealed a planet with a heavily cratered surface, much like that of Earth's moon. A new spacecraft called MESSENGER made three flybys of Mercury in 2008 and 2009, taking impressively high-resolution images (Figure 13.6a) and measurements of the parts of the planet not covered by Mariner 10's cameras. MESSENGER is scheduled to go into orbit around Mercury in 2011 and begin a yearlong closeup study. Analysis shows that large areas have been flooded by lava and then cratered. The largest impact feature on Mercury is the Caloris Basin, a multiringed area 1,300 km (800 mi) in diameter (Figure 13.6b).

Mercury is quite dense, and models indicate that it must have a large metallic core. In fact, the metallic core occupies about 70 percent of the radius of the planet. Mercury is essentially a metal planet with a thin rock mantle. Mariner 10 photos revealed long curving ridges up to 3 km (2 mi) high and 500 km (300 mi) long, as shown in Figure 13.6c. The ridges even cut through

craters, indicating that they formed after most of the heavy bombardment. Planetary scientists understand these ridges as evidence that the entire crust of Mercury compressed and shrank long ago.

Micrometeorite: Meteorite of microscopic size.

A History of Mercury

Like Earth's moon, Mercury is small, and that fact has determined much of its history. Not only is Mercury too small to retain an atmosphere, it has also lost much of its internal heat and thus is not geologically active.

In the first stage of its formation, Mercury differentiated to form a metallic core and a rocky mantle. Mariner 10 discovered a magnetic field with a strength about 10^{-4} times that of Earth's—further evidence of a metallic core. In Chapter 12, you saw that the condensation sequence could explain the abundance of metals in Mercury, but careful model calculations show that Mercury contains even more iron than would be expected from its position in the solar system. Drawing on the large-impact hypothesis for the origin of Earth's moon, scientists have proposed that Mercury suffered a major impact soon after it differentiated, an impact so large it ejected much of the rocky mantle into space. Such catastrophic events are rare in nature, but they do occur, so astronomers must be prepared to consider such hypotheses.

In the second stage of planet formation, cratering battered the crust, and lava flows welled up to fill the lowlands, just as they did on the moon. As the small world lost internal heat, its large metal core contracted, and its crust compressed and broke to form the long ridges, much as the peel of a drying apple wrinkles.

Lacking an atmosphere to erode it, Mercury has changed little since the last lava hardened, and it is now a "one-plate" planet like Earth's moon.

Figure 13.6

(a) Mercury is an airless, cratered world. (b) The Caloris multiringed basin was half in shadow and half in sunlight when the Mariner 10 spacecraft flew past the planet. (c) Lobate scarps are distributed all around Mercury. (d) The origin of the spider formation photographed by MESSENGER is a puzzle.

The impact that formed the multiringed basin Caloris pushed up mountain ranges as high as 3 km (2 mi).

The surface of Mercury is heavily cratered.

Almost no detail is visible from Earth.

a Visual-wavelength images

Discovery Rupes, a long curved ridge, cuts through craters that must have formed first.

13.5 Venus

You might expect Venus to be much like Earth. It is 95 percent Earth's diameter, has a similar average density, and is only 30 percent closer to the sun. Your Celestial Profile Cards compare Earth and Venus's data more specifically. Unfortunately, the surface of Venus is perpetually hidden below thick clouds, and only in the last few decades have planetary scientists discovered that Venus is a hot desert world of volcanoes, lava flows, and impact craters lying at the bottom of a deep, hot atmosphere. No spacesuit will allow you to visit the surface of Venus.

The Atmosphere of Venus

In composition, temperature, and density, the atmosphere of Venus is more Hades than Heaven. The air is unbreathable, very hot, and very dense. How do planetary scientists know this? Because space probes have descended into the atmosphere and in some cases landed on the surface.

Figure 13.7

Notice how three radar maps show different things. The main radar map, made by the Magellan robot probe, shows elevation over most of the surface of Venus, omitting the polar areas. The detailed map of Maxwell Montes and Lakshmi Planum are colored according to roughness, with orange the roughest terrain. The map of volcano Sapas Mons also shows roughness but is given an orange color to mimic the color of sunlight reaching the surface through the thick atmosphere.

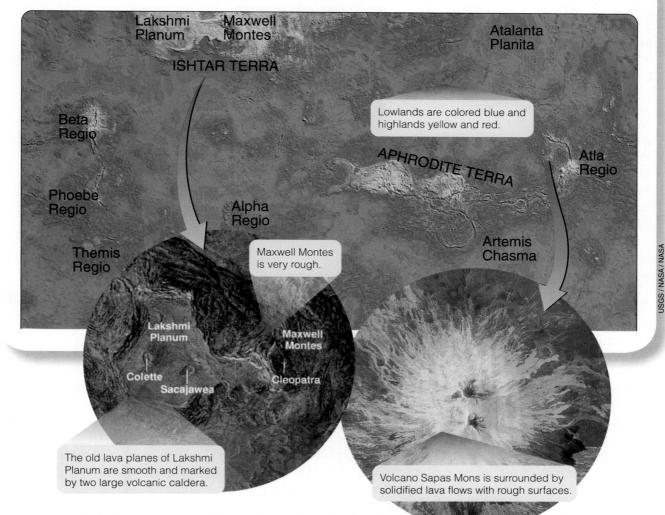

Lakshmi Planum — Maxwell Montes

Atalanta Planita

ISHTAR TERRA

Lowlands are colored blue and highlands yellow and red.

Beta Regio

APHRODITE TERRA

Atla Regio

Phoebe Regio

Alpha Regio

Artemis Chasma

Themis Regio

Maxwell Montes is very rough.

USGS / NASA / NASA

Lakshmi Planum

Maxwell Montes

Colette   Cleopatra
Sacajawea

Volcano Sapas Mons is surrounded by solidified lava flows with rough surfaces.

The old lava planes of Lakshmi Planum are smooth and marked by two large volcanic caldera.

In composition, the atmosphere of Venus is roughly 96 percent carbon dioxide. The rest is mostly nitrogen, with some argon, sulfur dioxide, and small amounts of sulfuric acid, hydrochloric acid, and hydrofluoric acid. There is only a tiny amount of water vapor. On the whole, the composition is deadly unpleasant, and most certainly smells bad too. Spectra show that the impenetrable clouds that hide the surface are made up of droplets of sulfuric acid and microscopic crystals of sulfur. This unbreathable atmosphere is 90 times denser than Earth's. The air you breathe is 1,000 times less dense than water, but on Venus the air is only 10 times less dense than water. If you could survive the unpleasant composition and intense heat, you could strap wings on your arms and fly.

The surface temperature on Venus is hot enough to melt lead, and you can understand that because the thick atmosphere creates a severe greenhouse effect. Sunlight filters down through the clouds and warms the surface, but infrared radiation cannot escape because the atmosphere is opaque to infrared. It is the overwhelming abundance of carbon dioxide that makes the greenhouse effect on Venus much more severe than on Earth.

The Surface of Venus

Although the thick clouds and atmosphere on Venus are opaque to visible and infrared light, they are transparent to radio waves. Orbiting spacecraft mapped Venus by radar, revealing details as small as 100 meters in diameter (Figure 13.7).

Radar maps of Venus are reproduced using arbitrary colors. Lowlands are colored blue in some maps, but this is misleading in a sense because there are no oceans on

Venus. In other maps the scientists have chosen instead to give Venus an overall orange tint because sunlight filtering down through the clouds bathes the landscape in a perpetual sunset glow. Some radar maps are colored gray, the natural color of the rocks.

Radar maps show that Venus is similar to Earth in one way but strangely different in other ways. Nearly 75 percent of Earth is covered by low-lying basaltic sea floors, and 85 percent of Venus is covered by basaltic lowlands. Of course on Venus the lowlands are not seafloors, and the remaining highlands are not the well-defined continents you see on Earth. Whereas Earth is dominated by plate tectonics, something different is happening on Venus.

The highland area Ishtar Terra, named for the Babylonian goddess of love, is about the size of Australia (Figure 13.7). At its eastern edge, the mountain called Maxwell Montes thrusts up 12 km (Everest, the tallest mountain on Earth, is 8.8 km high). The mountains bounding Ishtar Terra, including Maxwell, resemble folded mountain ranges, which suggests that some horizontal motion in the crust as well as volcanism has helped form the highlands.

Many features on Venus testify to its volcanic history. Long, narrow lava channels meander for thousands of kilometers. Radar maps reveal many smaller volcanoes, faults, and sunken regions produced when magma below the surface drained away. Other volcanic features include the **coronae**, circular bulges up to 2,100 km (1,300 mi) in diameter bordered by fractures, volcanoes, and lava flows (Figure 13.8). These appear to be produced by rising convection currents of molten magma that push up under the crust. When the magma withdraws, the crust sinks back, and the circular fractures mark the edge of the original upwelling.

Radar images also show that Venus is marked by numerous craters. Meteorites big enough to make craters larger than 3 km in diameter have no trouble penetrating the thick atmosphere but have formed only about 10 percent as many craters on Venus as on the maria of Earth's moon. The number of craters shows that the crust of Venus is younger than the lunar maria but older than most of Earth's active surface. The average age of the surface of Venus is estimated to be roughly half a billion

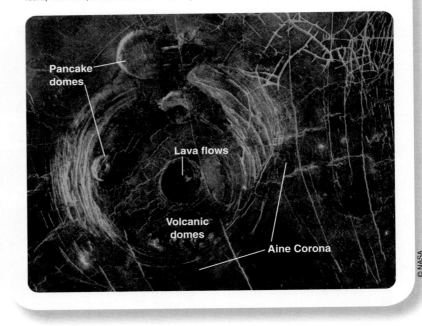

Figure 13.8

Radar map of surface features on Venus: Aine Corona, about 200 km in diameter, is marked by faults, lava flows, small volcanic domes, and pancake domes of solidified lava.

© NASA

years old. Clearly, geological processes cannot be renewing the surface as rapidly as they do on Earth.

Ten robot probes landed successfully on the surface of Venus in the 1970s and 80s and managed to survive the heat and pressure for a few minutes or hours, transmitting data to Earth. Some of those spacecraft analyzed the rock and snapped a few photographs (Figure 13.9). The surface rocks on Venus are dark gray basalts much like those in Earth's ocean floors. This evidence confirms that volcanism is important.

The History of Venus

To tell the story of Venus, you must draw together all the evidence and find hypotheses to explain two things, the thick carbon dioxide atmosphere and the peculiar geology.

Calculations show that Venus and Earth have outgassed about the same amount of carbon dioxide, but Earth's oceans have dissolved most of that and converted it to sediments such as limestone. If all of the carbon in Earth's sediments and crust were dug up and converted back to carbon dioxide, our atmosphere would be about as dense as the air on Venus, and with similar composition. This suggests that the main difference between Earth and Venus is the lack of water on Venus.

Venus may have had oceans when it was young and much more Earthlike than at present, but, being closer to the sun than Earth, it was warmer, and the carbon dioxide in the atmosphere created a greenhouse effect that made the planet even warmer. That process could have dried up any oceans that did exist and prevented Venus from purging its atmosphere of carbon dioxide. In fact, evidence from the composition of Venus's atmosphere indicates that an ocean's worth of water might have been vaporized and lost. As carbon dioxide continued to be outgassed, the greenhouse effect grew even more severe. Thus, planetary scientists conclude that Venus was trapped in a **runaway greenhouse effect**.

The intense heat at the surface may have affected the geology of Venus by making the crust drier and more flexible so that it was unable to break into moving plates as on Earth. There is no sign of plate tectonics on Venus, but there is evidence that convection currents below the crust are deforming the crust to make coronae, push up mountains such as Maxwell, and create some folded mountains like those around Ishtar Terra by minor horizontal crust motions.

As you learned earlier, the small number of craters on the surface of Venus hints that the entire crust has been replaced within the last half-billion years or so.

This may have occurred in a planetwide overturning as the old crust broke up and sank and lava flows created a new crust. Comparing Earth to Venus may eventually reveal more about how our own world's volcanism and tectonics work.

13.6 Mars

If you ever visit another world, Mars may be your best choice. You will need a heated, pressurized spacesuit, but Mars is not as inhospitable as the moon. It is also more interesting, with weather, complex geology, craters, volcanoes, and signs that water once flowed over its surface.

The Atmosphere of Mars

The Martian air contains 95 percent carbon dioxide, 3 percent nitrogen, and 2 percent argon. That is similar to the composition of air on Venus, but the Martian atmosphere is very thin, less than 1 percent as dense as Earth's atmosphere and 1/10,000 as dense as Venus's.

There is very little water in the Martian atmosphere. Liquid water cannot survive on the surface of Mars because the air pressure is too low: Any liquid water would immediately boil away. The polar caps appear to be composed of frozen water ice coated over by frozen carbon dioxide (dry ice). Whatever water is present on Mars is frozen either within the polar caps or as **permafrost** in the soil.

*Figure 13.9

The Venera 13 lander touched down on Venus in 1982 and carried a camera that swiveled from side to side to photograph the surface. The orange glow is produced by the thick atmosphere; when that is corrected digitally, you can see that the rocks are dark gray. Isotopic analysis suggests they are basalts.

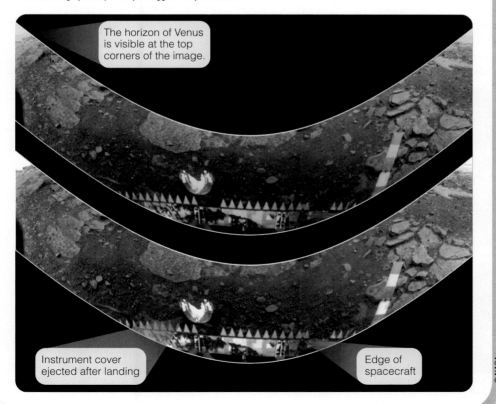

The horizon of Venus is visible at the top corners of the image.

Instrument cover ejected after landing

Edge of spacecraft

© NASA

Runaway greenhouse effect: A greenhouse effect so dramatic that it amplifies itself, becoming stronger with time.

Permafrost: Permanently frozen soil.

Although the present atmosphere of Mars is very thin, you will see evidence that the climate once permitted liquid water to flow over the surface, so Mars must have once had a thicker atmosphere. As a Terrestrial planet, it should have outgassed significant amounts of carbon dioxide, nitrogen, and water vapor; but because it was small it could not hold onto its atmosphere. The escape velocity on Mars is only 5 km/s, less than half of Earth's, so it was easier for rapidly moving gas molecules to escape into space. Furthermore, if Mars had been colder, the gas molecules in its atmosphere would have been traveling more slowly and would not have escaped as easily.

Another process that has reduced Mars's atmosphere is the planet's lack of an ozone layer to screen out UV radiation: Sunbathing on Mars would be a fatal mistake. Solar UV photons can break atmospheric molecules up into smaller fragments. Water, for example, can be broken by UV into hydrogen and oxygen. The hydrogen then escapes into space, and the oxygen easily combines chemically with rocks and soil. Mars may have had a substantial atmosphere when it was young, but it gradually lost much of that both by direct escape and by UV destruction. As a result, with a thin atmosphere that does not provide much greenhouse warming

nor permit liquid water to persist on the surface, Mars is now a cold, dry world.

Exploring the Surface of Mars

Data recorded by orbiting satellites show that the southern hemisphere of Mars is a heavily cratered highland region up to 4 billion years old. The northern hemisphere is mostly a much younger lowland plain with few craters, compared in Figure 13.10.

Volcanism on Mars is dramatically evident in the Tharsis region, a highland region of volcanoes and lava flows bulging 10 km (6 mi) above the surrounding surface. A similar uplifted volcanic plain, the Elysium region, is more heavily cratered and eroded and appears to be older than the Tharsis bulge. The lack of impact craters on the slopes of some volcanoes in both Tharsis and Elysium suggests that there has been volcanic activity within the past few hundred million years, and maybe even more recently.

All of the volcanoes on Mars are **shield volcanoes**, very broad mountains with gentle slopes that are produced on Earth by hot spots penetrating upward through the crust. Shield volcanoes are not related to plate tectonics; in fact, the large shield volcanoes on Mars, the largest of which is Olympus Mons, provide evidence that plate

Figure 13.10

These hemisphere maps of Mars are color coded to show elevation. The northern lowlands lie about 4 km (2.5 mi) below the southern highlands. Volcanoes are very high (white), and the giant impact basins, Hellas and Argyre, are low. Note the depth of the great canyon Valles Marineris.

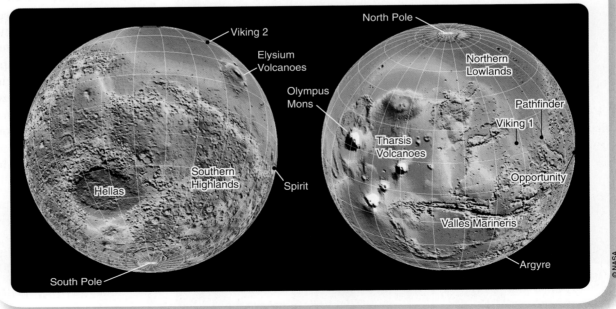

tectonics has not been significant on that planet. Olympus Mons is 600 km (almost 400 mi) in diameter at its base and rises 21 km (13 mi) high. In comparison, the largest volcano on Earth, also a shield volcano, is Mauna Loa in Hawaii, rising only 10 km (6 mi) above its base on the seafloor (Figure 13.11a). On Earth, volcanoes like those that formed the Hawaiian Islands occur over rising currents of hot material in the mantle. Because the crust plate moves horizontally, a chain of volcanoes is formed instead of a single large feature (Figure 13.11b). A lack of plate motion on Mars has allowed rising currents of magma to heat the crust repeatedly in the same places and build Olympus Mons plus other very large volcanic shields, especially in the Tharsis region (Figure 13.10).

When the crust of a planet is strained, it may break, producing faults and rift valleys. Near the Tharsis region is a great valley, Valles Marineris, which you can see in Figure 13.10, named after the Mariner spacecraft that first photographed it. The valley is a block of crust that has dropped downward along parallel faults. Erosion and landslides have further modified the valley into a great canyon. It is four times deeper, nearly ten times wider, and over ten times longer than the Grand Canyon. The number of craters in the valley indicates that it is 1 to 2 billion years old, placing its origin sometime before the end of the most active volcanism in the Tharsis region.

Searching for Water on Mars

Before you can tell the story of Mars, you must consider a difficult issue—water. How much water has Mars had, how much has been lost, and how much remains? As you learned earlier, liquid water cannot exist on the surface of Mars now because of its low atmospheric pressure and low temperature. Observations from orbiting spacecraft, however, have revealed landforms that suggest the effects of flowing water on Mars, and rovers on the surface have turned up positive proof of surface water.

In 1976, the two Viking spacecraft reached orbit around Mars and photographed its surface. Those photos revealed two kinds of water-related features. **Outflow channels** ap-

Figure 13.11

(a) Olympus Mons on Mars is much larger than Mauna Loa, the largest mountain on Earth. Mauna Loa is so heavy that it has sunk into Earth's crust, producing an undersea moat, but Olympus Mons has not, suggesting Mars's crust is much stronger than Earth's. (b) Volcanoes like Mauna Loa do not grow very large on Earth because the crust plates move horizontally and carry older volcanoes away from hot spots.

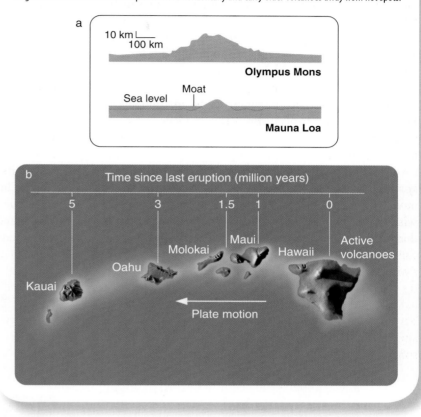

pear to have been cut by massive floods carrying as much as 10,000 times the water flowing down the Mississippi River. In a matter of hours or days, such floods swept away geological features and left scarred land such as that shown in the following images made by Viking orbiters and the Mars Global Surveyor. In contrast, **valley networks** look like meandering riverbeds with sandbars, deltas, and tributaries typical of streams that flowed for extended periods of time. The number of craters on top of these features reveals that they are quite old.

Spacecraft in orbit around Mars have used remote instruments to detect large amounts of water frozen in the soil. A radar study has found frozen water extending at least a kilometer beneath both polar caps. Images made from orbit also show regions of

Outflow channel: Geological features on Mars and Earth caused by flows of vast amounts of water released suddenly.

Valley network: A system of dry drainage channels on Mars that resembles the beds of rivers and tributary streams on Earth.

jumbled terrain and gullies leading down slopes, suggesting water has flowed onto the surface from underground sources, perhaps melting of subsurface ice. The terrain at the edges of the northern lowlands has been compared to shorelines, and some scientists suspect that the northern lowlands were filled with an ocean roughly 3 billion years ago. Look again at Figure 13.10, where the lowlands have been color-coded blue, and notice the major outflow channels leading from highlands into lowlands, like rivers flowing into an ocean.

Rovers named Spirit and Opportunity landed in January 2004 and carried sophisticated instruments to explore the rocky surface. Both rovers landed in areas suspected of having had water on their surfaces, and both made exciting discoveries. Using close-up cameras, the rovers found small spherical concretions of the mineral hematite (dubbed "blueberries") that must have formed in water. In other places, they found layers of sediments with ripple marks and crossed layers showing deposits that must have formed in moving water. You can see both

of these features in the images on the next page. Chemical analysis revealed minerals in the soil such as sulfates that are left behind when standing water evaporates.

In 2008, the space probe Phoenix landed in the north polar region of Mars and used its robotic arm to uncover water ice frozen in the Martian soil. Detailed chemical analysis of soil samples were completed confirming the presence of water and minerals necessary for any life that might exist there.

The History of Mars

Did Mars ever have plate tectonics? Where did the water go? These fundamental questions challenge you to assemble the evidence and hypotheses for Mars and tell the story of its evolution.

The history of Mars is a case of arrested development. The planet began by differentiating into a crust, mantle, and core. Studies of its rotation reveal that it has a dense core but no planetwide magnetic field. The core

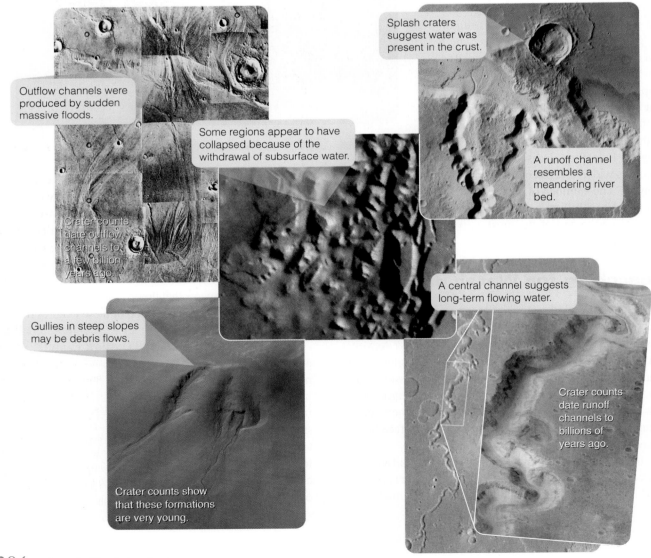

Outflow channels were produced by sudden massive floods.

Some regions appear to have collapsed because of the withdrawal of subsurface water.

Crater counts date outflow channels to a few billion years ago.

Gullies in steep slopes may be debris flows.

Crater counts show that these formations are very young.

Splash craters suggest water was present in the crust.

A runoff channel resembles a meandering river bed.

A central channel suggests long-term flowing water.

Crater counts date runoff channels to billions of years ago.

Marlin Space Science Systems/NASA

"blueberries"

flowing water deposits

0   1   2
cm

© NASA/JPL/Cornell/USGS

must have cooled quickly and shut off the dynamo effect that would have produced a field.

The crust of Mars is now quite thick, as shown by the mass of the Olympus Mons volcano (Figure 13.11), but it was probably thinner in the past. Cratering may have broken or at least weakened the crust, triggering lava flows that flooded some basins. Mantle convection may have pushed up the Tharsis and Elysium volcanic regions and broken the crust to form Valles Marineris, but moving crustal plates never dominated Mars. There are no folded mountain ranges on Mars and no sign of plate boundaries. As the planet cooled, its crust grew thick and immobile.

For Mars, the last stage of planetary development has been one of slow decline. Volcanoes may still occasionally erupt, but the little planet has lost much of its internal heat, and most volcanism occurred long ago. At some point in the history of Mars, water was abundant enough to flow over the surface in great floods and may have filled an ocean, but the age of liquid water must have ended over 3 billion years ago. The climate on Mars changed as the atmosphere gradually became thinner.

Atmospheric gases and water were lost to space, and the volcanic activity that could have replaced them had nearly stopped.

The water remaining on Mars today is frozen in the polar caps or in the soil. However, note that water is the first necessity of life, so its presence long ago on Mars is exciting. Someday an astronaut may scramble down an ancient Martian streambed, turn over a rock, and find a fossil.

The Moons of Mars

Unlike Mercury or Venus, Mars has moons. Small and irregular in shape, Phobos (28 × 23 × 20 km in diameter) and Deimos (16 × 12 × 10 km) are almost certainly captured asteroids (see Chapter 12). These moons are so small they cannot pull themselves into spherical shape. Phobos and Deimos are not just small; they are tiny. An athletic astronaut who could jump 2 m (6 ft) high on Earth could jump almost 3 km (2 mi) on Phobos. However they formed, these moons are so small that any interior heat would have leaked away very quickly, and there is

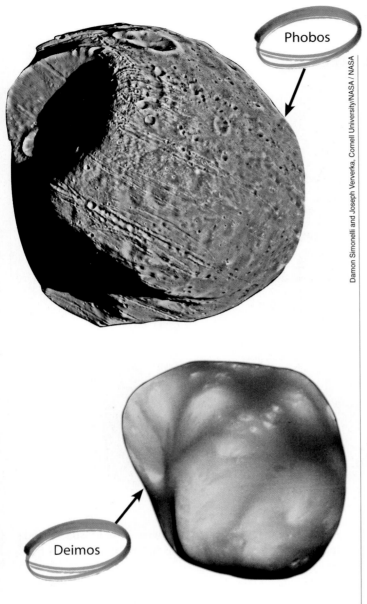

Phobos

Deimos

recap Humans have intense curiosity about their world. We want to know how things work and what is going to happen. We are looking for other worlds, but leaving Earth is difficult for two reasons. First, to achieve escape velocity you need a big, expensive rocket. Second, we face evolutionary challenges. If you leave Earth, you have to take the Earthly environment with you.

On the moon, astronauts' spacesuits provided them with air to breathe, water to drink, and climate control. Their spacesuits could not provide artificial gravity and medical tests show that low gravity conditions cause serious deterioration of bones and muscles unless difficult preventive measures are taken. It would be hard to stay healthy away from Earth. The Terrestrial planets are the only ones in our solar system that you might reasonably expect to visit, and not all of them are welcoming. Visiting Venus would require a deep-sea submersible; Mercury a superpower cooling system. Even Mars, the friendliest, is only somewhat more hospitable than the moon.

Your exploration should have guided you to the answers to these essential questions:

1. How can comparison help you understand the Terrestrial planets?

2. What are the main features of Earth when you view it as a planet?

3. How does size determine the geological activity and evolution of a planet?

4. How do distance from the sun and size of a planet affect the properties of its atmosphere?

5. What is the evidence that surface conditions on Mars were originally more Earth-like than they are at present?

no evidence of any internally driven geologic activity on either object.

Some futurists suggest that the first human missions to Mars may not land on the planet's surface but instead could build a colony on Phobos or Deimos. These plans involve speculation that these moon and other asteroids may have water in deep interior rocks that colonists could use.

Mars is a red desert planet, as shown in this true-color photograph by the Rover Opportunity. The rock outcrop is a meter-high crater wall. Dust suspended in the atmosphere colors the sky red-orange. Data from orbiters, landers, and rovers show that billions of years ago Mars was less of a desert and had water on its surface.

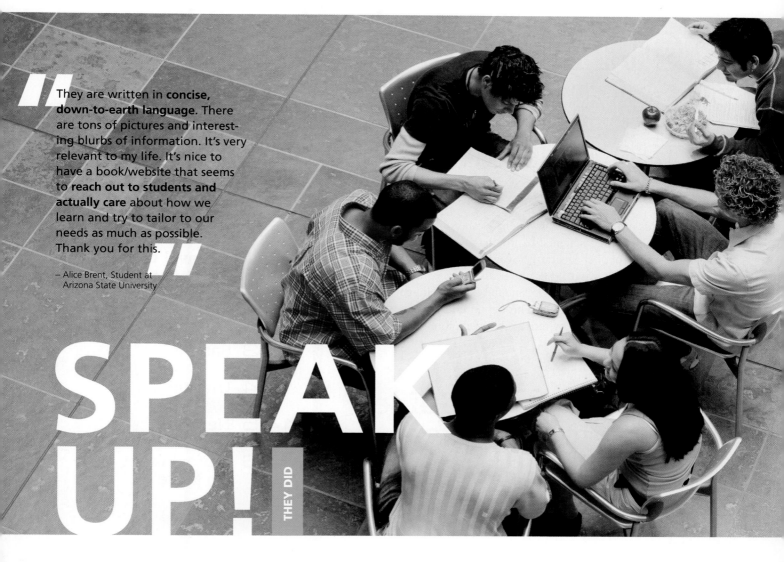

" They are written in **concise, down-to-earth language.** There are tons of pictures and interesting blurbs of information. It's very relevant to my life. It's nice to have a book/website that seems to **reach out to students and actually care** about how we learn and try to tailor to our needs as much as possible. Thank you for this. "

– Alice Brent, Student at Arizona State University

SPEAK UP!

THEY DID

ASTRO was built on a simple principle: to create a new teaching and learning solution that reflects the way today's faculty teach and the way you learn.

Through conversations, focus groups, surveys, and interviews, we collected data that drove the creation of the current version of **ASTRO** that you are using today. But it doesn't stop there – in order to make **ASTRO** an even better learning experience, we'd like you to SPEAK UP and tell us how **ASTRO** worked for you.

What did you like about it? What would you change? Are there additional ideas you have that would help us build a better product for next semester's astronomy students?

At **4ltrpress.cengage.com/astro** you'll find all of the resources you need to succeed in astronomy— **Glossary of Key Terms, animations, videos, flashcards,** and more!

Speak Up! Go to **4ltrpress.cengage.com/astro.**

Comparative
Planetology of the Jovian Planets

The worlds of the outer solar system can be studied from Earth, but much of what scientists know has been radioed back to Earth from robot spacecraft. Voyager 2 flew past each of the outer planets in the 1970s and 1980s, the Galileo spacecraft circled Jupiter dozens of times starting in 1995, and the Cassini/Huygens orbiter and probe arrived at Saturn in 2004. Throughout this chapter, you will find images and data returned by those robotic explorers.

14.1 A Travel Guide to the Outer Planets

You are about to visit four worlds that are truly unearthly. This travel guide will warn you about what to expect.

The Outer Planets

The outermost planets in our solar system are Jupiter, Saturn, Uranus, and Neptune. They are often called the "Jovian planets," meaning they resemble Jupiter. In fact, they have their own separate personalities. The illustration on page 290 compares these four worlds. Jupiter is the largest of the Jovian planets, over 11 times the diameter of Earth. Saturn is slightly smaller, and Uranus and Neptune are quite a bit smaller than Jupiter. Pluto, not included in the illustration, is smaller than Earth's moon but was considered a planet from the time of its discovery in 1930 until a decision by the International Astronomical Union (IAU) in 2006 that reclassified Pluto as a dwarf planet. You will learn about Pluto's characteristics, and reasons for the IAU decision, later in this chapter.

The other feature you will notice immediately is Saturn's rings. They are bright and beautiful and composed of billions of ice particles. Jupiter, Uranus, and Neptune also have rings, but they are not easily detected from Earth and are not visible in this figure. Nevertheless, as you visit these worlds you will be able to compare four different sets of planetary rings.

looking back

In the last two chapters, you watched our solar system form and then explored the Terrestrial planets. Along the way you learned some important principles of comparative planetology. Now it is time to explore the outer solar system, where the planets are bigger and the temperatures are colder. Comparative planetology will still be your basic strategy.

looking ahead

As you finish this chapter, you will have an astronomer's insight into your place in nature. You live on the surface of a planet. Are other planets inhabited? That is the subject of the next chapter.

false color image

Saturn

Jupiter, more than 11 times Earth's diameter, is the largest planet in our solar system.

The cloud belts and zones on Saturn are less distinct than those on Jupiter.

Uranus and Neptune are both about four times Earth's diameter.

Shadow of one of Jupiter's many moons

Earth is the largest of the Terrestrial worlds, but it is small compared with the Jovian planets.

Uranus and Neptune are green- and blue-colored because of small amounts of methane in their hydrogen-rich atmospheres.

Atmospheres and Interiors

The four Jovian worlds have hydrogen-rich atmospheres filled with clouds. The atmospheres of the Jovian planets are not very deep; for example, Jupiter's atmosphere makes up only about one percent of its radius. On Jupiter and Saturn, you can see that the clouds form stripes that circle each planet. You will find traces of these same types of features on Uranus and Neptune, but they are not very distinct.

Models based on observations indicate that below their atmospheres Jupiter and Saturn are mostly liquid, so the old-fashioned term for these planets, the *gas giants*, should probably be changed to the *liquid giants*. Uranus and Neptune are sometimes called the *ice giants* because they are rich in water in both solid and liquid forms. Near their centers the Jovian planets have cores of dense material with the composition of rock and metal. None of the Jovian worlds has a definite solid surface on which you could walk.

Recall from Chapter 12 that the Jovian planets have low density because they formed in the outer solar nebula where water vapor could freeze to form ice particles. The ice accumulated into proto-planets with density lower than the rocky Terrestrial planets and asteroids. Once the Jovian planets grew massive enough, they could draw in even lower-density hydrogen and helium gas directly from the nebula by gravitational collapse.

Satellite Systems

You can't really land your spaceship on the Jovian planets, but you might be able to land on their moons. All of the outer solar system planets have extensive moon systems. In many cases the moons interact gravitationally, mutually adjusting their orbits and also affecting the planetary ring systems. Some of the moons are geologically active now, while others show signs of past activity. Of course, geological activity requires heat flow from the interior, so you might wonder what could be heating the insides of these small objects.

14.2 Jupiter

Jupiter is named for the Roman king of the gods. It can be very bright in the night sky, and its cloud belts and four largest moons can be seen through even a small telescope. Jupiter is the largest and most massive of the Jovian planets, containing 71 percent of all the planetary matter in the entire solar system. Just as you used Earth, the largest of the Terrestrial planets, as the basis for comparison with the others, you can examine Jupiter in detail as a standard in your comparative study of the other Jovian planets.

The Interior

Jupiter is only 1.3 times denser than water. For comparison, Earth is more than 5.5 times denser than water. This gives astronomers a clue about the average composition of the planet's interior. Jupiter's shape also gives information about its interior. Jupiter and the other Jovian planets are all slightly flattened. A world with a large rocky core and mantle would not be flattened much by rotation, but an all-liquid planet would flatten significantly. Thus Jupiter's **oblateness**, the fraction by which its equatorial diameter exceeds its polar diameter, combined with its average density, helps astronomers model the interior. As with the other planets, you can refer to the Celestial Profile Cards for more data about Jupiter.

Models indicate that the interior of Jupiter is mostly liquid hydrogen. However, if you jumped into Jupiter carrying a rubber raft expecting an ocean, you would be disappointed. The base of the atmosphere is so hot and the pressure is so high that there is no sudden boundary between liquid and gas. As you fell deeper and deeper through the atmosphere, you would find the gas density increased around you until you were sinking through a liquid, but you would never splash into a distinct liquid surface.

Under very high pressure, liquid hydrogen becomes **liquid metallic hydrogen**—a material that is a very good conductor of electricity. Most of Jupiter's interior is composed of this material. That large mass of conducting liquid, stirred by convection currents and spun by the planet's rapid rotation, drives the dynamo effect and generates a powerful magnetic field. Jupiter's field is over 10 times stronger than Earth's. A planet's magnetic field deflects the solar wind and dominates a volume of space around the planet called the **magnetosphere**.

The strong magnetic field around Jupiter traps charged particles from the solar wind in radiation belts a billion times more intense than the Van Allen belts that surround Earth. The spacecraft that have flown through these regions received over 1,000 times the radiation that would have been lethal for a human.

At Jupiter's center, a so-called rocky core contains heavier elements, such as iron, nickel, silicon, and so on. With a temperature four times hotter than the surface of the sun and a pressure 50 million times Earth's air pressure at sea level, this material is unlike any rock on Earth. The term *rocky core* refers to the chemical composition, not to the properties of the material.

Careful infrared measurements of the heat flowing out of Jupiter reveal that the planet emits about twice as much energy as it absorbs from the sun. This energy is understood to be heat left over from the formation of the planet. In Chapter 12 you saw that Jupiter should have grown very hot when it formed, and some of that heat remains in its interior.

Jupiter's Complex Atmosphere

Study "Jupiter's Magnetosphere and Complex Atmosphere" on pages 292–293 and notice three important ideas:

1. Jupiter's extensive magnetosphere is responsible for auroras around the magnetic poles. Jupiter's rings, discovered in 1979 by the Voyager I space probe, are close to the planet.

2. The pattern of colored cloud bands circling the planet like stripes on a child's ball is called **belt–zone circulation**. This pattern is related to the high- and low-pressure areas found in Earth's atmosphere.

3. The positions of the cloud layers are at certain temperatures within the atmosphere where ammonia (NH_3), ammonium hydrosulfide (NH_4SH), and water (H_2O) can condense.

Oblateness: The flattening of a spherical body, usually caused by rotation.

Liquid metallic hydrogen: A form of liquid hydrogen that is a good electrical conductor, inferred to exist in the interiors of Jupiter and Saturn.

Magnetosphere: The volume of space around a planet within which the motion of charged particles is dominated by the planetary magnetic field rather than the solar wind.

Belt–zone circulation: The atmospheric circulation typical of Jovian planets in which dark belts and bright zones encircle the planet parallel to its equator.

Jupiter's Magnetosphere and Complex Atmosphere

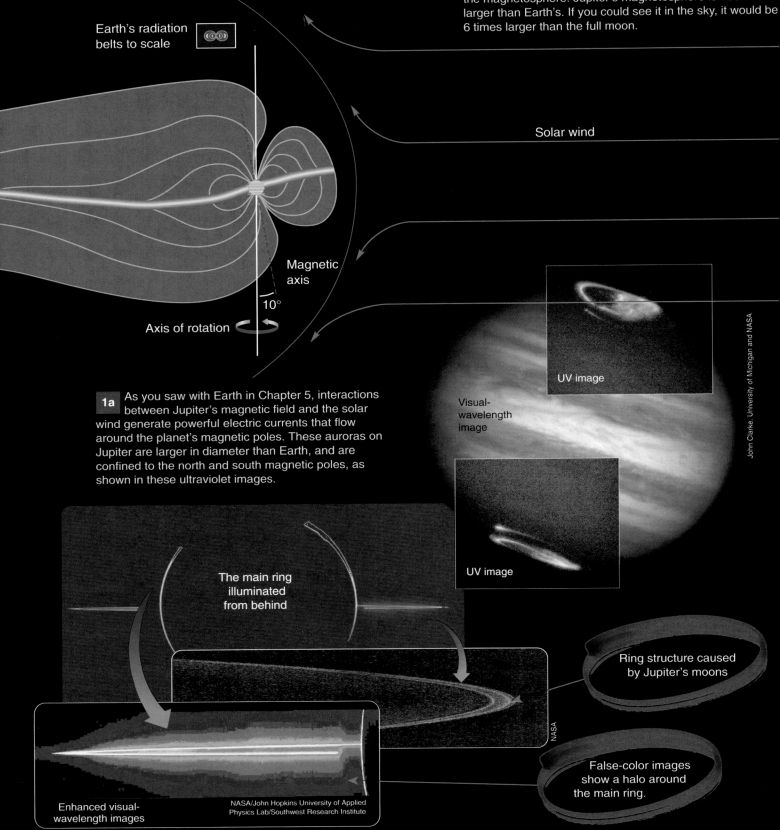

1 A planet's magnetic field deflects the solar wind and dominates a volume of space around the planet called the magnetosphere. Jupiter's magnetosphere is 100 times larger than Earth's. If you could see it in the sky, it would be 6 times larger than the full moon.

Earth's radiation belts to scale

Solar wind

Magnetic axis

10°

Axis of rotation

1a As you saw with Earth in Chapter 5, interactions between Jupiter's magnetic field and the solar wind generate powerful electric currents that flow around the planet's magnetic poles. These auroras on Jupiter are larger in diameter than Earth, and are confined to the north and south magnetic poles, as shown in these ultraviolet images.

UV image

Visual-wavelength image

UV image

John Clarke, University of Michigan and NASA

The main ring illuminated from behind

Ring structure caused by Jupiter's moons

NASA

False-color images show a halo around the main ring.

Enhanced visual-wavelength images

NASA/John Hopkins University of Applied Physics Lab/Southwest Research Institute

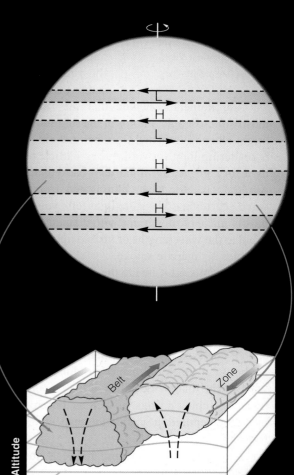

2 The visible clouds are not at the top of the atmosphere; an upper atmosphere of clear hydrogen and helium extends far above the cloud tops. Jupiter's atmosphere makes up only one percent of its radius. Major spots on Jupiter, like the Great Red Spot, are embedded circulating storms. They can remain stable for decades or even centuries.

Great Red Spot

2a Cloud belts and zones circle Jupiter like stripes on a ball. This form of atmospheric circulation is belt–zone circulation. The poles and equator on Jupiter are about the same temperature, perhaps because of heat rising from the interior. Because of that, and also because of Jupiter's rapid rotation, instead of wave-shaped winds as on Earth, Jupiter's high- and low-pressure areas are stretched into belts and zones parallel to the equator. Zones are brighter than belts because rising gas forms clouds high in the atmosphere where sunlight is stronger and reflected back to outside viewers.

North   Equator

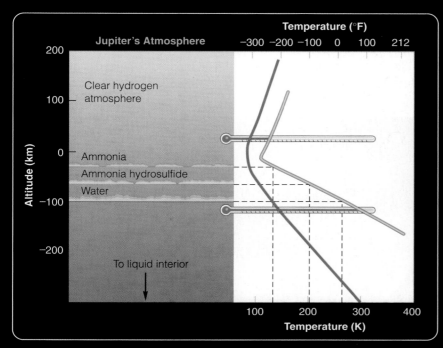

3 Jupiter's cloud layers lie at certain temperatures within the atmosphere where ammonia (NH_3), ammonium hydrosulfide (NH_4SH), and water (H_2O) can condense into droplets. If you could put thermometers into the different levels of the atmosphere, you would discover that the temperatures rise as you descend beneath the uppermost clouds, represented for Jupiter by the solid orange line. Because Saturn is further from the sun, its temperature is colder, shown by the green line. The cloud layers on Saturn form at the same temperature as the cloud layers on Jupiter, but they are deeper in Saturn's hazy atmosphere.

Jupiter's Ring

Astronomers have known for centuries that Saturn has rings, but Jupiter's ring was not discovered until 1979, when the Voyager 1 spacecraft sent back photos. Less than 1 percent as bright as Saturn's icy rings, Jupiter's ring is very dark and reddish, indicating that the ring is rocky material rather than icy.

Astronomers conclude that the ring particles are mostly microscopic. Photos of the ring show that it is very bright when illuminated from behind—in other words, it is scattering light forward. Large particles do not scatter light forward: A ring filled with basketball-size particles would look dark when illuminated from behind. **Forward scattering** of visible light tells you that the ring is mostly made of tiny grains with diameters approximately equal to the wavelengths of visible light, about the size of particles in cigarette smoke.

The rings orbit inside the **Roche limit**, the distance from a planet within which a moon cannot hold itself together by its own gravity. If a moon comes inside the Roche limit, tidal forces overcome the moon's gravity and pull the moon apart. Also, raw material for a moon cannot coalesce inside the Roche limit. The Roche limit is about 2.4 times the planet's radius, depending somewhat on the relative densities of the planet and the moon material. Jupiter's rings, as well as those of Saturn, Uranus, and Neptune, lie inside the respective Roche limits for each planet.

Now you can understand Jupiter's dusty rings. If a dust speck gets knocked loose from a larger rock inside the Roche limit, the rock's gravity cannot hold the dust speck. For that same reason, the billions of dust specks in the ring can't pull themselves together to make a moon because of tidal forces inside the Roche limit.

You can be sure that Jupiter's ring particles are not old. The pressure of sunlight and the planet's powerful magnetic field alter the orbits of the particles. Images show faint ring material extending down toward the cloud tops; this is evidently dust grains spiraling into the planet. Dust is also destroyed by the intense radiation around Jupiter that grinds the dust specks down to nothing in a century or so. The rings you see today therefore can't be material left over from the formation of Jupiter: The rings of Jupiter must be continuously resupplied with new dust. Observations made by the Galileo spacecraft provide evidence that the source of ring material is micrometeorites eroding small moons orbiting near or within the rings.

The rings around Saturn, Uranus, and Neptune are also known to be short lived, and they also must be resupplied by new material, probably eroded from nearby moons. Besides supplying the Jovian rings with particles, moons confine the rings, keep them from spreading outward, and alter their shapes. You will explore these processes in detail when you study the rings of the other planets later in this chapter.

Jupiter's Family of Moons

Jupiter has four large moons and at least 60 smaller moons. Larger telescopes and modern techniques are rapidly finding more small moons orbiting each of the Jovian planets. Most of the small moons are probably captured asteroids. In contrast, the four largest moons (Figure 14.1), called the Galilean moons after their discoverer, Galileo, are clearly related to each other and probably formed with Jupiter.

The outermost Galilean moons, Ganymede and Callisto, are about the size of Mercury, one and a half times

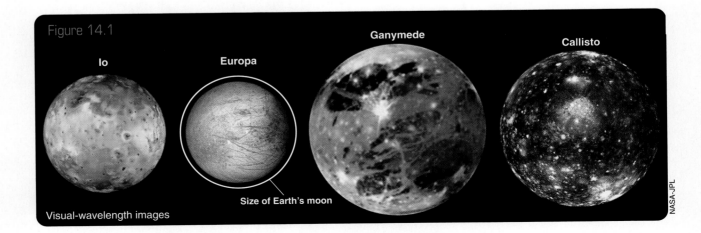

Figure 14.1

Io   Europa   Ganymede   Callisto

Size of Earth's moon

Visual-wavelength images

NASA-JPL

Figure 14.2

Like icebergs on an arctic ocean, blocks of crust on Europa appear to have floated apart. The blue icy surface is stained brown by mineral-rich water venting from below the crust. White areas are ejecta from the impact that formed nearby crater Pwyll.

© NASA

the size of Earth's moon. In fact, Ganymede is the largest moon in the solar system. Ganymede and Callisto have low densities of only 1.9 and 1.8 g/cm³, respectively, meaning they must consist roughly of half rock and half ice. Observations of their gravitational fields by the Galileo spacecraft reveal that both moons have rocky and metallic cores plus lower-density icy exteriors, so they have both differentiated. Both moons interact with Jupiter's magnetic field in a way that shows they probably have mineral-rich layers of liquid water 100 km or more below their icy crusts.

Callisto's surface and most of Ganymede's surface appear old because they are heavily cratered and very dark. The continuous blast of micrometeorites evaporates surface ice, leaving behind embedded minerals to form a dark skin like the grimy crust on an old snowbank, so surfaces get darker with age. More recent impacts dig up cleaner ice and leave bright craters, as you can see on Callisto in Figure 14.1. Ganymede has some younger, brighter grooved terrain believed to be systems of faults in the brittle crust. Some sets of grooves overlap other sets of grooves, suggesting extended episodes of geological activity.

The density of the next moon inward, Europa, is 3 g/cm³, high enough to mean that the moon is mostly rock with a thin icy crust. The visible surface is very clean ice, contains very few craters, has long cracks in the icy crust, and has complicated terrain that resembles blocks of ice in Earth's Arctic Ocean (Figure 14.2). The lack of craters tells you that Europa is an active world where craters are quickly erased. The pattern of folds on its surface suggests that the icy crust breaks as the moon is flexed by tides (see Chapter 3). Europa's gravitational influence on the Galileo spacecraft reveals that a liquid-water ocean perhaps 200 km deep lies below the 10- to 100-km-thick crust.

Images from spacecraft reveal that Io, the innermost of the four Galilean moons, has over 150 volcanic vents on its surface (Figure 14.3 on the next page). The active volcanoes throw sulfur-rich gas and ash high above the surface; the ash falls back to bury the surface at a rate of a few millimeters a year. This explains why you see no impact craters on Io—they are covered up as fast as they form. Io's density is 3.6 g/cm³, showing that it is not ice but rather rock and metal. Its gravitational influence on the passing Galileo spacecraft revealed that it is differentiated into a large metallic core, a rocky mantle, and a low-density crust.

The activity you see in the Galilean moons must be driven by energy flowing outward, yet these objects are too small to have remained hot from the time of their formation. Io's volcanism seems to be driven by **tidal heating**. Io follows a slightly eccentric orbit caused by its interactions with the other moons. Jupiter's gravitational field flexes Io with tides, and the resulting friction heats its interior. That heat flowing outward causes the volcanism. Europa is not as active as Io, but it too must have a heat source, presumably tidal heating. Ganymede is no longer active, but when it was younger it must have had internal heat to break the crust and produce the grooved terrain.

A History of Jupiter

Can you put all of the evidence together and tell the story of Jupiter? Creating such a logical argument from evidence and hypotheses is the ultimate goal of planetary astronomy.

Jupiter formed far enough from the sun to incorporate large numbers of icy planetesimals, and it must have grown rapidly. Once it became about 10 to 15 times more massive than Earth, it could begin to grow further by gravitational collapse

Tidal heating: The heating of a planet or satellite because of friction caused by tides.

(see Chapter 12), capturing gas directly from the solar nebula. Thus, it grew rich in hydrogen and helium from the solar nebula. Its present composition resembles the composition of the solar nebula and is also quite sunlike. Jupiter's gravity is strong enough to hold on to all of its gases, even hydrogen.

The large family of moons may be mostly captured asteroids, and Jupiter may still encounter a wandering asteroid or comet now and then. Some of these are deflected, some captured into orbit, and some, like comet fragments in 1994 (Figure 12.10) and an unidentified object in 2009, actually strike the planet. Dust blasted off of the inner moons by micrometeorites settles into the equatorial plane to form Jupiter's rings.

The four Galilean moons are large and seem to have formed like a mini-solar system in a disk of gas and dust around the forming planet. The innermost, Io, is densest, and the densities of the others decrease as you move away from Jupiter, similar to the way the densities of the planets decrease with distance from the sun. Perhaps the inner moons incorporated less ice because they formed closer to the heat of the growing planet. You can recognize that tidal heating also has been important, and the intense warming of the inner moons could have driven

off much of their ices. Thus two processes together are probably responsible for the differences in compositions of the Galilean moons.

14.3 Saturn

The Roman god Saturn, protector of the sowing of seed, was celebrated in a weeklong Saturnalia at the time of the winter solstice in late December. Early Christians took over the holiday to celebrate Christmas.

Saturn is most famous for its beautiful rings, which are easily visible through the telescopes of modern amateur astronomers. Large Earth-based telescopes have explored the planet's atmosphere, rings, and moons. The two Voyager spacecraft flew past Saturn in 1979, and the Cassini spacecraft went into orbit around Saturn in 2004 on an extended exploration of the planet, its rings, and its moons. Don't forget to look at the Celestial Profile Card for specific data on Saturn.

Figure 14.3

These color images of volcanic features on Io were produced by combining visual and near-infrared images and then digitally enhancing the color. To human eyes, most of Io would look pale yellow and light orange.

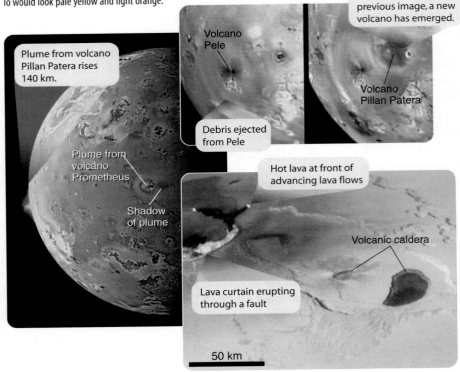

Plume from volcano Pillan Patera rises 140 km.

Plume from volcano Prometheus

Shadow of plume

Volcano Pele

Debris ejected from Pele

Five months after the previous image, a new volcano has emerged.

Volcano Pillan Patera

Hot lava at front of advancing lava flows

Volcanic caldera

Lava curtain erupting through a fault

50 km

© NASA

Saturn the Planet

As you can see in the opening illustration, Saturn has only faint belt–zone circulation, but Voyager, Hubble Space Telescope, and Cassini images show that belts and zones are present and that the associated winds blow up to three times faster than on Jupiter. Belts and zones on Saturn are less visible than on Jupiter because they occur deeper in the cold atmosphere, below a layer of methane haze.

Saturn is less dense than water (it would float!), suggesting that it is, like Jupiter, rich in hydrogen and helium. As photographs show, Saturn is the most oblate of the planets, and that evidence tells you that its interior is mostly liquid with a small core of heavy elements. Because its

who pays for science?

Science is an expensive enterprise, and that raises the question of payment. Some science has direct technological applications with immediate practical use, so the business community funds much of this type of research. For example, auto manufacturers need inexpensive, durable, quick-drying paint for their cars, and they find it worth the cost to hire chemists to study the way paint dries. Many industries have large research budgets, and some industries, such as pharmaceutical manufacturers, depend exclusively on scientific research to discover and develop new products.

If a field of research has immediate potential to help society, it is likely that government will supply funds. Research on public health is funded mostly by government institutions such as the National Institutes of Health and the National Science Foundation. The practical benefit of finding new ways to prevent disease, for example, is well worth the tax dollars.

Basic science, however, has no immediate practical use. That doesn't mean it is useless, but it does mean that the practical-minded stockholders of a company will not approve major investments in such research. Digging up dinosaur bones, for instance, is very poorly funded because no industry can make a profit from the discovery of a new dinosaur. Astronomy is another field of science that has few direct applications, and thus very little astronomical research is funded by industry.

The value of basic research is twofold. Discoveries that have no known practical use today may be critically important years from now. Thus, society needs to continue basic research to protect its own future. But basic research, such as studying Jupiter's atmosphere or Saturn's rings, is also of cultural value because it tells us something about Earth and indirectly about what we are. Each of us benefits in intangible ways from such research, and thus society needs to fund basic research for the same reason it funds art galleries and national parks—to make our lives richer and more fulfilling.

Because there is no immediate financial return from this kind of research, it falls to government institutions and private foundations to pay the bill. The Keck Foundation has built two giant telescopes with no expectation of financial return, and the National Science Foundation has funded thousands of astronomy research projects for the benefit of society. Debates rage as to how much money is enough and how much is too much, but, ultimately, funding basic scientific research is a public responsibility that society must balance against other needs. There isn't anyone else to pick up the tab.

Figure 14.4

As the Huygens probe descended by parachute through Titan's smoggy atmosphere, it photographed the surface from an altitude of 8 km (5 mi). Although no liquid was present, dark drainage channels lead into the lowlands. Radar images reveal lakes of liquid methane and ethane around the poles. Once the Huygens probe landed on the surface, it radioed back photos showing a level plain and chunks of ice rounded by a moving liquid.

Visual-wavelength images

Image recorded from 8 km above surface

Drainage channels were cut by flowing liquid.

Lakes of liquid methane and ethane look dark because they do not reflect radar waves.

At visual wavelengths, Titan's heavy atmosphere hides its surface.

False color Radar map

Icy grapefruit-size "rocks" on Titan are bathed in orange light from its hazy atmosphere.

Visual

© ESA/NASA/JPL/USGS/University of Arizona

internal pressure is lower, Saturn has less liquid metallic hydrogen than Jupiter. Perhaps that is why Saturn's magnetic field is 20 times weaker than Jupiter's. Like Jupiter, Saturn radiates more energy than it receives from the sun, and models predict that it has a very hot interior.

Saturn's Rings

Study "The Ice Rings of Saturn" on pages 300–301 and notice three things:

1. The rings are made up of billions of ice particles, each in its own orbit around the planet. The ring particles you observe now can't be as old as Saturn. The rings must be replenished now and then by impacts on Saturn's moons or other processes. The same is true of the rings around the other Jovian planets.

2. The gravitational effects of small moons can confine some rings in narrow strands or keep the edges of rings sharp. Moons can also produce waves in the rings that are visible as tightly wound ringlets.

3. The ring particles are confined in a thin layer in Saturn's equatorial plane, spread among small moons and confined by gravitational interactions with larger moons. The rings of Saturn, and the rings of the other Jovian worlds, are created by and controlled by the planet's moons. Without the moons, there would be no rings.

Saturn's Family of Moons

Saturn has more than 60 known moons, many of which are small and all of which contain mixtures of ice and rock. Many are probably captured objects.

The largest of Saturn's moons, called Titan, is a bit larger than the planet Mercury. Its density suggests that it must contain a rocky core under a thick mantle of ices. Titan is so cold that its gas molecules do not travel fast enough to escape, so it has an atmosphere composed mostly of nitrogen with traces of argon and methane. Sunlight converts some of the methane into complex carbon-rich molecules that collect into small particles filling the atmosphere with orange smog, visible in Figure 14.4.

These particles are understood to slowly settle downward to the surface in the form of a dark organic goo, meaning it is composed of carbon-rich molecules.

Titan's surface is mainly composed of ices of water and methane at −180°C (−290°F). The Cassini spacecraft dropped the Huygens probe into the atmosphere of Titan, and it photographed dark drainage channels suggesting that liquid methane falls as rain, washes the dark goo off the higher terrain, and drains into the lowlands. Such methane downpours may be rare, however. No direct evidence of liquid methane was detected as the probe descended, but later radar images made by the Cassini orbiter have detected what appear to be lakes presumably containing liquid methane. Infrared images suggest the presence of methane volcanoes that replenish the methane in the atmosphere, so Titan must have some internal heat source to power the activity.

Most of the remaining moons of Saturn are small and icy, have no atmospheres, and are heavily cratered. Most have dark, ancient surfaces. The moon Enceladus, however, shows signs of recent geological activity. Some parts of its surface contain 1,000 times fewer craters than other regions, and infrared observations show that its south polar region is unusually warm and venting water and ice geysers (Figure 14.5). Evidently, a reservoir of liquid water lies only tens of meters below the surface. At some point in its history, this moon must have been caught in a resonance with another moon and was warmed by tidal heating. Enceladus appears to maintain the faint E ring that extends far beyond the visible rings. In 2009 astronomers detected infrared radiation from a dark ring 13 million km (8 million mi) in radius, beyond the orbits of most of Saturn's moons.

14.4 Uranus

Now that you are familiar with the gas giants in our solar system, you will be able to appreciate how weird the ice giants, Uranus and Neptune, are. Uranus, especially, seems to have forgotten how to behave like a planet.

Uranus was discovered in 1781 by the scientist William Herschel, a German expatriate living in England. He named it *Georgium Sidus*, George's Star, after the English King George III. European astronomers, especially the French, refused to accept a planet named after an English king. They called it Herschel. Years later, German astronomer J. E. Bode suggested Uranus, the oldest of the Greek gods.

Planet Uranus

Uranus is only one-third the diameter of Jupiter and only one-twentieth as massive, and, being four times

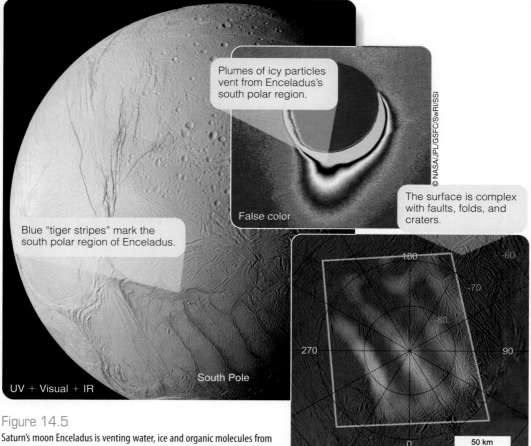

Blue "tiger stripes" mark the south polar region of Enceladus.

Plumes of icy particles vent from Enceladus's south polar region.

False color

The surface is complex with faults, folds, and craters.

UV + Visual + IR

South Pole

© NASA/JPL/GSFC/SwRI/SSI

IR image

50 km

Figure 14.5

Saturn's moon Enceladus is venting water, ice and organic molecules from geysers near its south pole. A thermal infrared image reveals internal heat leaking to space from the "tiger stripe" cracks where the geysers are located.

The Ice Rings of Saturn

1 The brilliant rings of Saturn are made up of billions of ice particles ranging from microscopic specks to chunks bigger than a house. Each particle orbits Saturn in its own circular orbit. Much of what astronomers know about the rings was learned when the Voyager 1 spacecraft flew past Saturn in 1980, followed by the Voyager 2 spacecraft in 1981. The Cassini Spacecraft reached orbit around Saturn in 2004. From Earth, astronomers see three rings labeled A, B, and C. Voyager and Cassini images reveal over a thousand ringlets within the rings.

Saturn's rings can't be leftover material from the formation of Saturn. The rings are made of ice particles, and the planet would have been so hot when it formed that it would have vaporized and driven away any icy material. Rather, the rings must be debris from collisions between passing comets, or other objects, and Saturn's icy moons. Such impacts should occur every 100 million years or so, and they would scatter ice throughout Saturn's system of moons. The ice would quickly settle into the equatorial plane, and some would become trapped in rings. Although the ice may waste away due to meteorite impacts and damage from radiation in Saturn's magnetosphere, new impacts could replenish the rings with fresh ice. The bright, beautiful rings you see today may be only a temporary enhancement caused by an impact that occurred since the extinction of the dinosaurs.

Encke Gap

Cassini Division

A ring

B ring

C ring

Earth to scale

Visual-wavelength image

As in the case of Jupiter's ring, Saturn's rings lie inside the planet's Roché limit where the ring particles cannot pull themselves together to form a moon.

Because it is so dark, the C ring was once called the crepe ring.

1a An astronaut could swim through the rings. Although the particles orbit Saturn at high velocity, all particles at the same distance from the planet orbit at about the same speed, so they collide gently at low velocities. If you could visit the rings, you could push your way from one icy particle to the next. This artwork is based on a model of particle sizes in the A ring.

The C ring contains boulder-size chunks of ice, whereas most particles in the A and B rings are more like golf balls, down to dust-size ice crystals. Further, C ring particles are less than half as bright as particles in the A and B rings. Cassini observations show that the C ring particles contain less ice and more minerals.

NASA

2 Because of collisions among ring particles, planetary rings should spread outward. The sharp outer edge of the A ring and the narrow F ring are confined by **shepherd satellites** that gravitationally usher straying particles back into the rings.

Some gaps in the rings, such as the Cassini Division, are caused by resonances with moons. A particle in the Cassini Division orbits Saturn twice for each orbit of the moon Mimas. On every other orbit, the particle feels a gravitational tug from Mimas. These tugs always occur at the same places in the orbit and force the orbit to become slightly elliptical. Such an orbit crosses the orbits of other particles, which results in collisions, and that removes the particle from the gap.

This image was recorded by the Cassini spacecraft looking up at the rings as they were illuminated by sunlight from below. Saturn's shadow falls across the upper side of the rings.

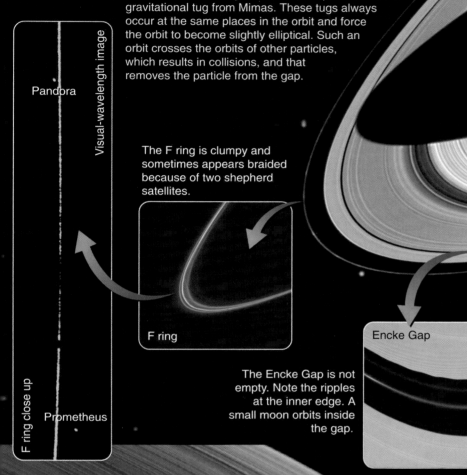

Visual-wavelength image

Pandora

F ring close up

Prometheus

The F ring is clumpy and sometimes appears braided because of two shepherd satellites.

F ring

The Encke Gap is not empty. Note the ripples at the inner edge. A small moon orbits inside the gap.

Encke Gap

Visual-wavelength images

Waves in the A ring

Saturn does not have enough moons to produce all of its ringlets by resonances. Many are produced by tightly wound waves, much like the spiral arms found in disk galaxies.

Cassini Division

A ring

Encke Gap

This combination of UV images has been given false color to show the ratio of mineral material to pure ice. Blue regions such as the A ring are the purest ice, and red regions such as the Cassini Division are the dirtiest ice. How the particles become sorted by composition is unknown.

Ultraviolet image

3 How do moons happen to be at just the right places to confine the rings? That puts the cosmic cart before the horse. The ring particles get caught in the most stable orbits among Saturn's innermost moons. The rings push against the inner moons, but those moons are locked in place by resonances with larger, outer moons. Without the moons, the rings would spread and dissipate.

Saturn's rings are a very thin layer of particles and nearly vanish when the rings turn edge-on to Earth. Although ripples in the rings caused by waves may be hundreds of meters high, the sheet of particles may be only about ten meters thick.

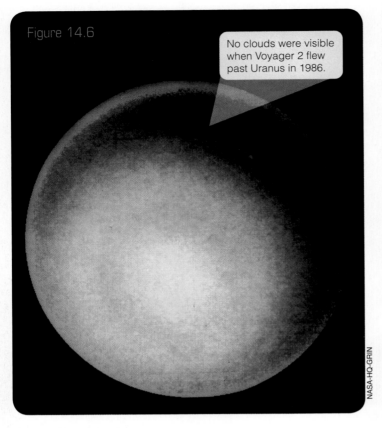

Figure 14.6

No clouds were visible when Voyager 2 flew past Uranus in 1986.

NASA-HQ-GRIN

ter 2. Uranus's odd rotation may have been produced when it was struck by a very large planetesimal late in its formation, or by tidal interactions with the other giant planets as it migrated outward early in the history of the solar system (see Chapter 12).

Voyager 2 photos, like the one in Figure 14.6, show a nearly featureless ball. The atmosphere is mostly hydrogen and helium, but traces of methane absorb red light and thus make the atmosphere look blue or teal, depending on the image. There is no belt–zone circulation visible in the Voyager photographs, although extreme computer enhancement revealed a few clouds and bands around the south pole. In the decades since Voyager 2 flew past Uranus, spring has come to the northern hemisphere of Uranus and autumn to the southern hemisphere. Images made by the Hubble Space Telescope as well as the most powerful Earth-based telescopes reveal changing clouds and cloud bands (Figure 14.7).

Infrared measurements show that Uranus is radiating about the same amount of energy that it receives from the sun, meaning it has much less heat flowing out of its interior than Jupiter or Saturn (or Neptune). This may account for its limited atmospheric activity. Astronomers are not sure why Uranus differs in this respect from the other Jovian worlds.

farther from the sun, its atmosphere is almost 100°C colder than Jupiter's.

Uranus never grew massive enough to capture large amounts of gas from the nebula as Jupiter and Saturn did, so it has much less hydrogen and helium. Its internal pressure is enough lower than Jupiter's that it should not contain any liquid metallic hydrogen. Models of Uranus based in part on its density and oblateness suggest that it has a small core of heavy elements and a deep mantle of partly solid water. Although referred to as ice, this material would not be anything like ice on Earth, given the temperatures and pressures inside Uranus. The mantle also contains rocky-composition material and dissolved ammonia and methane. Circulation in this electrically conducting mantle may generate the planet's peculiar magnetic field, which is highly inclined to its axis of rotation. Above this mantle lies a deep hydrogen and helium atmosphere.

Uranus rotates on its side, with its equator inclined about 98° to its orbit. With an orbital period of 84 years, each of the four seasons lasts 21 years, and the winter–summer contrast is extreme. During a season when one of its poles is pointed nearly at the sun (a solstice), inhabitants of Uranus would never see the sun rise or set. Compare this with seasons on Earth discussed in Chap-

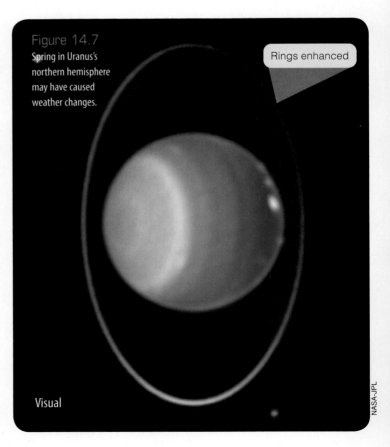

Figure 14.7

Spring in Uranus's northern hemisphere may have caused weather changes.

Rings enhanced

Visual

NASA-JPL

The Uranian Moons

Until recently, astronomers could see only five moons orbiting Uranus. Voyager 2 discovered 10 more small moons in 1986, and yet more have been found in images recorded by new, giant telescopes on Earth.

The five major moons of Uranus are smaller than Earth's moon and have old, dark, cratered surfaces. A few have deep cracks, produced, perhaps, when the interior froze and expanded. In some cases, liquid water "lava" appears to have erupted and smoothed over some regions. Miranda, the innermost moon, is only 14 percent the diameter of Earth's moon, but its surface is marked by grooves called **ovoids** (Figure 14.8). These may have been caused by internal heat driving convection in the icy mantle. By counting craters on the ovoids, astronomers conclude that the entire surface is old, and the moon is no longer active.

The Uranian Rings

The rings of Uranus are dark and faint, contain little dust, are confined by shepherd satellites, and must be continuously resupplied with material from the moons. They are not easily visible from Earth: The first hint that Uranus had rings came from **occultations**, the passage of the planet in front of a star. Most of what astronomers know about these rings comes from the observations of the Voyager 2 spacecraft. Their composition appears to be water ice mixed with methane that has been darkened by exposure to radiation.

In 2006, astronomers found two new, very faint rings orbiting far outside the previously known rings of Uranus. The newly discovered satellite Mab appears to be the source of particles for the larger ring, and the smaller of the new rings is confined between the orbits of the moons Portia and Rosalind. Note that the International Astronomical Union (IAU) has declared that the newly discovered moons of Uranus are to be named after characters in Shakespeare's plays.

14.5 Neptune

A British astronomer and a French astronomer independently calculated the existence and location of Neptune from irregularities in the motion of Uranus. British observers were too slow to respond; Neptune was discovered in 1846, and the French astronomer got the credit. Neptune looks like a tiny blue dot with no visible cloud features. Because of its blue color, astronomers named it after the god of the sea. In 1989, Voyager 2 flew past and revealed some of Neptune's secrets.

Planet Neptune

Almost exactly the same size as Uranus, Neptune is calculated to have a similar interior: a small core of heavy elements lies within a slushy mantle of water, ices, and minerals (rock) below a hydrogen-rich atmosphere. Yet Neptune looks quite different; it is dramatically blue and has active cloud formations. Neptune has a dark blue tint because its atmosphere contains one and a half times more methane than Uranus. Methane absorbs red photons better than blue and scatters blue photons better than red, giving Neptune a blue color and Uranus a green-blue color.

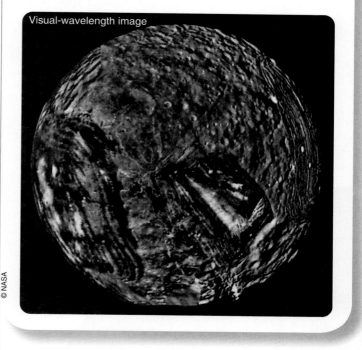

Figure 14.8

Geological activity on Uranus's moon Miranda. The face of Miranda is marked by ovoids, which are believed to have formed when internal heating caused slow convection in the ice of the moon's mantle. Note the 5-km-high cliff at the lower right edge of the moon.

Visual-wavelength image

© NASA

Ovoid: The oval features found on Miranda, a satellite of Uranus.

Occultation: The passage of a larger body in front of a smaller body.

Atmospheric circulation on Neptune is much more dramatic than on Uranus. When Voyager 2 flew by Neptune in 1989, the largest feature was the Great Dark Spot, which you can see in Figure 14.9. Roughly the size of Earth, the spot seemed to be an atmospheric circulation similar to Jupiter's Great Red Spot. Smaller spots were visible in Neptune's atmosphere, and photos showed they were circulating like hurricanes. More recently, the Hubble Space Telescope has photographed Neptune and found that the Great Dark Spot is gone, and new cloud formations have appeared. Evidently, the weather on Neptune is surprisingly changeable.

The atmospheric activity on Neptune is apparently driven by heat flowing from the interior plus some contribution by dim light from the sun 30 AU away. Neptune may have more atmospheric activity than Uranus because it has more heat flowing out of its interior, for reasons that are unclear.

Like Uranus, Neptune has a highly inclined magnetic field that must be linked to circulation in the interior. In both cases, astronomers suspect that ammonia dissolved in the liquid water mantle makes the mantle a good electrical conductor and that convection in the water, coupled with the rotation of the planet, drives the dynamo effect and generates the magnetic field.

The Neptunian Moons

Neptune has two moons that were discovered from Earth before Voyager 2 flew past in 1989. The passing spacecraft discovered six more very small moons. Since then a few more small moons have been found by astronomers using large Earth-based telescopes.

The two largest moons have peculiar orbits. Nereid, about one-tenth the size of Earth's moon, follows a large, elliptical orbit, taking nearly an Earth year to circle Neptune once. Triton, nearly 80 percent the size of Earth's moon, orbits Neptune backward—clockwise as seen from the north. These odd orbits suggest that the system was disturbed long ago in an interaction with some other body, such as a massive planetesimal.

Triton has an atmosphere of nitrogen and methane about 10^5 times less dense than Earth's, and a temperature of 37 K (–393°F). A significant part of Triton is ice, and deposits of nitrogen frost are visible at the southern pole (Figure 14.10). Many features on Triton suggest it has had an active past. It has few craters on its surface, but it does have long faults that appear to have formed when the icy crust broke, plus large basins that seem to have been flooded repeatedly by liquids from the interior. Even more interesting are the dark smudges visible in the southern polar cap that are interpreted as sunlight-darkened deposits of methane erupted out of liquid nitrogen geysers.

The Neptunian Rings

Neptune's rings are faint and very hard to detect from Earth, but they illustrate some interesting processes of comparative planetology (Figure 14.11). Neptune's rings are similar to those of Uranus but contain more small dust particles. One of Neptune's moons is producing short arcs in the outermost ring. Neptune's ring system, like the others, is apparently resupplied by impacts on moons scattering debris that fall into the most stable places among the orbits of the moons.

Figure 14.9

Neptune's equator is inclined almost 29 degrees in its orbit, and it experiences seasons that each last about 40 years. Since 1989, spring has come to the southern hemisphere, and the weather has changed, which is surprising because sunlight on Neptune is 900 times dimmer than on Earth.

Great Dark Spot

© NASA-JPL

Figure 14.10

Visible-wavelength image of Triton. Triton's southern polar cap is formed of nitrogen frost.

© NASA-JPL

14.6 Pluto: Planet No More

Out on the edge of the solar system orbits a family of small, icy worlds. Pluto was the first to be discovered, in 1930, but modern telescopes have found more.

You may have learned in school that there are nine planets in our solar system, but in 2006 the International Astronomical Union voted to remove Pluto from the list of planets. Pluto is a very small, icy world: It isn't Jovian, and it isn't Terrestrial. Its orbit is highly inclined and so elliptical that Pluto actually comes closer to the sun that Neptune at times. To understand Pluto's status, you must use comparative planetology to analyze Pluto and then compare it with its neighbors.

Pluto is very difficult to observe from Earth. It has only 65 percent the diameter of Earth's moon. In Earth-based telescopes it never looks like more than a faint point of light and even in Space Telescope images it shows little detail. Orbiting so far from the sun, Pluto is cold enough to freeze most compounds you think of as gases, and spectroscopic observations have found evidence of nitrogen ice. It has a thin atmosphere of nitrogen and carbon monoxide with small amounts of methane.

Pluto has three moons. Two, named Nix and Hydra, are quite small, but Charon is relatively large, with half of Pluto's diameter. Charon orbits Pluto with a period of 6.4 days in the same plane as Nix and Hydra, highly inclined to the ecliptic (Figure 14.12, on the next page). Pluto and Charon are tidally locked to face each other, so Pluto's rotation is also highly inclined.

Charon's orbit size and period plus Kepler's third law reveal that the mass of the system is only about 0.002 Earth mass. Most of that mass is Pluto, which has about 12 times the mass of Charon. Knowing the diameters and masses of Pluto and Charon allows astronomers to calculate that their densities are both about 2 g/cm^3. This indicates that Pluto and Charon must each contain about 35 percent ice and 65 percent rock.

The best photos by the Hubble Space Telescope reveal almost no surface detail, but you know enough about icy moons to guess that Pluto has craters and probably shows signs of tidal heating caused by interaction with its large moon Charon. The New Horizons spacecraft will fly past Pluto in July 2015, and the images radioed back to Earth will certainly show that Pluto is an interesting world.

Figure 14.11

The bright disk of Neptune is hidden behind the black bar in this Voyager 2 image. Two narrow rings are visible, and a wider, fainter ring lies closer to the planet. More ring material is visible between the two narrow rings. The rings are bright in forward-scattered light, indicating that the rings contain significant amounts of very small dust particles with short lifetimes. Like the rings of Uranus, the rings of Neptune are dark (reflect very little light) and probably contain methane-rich ice darkened by radiation.

Disk of Neptune

Visual-wavelength image

© NASA

What Defines a Planet?

To understand why Pluto is no longer considered a planet, you should recall the Kuiper belt (see Chapter 12). Since 1992 astronomers have discovered roughly a thousand icy bodies orbiting beyond Neptune. There may be as many as 100 million objects in the Kuiper belt larger than 1 km in diameter. They are understood to be icy bodies left over from the formation of the outer solar system. Some of the Kuiper belt objects are quite large, and one, named Eris, is 5 percent larger in diameter than Pluto. Three other Kuiper belt objects found so far, Sedna, Orcus, and Quaoar (pronounced kwah-o-wahr), are half the size of Pluto or larger. Some of these objects have moons of their own. In that way, they resemble Pluto and its three moons.

A bit of comparative planetology shows that Pluto is not related to the Jovian or Terrestrial planets; it is obviously a member of a newfound family of icy worlds that orbit beyond Neptune. These bodies must have formed at about the same time as the eight classical planets of the solar system, but they did not grow massive to clear their orbital zones of remnant planetesimals and remain embedded among a swarm of other objects in the Kuiper belt.

One of the IAU's criteria for planet status is that an object must be large enough to dominate and gravitationally clear its orbital region of most or all other objects. Eris and Pluto, the largest objects found so far in the Kuiper belt, and Ceres, the largest object in the asteroid belt, do not meet that standard. On the other hand, all three are large enough for their gravities to have pulled them into spherical shapes, so they are the prototypes of a new class of objects defined by the IAU as **dwarf planets**.

Pluto and the Plutinos

No, this section is not about a 1950s rock and roll band. It is about the history of the dwarf planets, and it will take you back 4.6 billion years to watch the outer planets form.

Over a hundred Kuiper belt objects are known that are caught with Pluto in a 3:2 resonance with Neptune.

Dwarf planet: A body that orbits the sun, is not a satellite of a planet, is massive enough to pull itself into a spherical shape but not massive enough to clear out other bodies in and near its orbit. For example, Pluto, Eris, and Ceres.

Plutino: One of the icy Kuiper belt objects that, like Pluto, is caught in a 3:2 orbital resonance with Neptune.

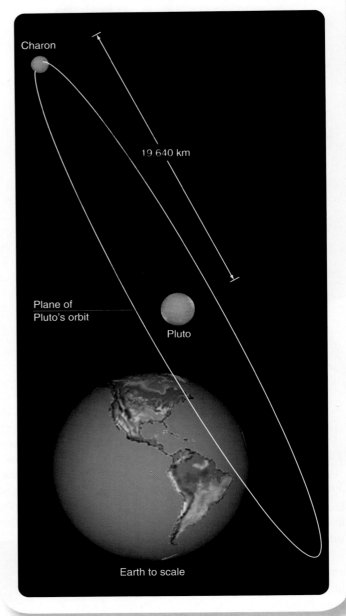

Figure 14.12
The circular orbit of Pluto's moon Charon is seen here at an angle. The orbit is only a few times bigger than Earth and is tipped 118° to the plane of Pluto's orbit around the sun.

Charon

19,640 km

Plane of Pluto's orbit

Pluto

Earth to scale

That is, they orbit the sun twice while Neptune orbits three times. These Kuiper belt objects have been named **plutinos**. The plutinos formed in the outer solar nebula, but how did they get caught in resonances with Neptune? In Chapter 12, you learned that models of the formation of the planets suggest that Uranus and Neptune may have formed closer to the sun and that sometime later, gravitational interactions among the Jovian planets could have gradually shifted Uranus and Neptune outward. As Neptune migrated outward, its orbital resonances could have swept up small objects like a strange kind of snowplow. The Plutinos are caught in the 3:2 resonance, and other Kuiper belt objects are caught in other resonances. This appears to support the models that predict that Uranus and Neptune migrated outward.

The migration of the outer planets would have dramatically upset the motion of some of these Kuiper belt objects, and some could have been thrown inward where they could interact with the Jovian planets. Some of those objects may have been captured as moons, and astronomers wonder if moons such as Neptune's Triton could have started life as Kuiper belt objects. Other objects may have impacted bodies in the inner solar system and caused the late heavy bombardment episode especially evident on the surface of Earth's moon. The small frozen worlds on the fringes of the solar system hold clues to the formation of Earth and the other planets 4.6 billion years ago.

recap

No one has ever been further from Earth than the moon. Humans have sent robotic spacecraft to visit most of the larger worlds in our solar system, and we have found them to be strange and wonderful places. We are trapped on Earth.

We lack the technology to leave Earth. Getting away from Earth's gravitational field is difficult and calls for very large rockets. America built such rockets in the 1960s and early 1970s. They could send astronauts to the moon, but such rockets no longer exist. The best technology today can carry astronauts just a few hundred kilometers above Earth's surface to orbit above the atmosphere. The United States is beginning an ambitious plan to build a new generation of human-piloted rockets meant to carry people back to the moon, and eventually to Mars. We'll have to wait and see whether or not Earth's civilization has the resources to build spacecraft capable of carrying human explorers to other worlds.

In the previous chapter, you discovered another reason we Earthlings are trapped on Earth. We have evolved to fit the environment on Earth. None of the planets or moons you explored in this chapter would welcome you. Extreme temperatures and lack of air are obvious problems, but in addition, Earthlings have evolved to live with Earth's gravity. Astronauts in space for just a few weeks suffer biomedical problems because they are no longer in Earth's gravity. Living in a colony on Mars or the moon might raise similar problems. Just getting to the outer planets would take decades of space travel; living for years in a colony on one of the Jovian moons under low gravity and exposed to the planet's radiation belts may be beyond the capability of the human body. We may be trapped on Earth not because we lack large enough rockets but because we need Earth's protection. It seems likely that we need Earth more than it needs us. The human race is changing the world we live on at a startling pace, and some of those changes could make Earth less hospitable to human life.

All of your exploring of un-Earthly worlds serves to remind you of the nurturing beauty of our home planet, and enables you to answer the following questions from this chapter:

1. What are the properties of the Jovian (Jupiter-like) planets?

2. How do you know that some moons have been geologically active?

3. How are planetary rings formed and maintained?

4. How do the two smaller Jovian worlds—Uranus and Neptune—differ from their more massive siblings?

5. Why did astronomers redefine Pluto as a "dwarf planet"?

Part 5

Chapter 15

Life on Other Worlds

"Did I solicit thee from darkness to promote me?"

ADAM, TO GOD
JOHN MILTON, *PARADISE LOST*

As a living thing, you have been promoted from darkness. The atoms of carbon, oxygen, and other heavy elements that are necessary components of your body did not exist at the beginning of the universe but were cooked up by successive generations of stars. The elements from which you are made are common everywhere in the observable universe, so it is possible that life began on other worlds and evolved to intelligence there as well. If so, perhaps those other civilizations will be detected from Earth. Future astronomers may discover distant alien species completely different from any life on Earth. Your goal in this chapter is to try to understand truly intriguing puzzles—the origin and evolution of life on Earth and possibly on other worlds.

15.1 The Nature of Life

What is life? Philosophers have struggled with that question for thousands of years, and it is not possible to answer it in a single chapter. An attempt at a general definition of what living things do, distinguishing them from nonliving things, might be: Life is a *process* by which an organism extracts energy from the surroundings, maintains itself, and modifies the surroundings to foster its own survival and reproduction.

One very important observation is that all living things on Earth, no matter how apparently different, share certain characteristics in how they perform the process of life.

The Physical Basis of Life

The physical basis of life on Earth is the element carbon. Because of the way carbon atoms bond to each other and to other atoms, they can form long, complex, stable chains that are

looking back

This chapter is either unnecessary or vital, depending on your point of view. If you believe that astronomy is the study of the physical universe above the clouds, then you are done; the previous 14 chapters completed your basic study of astronomy. But if you believe that astronomy is the study of your position in the universe—not just your physical location but also your role as a living being in the evolution of the universe—then everything you have learned so far from this book was preparation for this final chapter.

looking ahead

You have explored the universe from the phases of the moon to the big bang, from the origin of Earth to the death of the sun. Astronomy is important, not because it is about stars and galaxies, but because it is about you: It can tell you your role in the universe. Now that you know astronomy, you see yourself and your world in a different way. Astronomy has changed you.

The Nature of Scientific Explanation

Science is a way of understanding the world around you, and at the heart of that understanding are explanations that science gives you for natural phenomena. Whether you call these explanations stories, histories, hypotheses, or theories, they are attempts to describe how nature works based on evidence and intellectual honesty. While you may take these explanations as factual truth, you should understand that they are not the only explanations that describe the natural world.

A separate class of explanations involves religion. The Old Testament description of the creation of the world does not fit well with scientific observations, but it is a way of understanding the world nonetheless. Religious explanations are based partly on faith rather than on strict rules of logic and evidence, and it is wrong to demand that they follow the same rules as scientific explanations. In the same way, it is wrong to demand that scientific explanations take into account religious beliefs. The so-called conflict between science and religion arises when people fail to recognize that science and religion are different ways of knowing about the world.

Scientific explanations are compelling because they provide tremendous insights into the workings of nature, and have been very successful at yielding technological innovations that have changed the world you live in. From new vaccines to digital music players to telescopes that can observe the most distant galaxies, the products of the scientific process are all around you. Scientific explanations have provided tremendous insights into the workings of nature, and it is easy to forget that there are other explanations. Many people are attracted to the suggestion, made by evolutionary biologist Stephen Jay Gould and others, that religious explanations and scientific explanations should be considered as "separate magisteria." In other words, religion and science are devoted to different realms of the mystery of existence.

Science and religion both have a lot to offer by their differing ways of explaining the world, but the two ways follow separate rules and cannot be judged by each other's standards. The trial of Galileo (see Chapter 3) can be understood as a conflict between these two ways of knowing.

capable of storing and transmitting information. A large amount of information of some sort is necessary to maintain the forms and control the functions of living things.

Carbon may not be crucial to life. Science fiction authors have speculated that silicon could be substituted for carbon because the two elements share some chemical properties, but that seems unlikely because silicon chains are harder to assemble and disassemble than their carbon counterparts and can't be as lengthy. Even stranger life forms have been proposed, based on electromagnetic fields and ionized gas, and none of these possibilities can

be ruled out. These hypothetical life forms make for fascinating speculation, but they can't be studied as systematically in the way that life on Earth can. This chapter is concerned with the origin and evolution of life as it is on Earth, based on carbon, not because of lack of imagination, but because it is the only form of life about which we know anything.

Even carbon-based life has its mysteries. What makes a lump of carbon-based molecules a living thing? An important part of the answer lies in the transmission of information from one molecule to another.

MUTATION: ALBINISM

Information Storage and Duplication

Almost every action performed by a living cell is carried out by chemicals it manufactures. Cells must store recipes for all these chemicals, use them when they need them, and pass them on to their offspring.

Study "DNA: The Code of Life" on pages 312–313 and notice three important points:

1. The chemical recipes of life are stored in each cell as information on DNA molecules that resemble a ladder with rungs that are composed of chemical bases. The recipe information is expressed by the sequence of ladder rungs, providing instructions to guide chemical reactions within the cell.

2. DNA instructions normally are expressed by being copied into a messenger molecule called RNA that causes molecular units called amino acids to be connected into large molecules called proteins. Proteins serve as the cell's basic structural molecules or as enzymes that control chemical reactions.

3. The instructions stored in DNA are genetic information passed along to offspring. The DNA molecule reproduces itself when a cell divides so that each new cell contains a copy of the original information.

To produce viable offspring, a cell must be able to make copies of its DNA. Surprisingly, it is important for the continued existence of all life that not all the copies be exact duplicates.

Modifying the Information

Earth's environment changes continuously. To survive, species must change as their food supply, climate, or home terrain changes. If the information stored in DNA could not change, then life would quickly go extinct. The process by which life adjusts itself to its changing environment is called **biological evolution**.

When an organism reproduces, its offspring receives a copy of its DNA. Sometimes external effects such as radiation alter the DNA during the parent organism's lifetime, and sometimes mistakes occur in the copying process, so that occasionally the copy is slightly different from the original. Offspring born with random alterations to their DNA are called **mutants**. Most mutations make no difference, but some mutations are fatal, killing the afflicted organisms before they can reproduce. In rare but vitally important cases, a mutation can actually help an organism survive.

These changes produce variation among the members of a species. All of the squirrels in the park may look the same, but they carry a range of genetic variation. Some may have slightly longer tails or faster-growing claws. These variations make almost no difference until the environment changes. For example, if the environment becomes colder, a squirrel with a heavier coat of fur will, on average, survive longer and produce more offspring than its normal contemporaries. Likewise, the offspring that inherit this beneficial variation will also live longer and have more offspring of their own. These differing rates of survival and reproduction are examples of **natural selection**. Over time, the beneficial variation becomes more common and a species can evolve until the entire population shares the trait. In this way, natural selection adapts species to their changing environments by selecting, from the huge array of random variations, those that would most benefit the survival of the species.

It is commonly believed that evolution is random, but that is not true. The underlying variation within species is random, but natural selection is not random because progressive changes in a species are directed by changes in the environment.

Biological evolution: The process of mutation, variation, and natural selection by which life adjusts itself to its changing environment.

Mutant: Offspring born with DNA that is altered relative to parental DNA.

Natural selection: The process by which the best genetic traits are preserved and accumulated, allowing the fittest organisms and species to survive and proliferate.

DNA: The Code of Life

1 The key to understanding life is information—the information that guides all of the processes in an organism. In most living things on Earth, that information is stored on a long spiral molecule called **DNA (deoxyribonucleic acid)**.

1a The DNA molecule looks like a spiral ladder with rails made of phosphates and sugars. The rungs of the ladder are made of four chemical bases arranged in pairs. The bases always pair the same way. That is, base A always pairs with base T, and base G always pairs with base C.

1b Information is coded on the DNA molecule by the order in which the base pairs occur. To read that code, molecular biologists have to "sequence the DNA." That is, they must determine the order in which the base pairs occur along the DNA ladder.

The Four Bases

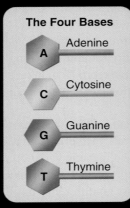

- **A** Adenine
- **C** Cytosine
- **G** Guanine
- **T** Thymine

2 DNA automatically combines raw materials to form important chemical compounds. The building blocks of these compounds are relatively simple **amino acids**. Segments of DNA act as templates that guide the amino acids to join together in the correct order to build specific **proteins**, chemical compounds important to the structure and function of organisms. Some proteins called **enzymes** regulate other processes. In this way, DNA recipes regulate the production of the compounds of life.

Jamie Backman

The traits you inherit from your parents, the chemical processes that animate you, and the structure of your body are all encoded in your DNA. When people say "you have your mother's eyes" they are talking about DNA codes.

Nucleus
(information
storage)

Cell membrane
(transport of raw
materials and
finished product)

Material
storage

Manufacture
of proteins
and enzymes

Energy
production

2a A cell is a tiny factory that uses the DNA code to manufacture chemicals. Most of the DNA remains safe in the nucleus of a cell, and the code is copied to create a molecule of **RNA (ribonucleic acid)**. Like a messenger carrying blueprints, the RNA carries the code out of the nucleus to the work site where the proteins and enzymes are made.

2b A single cell from a human being contains about 1.5 meters of DNA containing about 4.5 billion base pairs—enough to record the entire works of Shakespeare 200 times. A typical human contains a total of about 600 AU of DNA. Yet the DNA in each cell, only 1.5 meters in length, contains all of the information to create a new human. A clone is a new creature created from the DNA code found in a single cell.

Original DNA

Copy DNA

3 DNA, coiled into a tight spiral, makes up the **chromosomes** that are the genetic material in a cell. A **gene** is a segment of a chromosome that controls a certain function. When a cell divides, each of the new cells receives a copy of the chromosomes, as genetic information is handed down to new generations.

Copy DNA

3a To divide, a cell must duplicate its DNA. The DNA ladder splits, and new bases match to the exposed bases of the ladder to build two copies of the original DNA code. Because the base pairs almost always match correctly, errors in copying are rare. One set of the DNA code goes to each of the two new cells.

Cell Reproduction by Division

As a cell begins to divide, its DNA duplicates itself.

The duplicated chromosomes move to the middle.

The two sets of chromosomes separate, and . . .

the cell divides to produce . . .

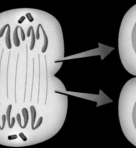

two cells, each containing a full set of the DNA code.

15.2 Life in the Universe

It is obvious that the 4.5 billion chemical bases that make up human DNA did not just come together in the right order by chance. The key to understanding the origin of life lies in the processes of evolution. The complex interplay of environmental factors with the DNA of generation after generation of organisms drove some life forms to become more sophisticated over time, until they became the unique and specialized creatures on Earth today.

This means that life on Earth could have begun very simply, even as simple a form as carbon chain molecules able to copy themselves. Of course, this is a hypothesis for which you can seek evidence. What evidence exists regarding the origin of life on Earth?

The Origin of Life on Earth

The oldest fossils are all the remains of sea creatures, and this indicates that life began in the sea. Identifying the oldest fossils is not easy, however. Fossils billions of years old are difficult to recognize because the earliest living things contained no easily preserved hard parts like bones or shells, and because the individual organisms were microscopic from western Australia that are more than 3 billion years old contain features that experts identify as fossil **stromatolites**, remains of colonies of single-celled organisms (Figure 15.1). The evidence,

though scarce, indicates that simple organisms lived in Earth's oceans 3.4 billion or more years ago, less than 1.2 billion years after Earth formed. Where did these simple organisms come from?

An important experiment performed by Stanley Miller and Harold Urey in 1952 sought to recreate the conditions in which life on Earth began. The **Miller experiment** consisted of a sterile, sealed glass container holding water, hydrogen, ammonia, and methane. An electric arc inside the apparatus created sparks to simulate the effects of lightning in Earth's early atmosphere (Figure 15.2).

Miller and Urey let the experiment run for a week and then analyzed the material inside. They found that the interaction between the electric arc and the simulated atmosphere had produced many organic molecules from the raw material of the experiment, including such important building blocks of life as amino acids. When the experiment was run again using different energy sources such as hot silica to represent molten lava spilling into the ocean, similar molecules were produced. Even the amount of UV radiation present in sunlight is sufficient to produce complex organic molecules.

According to updated models of the formation of the solar system and Earth (see Chapters 12 and 13), Earth's

Stromatolite: A layered formation caused by mats of algae or bacteria combined with sediments.

Miller experiment: An experiment that attempted to reproduce early Earth conditions and showed how easily amino acids and other organic compounds can form.

Figure 15.1

Cross section of layers in a fossilized stromatolite. Layers are made of mineralized mats of bacteria covered repeatedly by sediment. Fossilized stromatolites have been found within rocks more than 3.4 billion years. This sample measures 300 millimeters across.

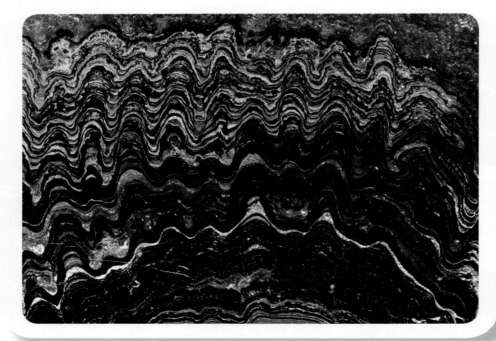

© Dirk Wiersma/Photo Researchers, Inc.

Figure 15.2

(a) The Miller experiment circulated gases through water in the presence of an electric arc. This simulation of primitive conditions on Earth produced many complex organic molecules, including amino acids, the building blocks of proteins. (b) Stanley Miller with a Miller apparatus.

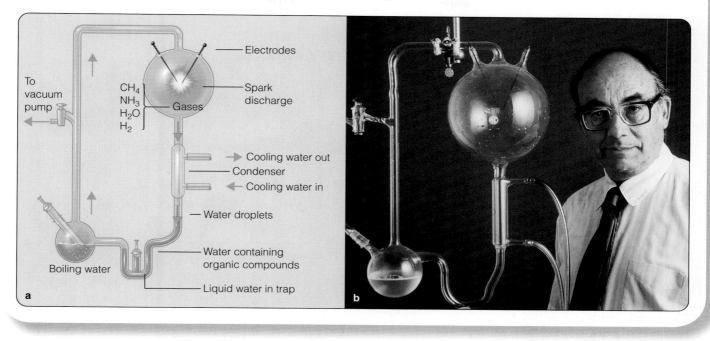

© Stanley Miller

early atmosphere probably consisted mostly of carbon dioxide, nitrogen, and water vapor instead of the mix of hydrogen, ammonia, and methane assumed by Miller and Urey. When gases corresponding to the newer understanding of the early Earth atmosphere are processed in a Miller apparatus, lesser but still significant numbers of organic molecules are created.

The Miller experiment is important because it shows that complex organic molecules form naturally in a wide variety of circumstances. Lightning, sunlight, and hot lava pouring into the oceans are just some of the energy sources that can naturally rearrange simple common molecules into the complex molecules that make life possible. If you could travel back in time, you would expect to find Earth's first oceans filled with a rich mixture of organic compounds called the **primordial soup**.

Many of these organic compounds would have been able to link up to form larger molecules. Amino acids, for example, can link together to form proteins by joining ends and releasing a water molecule (Figure 15.3). It was initially thought that this must have occurred in sun-warmed tidal pools where organic molecules were concentrated by evaporation. However, violent episodes of volcanism and catastrophic meteorite impacts would probably have destroyed complex molecules developing at the surface, so it is now thought that successful linkage of complex mol-

ecules might have taken place on the ocean floor, perhaps near the hot springs at midocean rises (see page 270).

These complex organic molecules were still not living things. Even though some proteins may have contained hundreds of amino acids, they did not reproduce but rather linked and broke apart at random. Because some molecules are more stable than others, and some bond together more easily than others, scientists hypothesize that a process of **chemical evolution** eventually concentrated the various smaller molecules into the most stable larger forms. Eventually, according to the hypothesis, somewhere in the oceans, after sufficient time, a molecule formed that could copy itself. At that point, the chemical evolution of molecules became the biological evolution of living things.

An alternate theory for the origin of life holds that reproducing molecules may have arrived here from space. Radio astronomers have found a wide variety of organic molecules in the interstellar medium, and similar compounds have been found inside meteorites, such as the Murchison

Primordial soup: The rich solution of organic molecules in Earth's first oceans.

Chemical evolution: The chemical process that led to the growth of complex molecules on primitive Earth. This did not involve the reproduction of exact molecules.

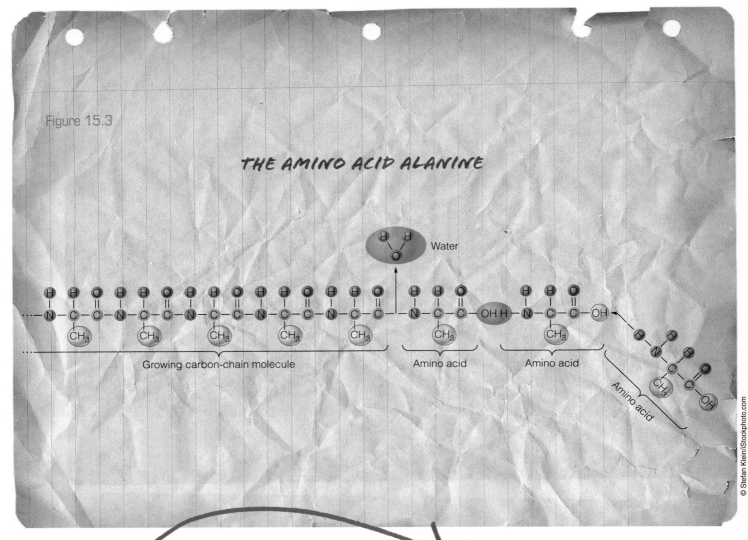

Figure 15.3

THE AMINO ACID ALANINE

Water

Growing carbon-chain molecule

Amino acid

Amino acid

Amino acid

meteorite pictured here. The Miller experiment showed how easy it is to create organic molecules in a hydrogen-right environment, so it is not surprising to find them in space. Although speculation is fun, the hypothesis that life arrived on Earth from space is presently more difficult to test than the hypothesis that Earth life originated on Earth.

Whether the first reproducing molecules formed here on Earth or in space, the important thing is that they could have formed by natural processes. Scientists know enough about these processes to feel confident about them, even though some of the steps remain unknown.

The details of the evolution of the first cells are unknown, but the first reproducing molecule to surround itself with a protective membrane must have gained an important survival advantage. Experiments have shown that microscopic spheres the size of cells containing organic molecules form relatively easily in water, as in Figure 15.4, so the evolution of cell-like structures is not surprising.

The first cells must have been simple single-celled organisms much like modern bacteria. As you learned earlier, these kinds of cells are preserved in stromatolites

Figure 15.4

Single amino acids can be assembled into long proteinlike molecules. When such material cools in water, it can form microspheres, microscopic spheres with double-layered boundaries similar to cell membranes. Microspheres may have been an intermediate stage in the evolution of life from complex molecules to living cells holding molecules reproducing genetic information.

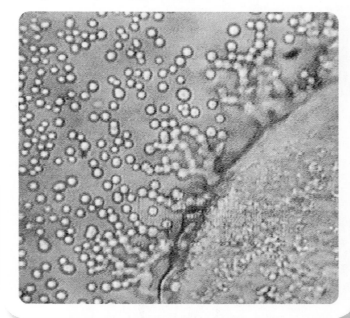

© Sidney Fox and Randall Grubbs

(Figure 15.1), mineral formations formed by layers of photosynthetic bacteria and shallow ocean sediments. Stromatolites formed in rocks with radioactive ages of 3.4 billion years, and they also still form in some places today. Stromatolites and other photosynthetic organisms would have begun adding oxygen, a product of photosynthesis, to Earth's early atmosphere (Figure 15.5). An oxygen abundance of only 0.1 percent would have created an ozone screen, protecting organisms from the sun's ultraviolet radiation and later allowing life to colonize the land.

Over the course of eons, the natural processes of evolution gave rise to stunningly complex **multicellular** life forms with their own widely differing ways of life. It is a common misconception to imagine that life is too complex to have evolved from such simple beginnings. It is possible because small variations can accumulate, but that accumulation requires huge amounts of time.

Geologic Time

Life has existed on Earth for at least 3.4 billion years, but there is no evidence of anything more than simple organisms until about 540 million years ago, when life suddenly branched into a wide variety of complex forms. This sudden increase in complexity is known as the **Cambrian explosion**, and it marks the beginning of the Cambrian period.

If you represented the entire history of Earth on a scale diagram, the Cambrian explosion would be near the top of the column, as shown at the left of Figure 15.6 on page 319. The emergence of most animals familiar to you today, including fishes, amphibians, reptiles, birds, and mammals, would be crammed into the topmost part of the chart, above the Cambrian explosion.

If you magnify this portion of the diagram, as shown on the right side of Figure 15.6, you can get a better idea of when these events occurred in the history of life. Humanoid creatures have walked the Earth for about 4 million years. This is a long time by the standard of a human lifetime, but it makes only a narrow red line at the top of the diagram. All of recorded history would be a microscopically thin line at the very top of the column.

To understand just how thin that line is, imagine that the entire 4.6-billion-year history of the Earth has been compressed onto a yearlong video and that you began watching this video on January 1. You would not see any signs of life until March or early April, and the slow evolution of the first simple forms would take up the next 6 or 7 months. Suddenly, in mid-November, you would see complex organisms of the Cambrian explosion.

You would see no life of any kind on land until November 28, but once it appeared it would diversify quickly, and by December 12 you would see dinosaurs walking the continents. By the day after Christmas they would be gone, and mammals and birds would be on the rise.

If you watched closely, you might see the first humanoid forms by suppertime on New Year's Eve, and by late evening you could see humans making the first stone tools. The Stone Age would last until 11:59 pm, after which the first towns, then cities, would appear. Suddenly things would begin to happen at lightning speed. Babylon would flourish, the Pyramids would rise, and Troy would fall. The Christian era would begin 14 seconds before the New Year. Rome would fall, the Middle Ages and the Renaissance would flicker past. The American

Multicellular: An organism composed of more than one cell.

Cambrian explosion: A geologically brief period about 540 million years ago during which fossil evidence indicates Earth life became complex and diverse. Cambrian rocks contain the oldest easily identifiable fossils.

Figure 15.5
Artist's conception of a scene on the young Earth, 3.4 billion years ago, with stromatolite bacterial mats growing near the shores of an ocean.

Illustration by Peter Sawyer ©Smithsonian Institution

and French revolutions would occur one second before the end of the video.

By imagining the history of Earth as a yearlong video, you have gained some perspective on the rise of life. Tremendous amounts of time were needed for the first simple living things to evolve in the oceans; but, as life became more complex, new forms arose more and more quickly as the hardest problems—how to reproduce, how to take energy efficiently from the environment, how to move around—were solved. The easier problems, like what to eat, where to live, and how to raise offspring, were solved in different ways by different organisms, leading to the diversity that is seen today.

Even intelligence—that which appears to set humans apart from other animals—may be a unique solution to an evolutionary problem posed to humanity's ancient ancestors. A smart animal is better able to escape predators, outwit its prey, and feed and shelter itself and its offspring, so under certain conditions evolution is likely to select for intelligence. Could intelligent life arise on other worlds? To try to answer this question, you can estimate the chances of any type of life arising on other worlds, then assess the likelihood of that life developing intelligence.

Life in Our Solar System

Could there be carbon-based life elsewhere in our solar system? Liquid water seems to be a requirement of carbon-based life, necessary both for vital chemical reactions

© Frederick Kaselow/iStockphoto.com

Figure 15.6

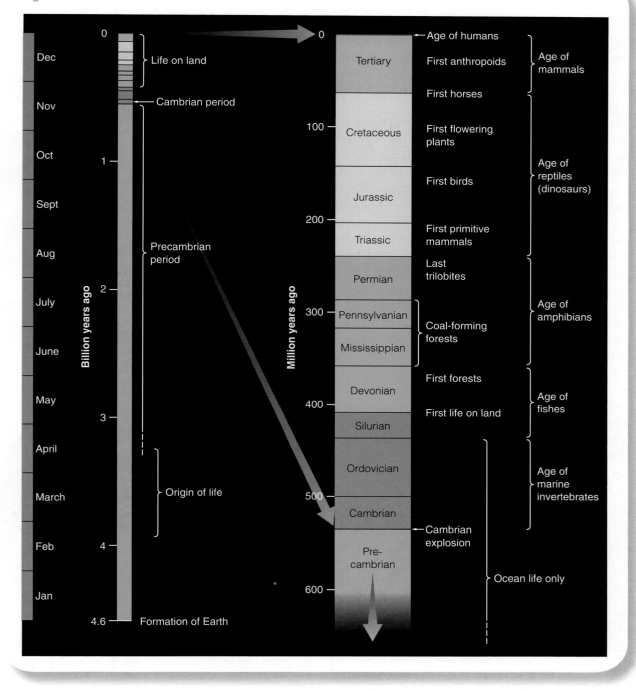

and as a medium to transport nutrients and wastes. It is not surprising that life developed in Earth's oceans and stayed there for billions of years before it was able to colonize the land.

Scientists are in agreement that any world harboring living things must have significant quantities of some type of liquid. Water is a cosmically abundant substance with properties such as high heat capacity that set it apart from other common molecules that are liquid at the temperatures of planetary surfaces.

Many worlds in the solar system can be eliminated immediately as hosts for life because liquid water is not possible there. The moon and Mercury are airless, and water would boil away into space immediately. Venus has traces of water vapor in its atmosphere, but it is too hot for liquid water to survive on the surface. The Jovian planets have deep atmospheres, and at a certain level water condenses into liquid droplets. However, it seems unlikely that life could have originated there; the Jovian planets do not have solid surfaces (see Chapter 14).

Isolated water droplets could never mingle to mimic the rich primordial oceans of Earth, where organic molecules grew and interacted. Additionally, powerful currents in the gas giants' atmospheres would quickly carry any reproducing molecules that did form there down into inhospitably hot regions of the atmosphere.

As you learned in Chapter 14, at least one of the Jovian satellites could potentially support life. Jupiter's moon Europa appears to have a liquid-water ocean below its icy crust, and minerals dissolved in the water could provide a source of raw material for chemical evolution. Europa's ocean is kept warm and liquid now by tidal heating. There also may be liquid water layers under the surfaces of Ganymede and Callisto. That can change as the orbits of the moons interact; Europa, Ganymede and Callisto may have been frozen solid at other times in their histories, destroying organisms that had developed there.

Saturn's moon Titan is rich in organic molecules. Chapter 14 described how sunlight converts the methane in Titan's atmosphere into organic smog particles that settle to the surface. The chemistry of any life that could have evolved from those molecules and survived in Titan's lakes of methane is unknown. It is fascinating to consider possibilities, but Titan's extremely low temperature of –180°C (–290°F) could make chemical reactions slow to the point where life processes seem unlikely.

Observations of water venting from the south-polar region of Saturn's moon Enceladus show that it has liquid water below its crust. It is possible that life could exist in that water, but the moon is very small and has been warmed by tidal heating that may operate only occasionally. Enceladus may not have had plentiful liquid water for the extended time thought necessary for the rise of life.

Mars is the most likely place for life to exist in the solar system because, as you learned in Chapter 13, there is a great deal of evidence that liquid water once flowed on its surface. Even so, results from searches for signs of life on Mars are not encouraging. The robotic spacecraft Viking 1 and Viking 2 landed on Mars in 1976 and tested soil samples for living organisms. Some of the tests had puzzling semipositive results that scientists hypothesize were caused by nonbiological chemical reactions in the soil. There is no clear evidence of life or even of organic molecules in the Martian soil.

In 1996 there were prominent news stories regarding chemical and physical traces of life on Mars discovered inside a Martian meteorite found in Antarctica (Figure 15.7). Scientists were excited by the announcement, but they immediately began testing the evidence. Their results suggest that the unusual chemical signatures in the rock may have formed by processes that did not in-

Figure 15.7

(a) Meteorite ALH84001 is one of a dozen meteorites known to have originated on Mars. A research group claimed that the meteorite contained chemical and physical traces of ancient life on Mars, including (b) what appear to be fossils of microscopic organisms. That evidence has not been confirmed, and the claim continues to be debated.

volve life. Tiny features in the rock that were originally thought to be fossils of ancient Martian microorganisms could be nonbiological mineral formations. This is the only direct evidence yet found regarding potential life on Mars, but it remains highly controversial. Conclusive evidence of life on Mars may have to wait until a geologist from Earth can scramble down dry Martian streambeds and crack open rocks looking for fossils.

There is no strong evidence for the existence of life elsewhere in the solar system. Now your search will take you to distant planetary systems.

Life in Other Planetary Systems

Could life exist in other planetary systems? You already know that there are many different kinds of stars and that many of these stars have planetary systems. As a first step toward answering this question, you can try to identify the kinds of stars that seem most likely to have stable planetary systems where life could evolve.

If a planet is to be a suitable home for living things, it must be in a stable orbit around its sun. That is simple in a planetary system like our own, but most planet orbits in binary star systems are unstable unless the component stars are very close together or very far apart. Astronomers can calculate that in binary systems with stars separated by distances of a few AU, the planets should eventually be swallowed up by one of the stars or ejected from the system. Half the stars in the galaxy are members of binary systems, and many of them are unlikely to support life on planets.

Moreover, just because a star is single does not necessarily make it a good candidate for sustaining life. Earth required between 0.5 and 1 billion years to produce the first cells and 4.6 billion years for intelligence to emerge. Massive stars that live only a few million years do not meet this criterion. If the history of life on Earth is at all representative, then stars more massive and luminous than about spectral type F5 are too short lived for complex life to develop. Main-sequence stars of types G and K, and possibly some of the M stars, are the best candidates.

The temperature of a planet is also important, and that depends on the type of star it orbits and its distance from the star. Astronomers have defined a **habitable zone** around a star as a region within which planets have temperatures that permit the existence of liquid water. The sun's habitable zone extends from around the orbit of Venus to the orbit of Mars, with Earth right in the middle. A low-luminosity star has a small habitable zone, and a high-luminosity star has a large one.

Scientists on Earth are finding life in places previously judged inhospitable, such as the bottoms of ice-covered lakes in Antarctica and far underground inside solid rock. Life has also been found in boiling hot springs with highly acidic water. As a result, it is difficult for

scientists to pin down a range of environments and state with certainty that life cannot exist outside those conditions. You should also note that three of the environments listed as possible havens for life—Europa, Titan, and Enceladus—are in the outer solar system and lie far outside the sun's conventional habitable zone. Stable planets inside the habitable zones of long-lived stars are the places where life seems most likely, but, given the tenacity and resilience of Earth's life forms, there might be other, seemingly inhospitable, places in the universe where life exists.

15.3 Intelligent Life in the Universe

Visiting extrasolar planets is, for now, impossible. Nevertheless, if other civilizations exist, it is possible humans can communicate with them. Nature puts restrictions on such conversations, but the main question lies in the unknown life expectancy of civilizations.

Travel between the Stars

The distances between stars are almost beyond comprehension. The space shuttle would take about 150,000 years to reach the nearest star. The obvious way to overcome these huge distances is with tremendously fast spaceships, but even the closest stars are many light-years away.

Nothing can exceed the speed of light, and accelerating a spaceship close to the speed of light takes huge amounts of energy. Even if you travel more slowly, your rocket would require massive amounts of fuel. If you were piloting a

Habitable zone: A region around a star within which planets have temperatures that permit the existence of liquid water on their surfaces.

spaceship with the mass of 100 tons (the size of a yacht) to the nearest star, 4 light years away, and you wanted to travel at half the speed of light so as to arrive in 8 years, the trip would require 400 times as much energy as the entire United States consumes in a year.

These limitations not only make it difficult for humans to leave the solar system, but they would also make it difficult for aliens to visit Earth. Reputable scientists have studied "unidentified flying objects" (UFOs) and have never found any evidence that Earth is being visited or has ever been visited by aliens. Humans are unlikely ever to meet aliens face to face. However, communication by radio across interstellar distances takes relatively little energy.

Radio Communication

Nature puts restrictions on travel through space, and it also restricts astronomers' ability to communicate with distant civilizations by radio. One restriction is based on simple physics. Radio signals are electromagnetic waves and travel at the speed of light. Due to the distances between the stars, the speed of radio waves would severely limit astronomers' ability to carry on normal conversations with distant civilizations. Decades could elapse between asking a question and getting an answer.

So, rather than try to begin a conversation, one group of astronomers decided in 1974 to broadcast a simple message of greeting toward the globular cluster M13, 26,000 light years away, using the Arecibo radio telescope (see below). When the signal arrives 26,000 years in the future, alien astronomers may be able to decode it.

The Arecibo beacon is an anticoded message, meaning that it is intended to be decoded by beings about whom we know nothing except that they build radio telescopes. The message is a string of 1,679 pulses and gaps. Pulses represent 1s, and gaps represent 0s. The string can be arranged in only two possible ways: as 23 rows of 73 or as 73 rows of 23. The second arrangement forms a picture containing information about life on Earth, as you can see from panel b of the following figure.

a A simple example of an anticoded message

5 rows of 7

7 rows of 5

b The Arecibo anticoded message

Start of number markers — Binary numbers 1 to 10

Atomic numbers of hydrogen, carbon, nitrogen, oxygen, and phosphorus

Formulas for sugars and bases in DNA

DNA double helix

Number of units in DNA

Start of number markers

Human figure

Start of number marker

Population of Earth

Height of human in wavelengths

Arecibo radio dish transmitting signal

Sun and planets with Earth offset

Diameter of dish in wavelengths

Start of number marker

© NASA

Although the 1974 Arecibo beacon was the only powerful signal sent purposely from Earth to other solar systems, Earth is sending out many other signals. Shortwave radio signals, such as TV and FM, have been leaking into space for the last 50 years or so. Any civilization within 50 light years could already have detected Earth's civilization. That works both ways: Alien signals, whether intentional messages of friendship or the blather of their equivalent to daytime TV, could be arriving at Earth now. Astronomers all over the world are pointing radio telescopes at the most likely stars and listening for alien civilizations.

If you conclude that there is likely to be life on other worlds, then you might be tempted to use UFO sightings as evidence to test your hypothesis. Scientists don't do this for two reasons.

First, the reputation of UFO sightings and alien encounters does not inspire confidence that these data are reliable. Most people hear of such events in grocery store tabloids, daytime talk shows, or sensational "specials" on viewer-hungry cable networks. You should take note of the low reputation of the media that report UFOs and space aliens. Most of these reports, like the reports that Elvis is alive and well, are simply made up for the sake of sensation, and you cannot use them as reliable evidence.

Second, the few UFO sightings that are not made up do not survive careful examination. Most are mistakes and unintentional misinterpretations of natural events or human-made objects, committed by honest people. Over many decades, experts have studied these incidents and found none that are convincing. In short, despite false claims to the contrary on TV shows, there is no dependable evidence that Earth has ever been visited by aliens.

That's too bad. A confirmed visit by intelligent creatures from beyond our solar system would answer many questions. It would be exciting, enlightening, and, like any real adventure, a bit scary. But there is not yet any direct evidence of life on other worlds.

Which channels should astronomers monitor? Wavelengths longer than 100 cm would get lost in the background noise of the Milky Way Galaxy, while wavelengths shorter than about 1 cm are absorbed in Earth's atmosphere. Between those wavelengths is a radio window that is open for communication. Even this restricted window contains millions of possible radio-frequency bands and is too wide to monitor easily, but astronomers may have found a way to narrow the search. Within this window lie the 21-cm spectral line of neutral hydrogen and the 18-cm line of OH (Figure 15.8). The interval between those lines has low background interference and is named the **water hole** because H plus OH yields water. Any civilizations sophisticated enough to do radio astronomy research must know of these lines and might appreciate their significance in the same way as do Earthlings.

A number of searches for extraterrestrial radio signals have been made, and some are now under way. This field of study is known as **SETI**, Search for Extra-Terrestrial Intelligence, and it has generated heated debate among astronomers, philosophers, theologians, and politicians. Congress funded a NASA search for a short time but ended support in the early 1990s. In fact, the annual cost of a major search is only about as much as a single Air Force attack helicopter, but much of the reluctance to fund searches probably stems from issues other than cost. One point to keep in mind is that the discovery of alien intelligence would cause a huge change in humans' worldview, akin to Galileo's discovery that the moons of Jupiter do not go around the Earth. Some turmoil would inevitably result.

In spite of the controversy, the search continues. The NASA SETI project canceled by Congress was renamed Project Phoenix and completed using private funds. The SETI Institute, founded in 1984, managed Project Phoenix plus several other important

Water hole: The interval of the radio spectrum between the 21-cm hydrogen emission line and the 18-cm OH emission line; these are likely wavelengths to use in the search for extraterrestrial life.

SETI: The Search for Extra-Terrestrial Intelligence.

Figure 15.8

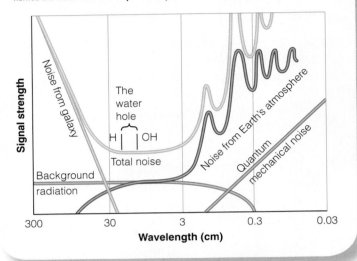

Radio noise from various astronomical sources makes it difficult to detect distant signals at wavelengths longer than 100 cm or shorter than 1 cm. In this range, radio emission lines from H atoms and from OH molecules mark a small wavelength range named the water hole that may be a likely channel for communication.

searches and is currently building a new radio telescope array in northern California, collaborating with the University of California, Berkeley, and partly funded by Paul Allen of Microsoft.

There is even a way for you to help with searches. The Berkeley SETI team (note: they are separate from the SETI Institute), with the support of the Planetary Society, has recruited about 4 million owners of personal computers that are connected to the Internet. Participants download a screen saver that searches data files from the Arecibo radio telescope for signals whenever the owner is not using the computer. For information, locate the seti@home project at http://setiathome.ssl.berkeley.edu/.

The search continues, but radio astronomers struggle to hear anything against the worsening babble of noise from human civilization. Wider and wider sections of the electromagnetic spectrum are being used for earthly communication, and this, combined with stray electromagnetic noise from electronic devices including everything from computers to refrigerators, makes hearing faint radio signals difficult. It would be ironic if humans fail to detect faint signals from another world because our own world has become too noisy. Ultimately, the chances of success depend on the number of inhabited worlds in the galaxy.

Drake equation: The equation that estimates the total number of communicative civilizations in our galaxy.

How Many Inhabited Worlds?

Given enough time, the searches will find other worlds with civilizations, assuming that there are at least a few out there. If intelligence is common, scientists should find signals relatively soon—within the next few decades—but if intelligence is rare, it may take much longer.

Simple arithmetic can give you an estimate of the number of technological civilizations in the Milky Way Galaxy with which you might communicate, N_c. The formula proposed for discussions about N_c is named the **Drake equation** after the radio astronomer Frank Drake, a pioneer in the search for extraterrestrial intelligence. The version of the Drake equation presented here is modified slightly from its original form:

$$N_c = N_* \times f_p \times n_{HZ} \times f_L \times f_I \times f_s$$

N_* is the number of stars in our galaxy, and f_p represents the fraction of stars that have planets. If all single stars have planets, f_p is about 0.5. The factor n_{HZ} is the average number of planets in each solar system suitably placed in the habitable zone. This factor actually means the number of planets possessing liquid water. Europa and Enceladus in our solar system show that liquid water can exist due to tidal heating outside the conventional habitable zone that in our system contains Earth's orbit. Thus, n_{HZ} may be larger than had been previously thought. The factor f_L is the fraction of suitable planets on which life begins, and f_I is the fraction of those planets where life evolved to intelligence.

The first five factors in the Drake equation can be roughly estimated, but the final factor, f_S, the fraction of a star's life during which an intelligent species is communicative, is extremely uncertain. If a society survives at a technological level for only 100 years, the chances of communicating with it are small. But a society that stabilizes and remains technological for a long time is much more likely to be detected. For a star with a life span of 10 billion years, f_S can range from 10^{-8} for extremely short-lived societies to 10^{-4} for societies that survive for a million years. Table 15.1 summarizes what many scientists consider a reasonable range of values for f_S and the other factors.

If the optimistic estimates are true, there could be a communicative civilization within a few tens of light-years of Earth. On the other hand, if the pessimistic estimates are true, Earth may be the only planet that is capable of communication within thousands of the nearest galaxies.

What Are We?

There are over four thousand religions around the world, and nearly all hold that humans have a dual nature: we are physical objects made of atoms, but we are also spiritual beings. Science is unable to examine the spiritual side of existence, but it can tell us about our physical nature.

You are made of very old atoms. The matter you are made of appeared in the big bang and was cooked into a wide range of elements inside stars. Your atoms may have been inside at least two or three generations of stars. Eventually, your atoms became part of a nebula that contracted to form the sun and the planets of the solar system.

Your atoms have been part of Earth for the last 4.6 billion years. They have been recycled many times through dinosaurs, stromatolites, fish, bacteria, grass, birds, worms, and other living things. Now you are using your atoms, but when you are done with them, they will be used again and again.

When the sun swells into a red giant star and dies in a few billion years, Earth's atmosphere and oceans will be driven away, and at least the outer few kilometers of Earth's crust will be vaporized and blown outward to become part of the nebula around the white-dwarf remains of the sun. Your atoms are destined to return to the interstel-

recap At the beginning of this chapter you tried to tackle a question that has troubled scientists and philosophers for centuries: What is life? You didn't get more than the beginning of an answer to that question here, but thinking about it will illuminate the problem and prepare you to search for life on other worlds.

You have explored difficult questions, and should be able to venture some answers to these four related questions:

1. What is life?

2. How did life originate on Earth?

3. Could life begin on other worlds?

4. Can Earthlings communicate with civilizations on other worlds?

But remember, often in science, asking a question is more important than getting an immediate answer.

lar medium and will become part of future generations of stars, planets, and perhaps living beings.

The message of astronomy is that humans are not just observers: We are participants, we are part of the universe. Among all of the galaxies, stars, planets, planetesimals, and bits of matter, humans are objects that can think, and that means we can understand what we are.

Is the human race the only thinking species? If so, we bear the sole responsibility to understand and admire the universe. The detection of signals from another civilization would demonstrate that we are not alone, and such communication would end the self-centered isolation of humanity and stimulate a re-evaluation of the meaning of human existence.

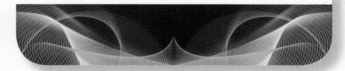

Table 15.1 The Number of Technological Civilizations per Galaxy

	Variables	Estimates	
		Pessimistic	Optimistic
N_*	Number of stars per galaxy	2×10^{11}	2×10^{11}
f_P	Fraction of stars with planets	0.1	0.5
n_{HZ}	Number of planets per star that lie in habitable zone for longer than 4 billion years	0.01	1
f_L	Fraction of suitable planets on which life begins	0.01	1
f_I	Fraction of planets where life forms evolve to intelligence	0.01	1
f_S	Fraction of star's existence during which a technological society survives	10^{-8}	10^{-4}
N_c	Number of communicative civilizations per galaxy	2×10^{-4}	1×10^7

Afterword

The aggregate of all our joys and sufferings, thousands of confident religions, ideologies and economic doctrines, every hunter and forager, every hero and coward, every creator and destroyer of civilizations, every king and peasant, every young couple in love, every hopeful child, every mother and father, every inventor and explorer, every teacher of morals, every corrupt politician, every superstar, every supreme leader, every saint and sinner in the history of our species, lived there on a mote of dust, suspended in a sunbeam.

CARL SAGAN (1934–1996)

Our journey together is over, but before we part company, let's ponder one final time the primary theme of this book—humanity's place in the universe. Astronomy gives us some comprehension of the workings of stars, galaxies, and planets, but its greatest value lies in what it teaches us about ourselves. Now that you have surveyed astronomical knowledge, you can better understand your own position in nature.

To some, the word nature conjures up visions of furry rabbits hopping about in a forest glade. To others, nature is the blue-green ocean depths, and still others think of nature as windswept mountaintops. As diverse as these images are, they are all Earthbound. Having studied astronomy, you can see nature as a beautiful mechanism composed of matter and energy interacting according to simple rules to form galaxies, stars, planets, mountaintops, ocean depths, forest glades, and people.

Perhaps the most important astronomical lesson is that humanity is a small but important part of the universe. Most of the universe is probably lifeless. The vast reaches between the galaxies appear to be empty of all

Earth photographed by Voyager 1 from the edge of the solar system. The vertical beams are sunlight reflected inside the camera.

Earth ⟶

but the thinnest gas, and stars are much too hot to preserve the chemical bonds that seem necessary for life to survive and develop. It seems that only on the surfaces of a few planets, where temperatures are moderate, can atoms link together in special ways to form living matter.

If life is special, then intelligence is very precious. The universe may contain many planets devoid of life, planets where the wind has blown unfelt for billions of years. There may also exist planets where life has developed but has not become complex, planets on which the wind stirs wide plains of grass and rustles through dark forests. On some planets, equivalents of Earth insects, fish, birds, and animals may watch the passing days at most dimly aware of their own existence. It is intelligence, human or non-human, that gives meaning to the landscape.

Science is the process by which Earth intelligence has tried to understand the physical universe. Science is not the invention of new devices or processes. It does not create home computers, cure the mumps, or manufacture plastic spoons—these are products of engineering and technology, the adaptation of scientific understanding for practical purposes. Science is the understanding of nature, and astronomy is that understanding on the grandest scale. Astronomy is the science by which the universe, through its intelligent lumps of matter, tries to understand its own existence.

As the primary intelligent species on this planet, we are the custodians of a priceless gift—a planet filled with living things. This is especially true if life is rare in the universe. In fact, if Earth is the only inhabited planet, our responsibility is overwhelming. We are the only creatures who can take action to preserve the existence of life on Earth, and, ironically, our own actions are the most serious hazards.

The future of humanity is not secure. We are trapped on a tiny planet with limited resources and a population growing faster than our ability to produce food. In our efforts to survive, we have already driven some creatures to extinction and now threaten others. If our civilization collapses because of starvation, or if our race destroys itself somehow, the only bright spot is that the rest of the creatures on Earth may be better off for our absence.

But even if we control our population and conserve and recycle our resources, life on Earth is doomed. In 5 billion years, the sun will leave the main sequence and swell into a red giant, incinerating Earth. However, Earth will be lifeless long before that. Within the next few billion years, the growing luminosity of the sun will first alter Earth's climate and then vaporize its oceans. Earth, like everything else in the universe, is only temporary.

To survive, humanity must eventually leave Earth and search for other planets. Colonizing the moon and other planets of our solar system will not save us because they will face the same fate as Earth when the sun dies. On the other hand, travel to other stars is tremendously difficult and may be impossible with the limited resources we have in our small solar system. We, and all of the living things on Earth, may be trapped.

This is a sad prospect, but a few factors are comforting. First, everything else in the universe is also temporary. Stars die, galaxies die; perhaps the entire universe will someday end. Our distant future is limited, and this assures us that we are a part of a much larger whole. Second, we have a few billion years to prepare, and a billion years is a very long time. Only a few million years ago, our ancestors were starting to walk upright and communicate. A billion years ago, our ancestors were microscopic organisms living in the oceans. To suppose that a billion years hence there will be beings who resemble today's humans, or that humans will still be the dominant intelligence on Earth, or that humans will even exist, are ultimately conceits.

Our responsibility is not to save our race for all eternity but to behave as dependable custodians of our planet, preserving it, admiring it, and trying to understand it. That calls for drastic changes in our behavior toward other living things and a revolution in our attitude toward our planet's resources. Whether we can change our ways is debatable—humanity is far from perfect in its understanding, abilities, or intentions. However, you must not imagine that we, and our civilization, are less than precious. We have the gift of intelligence, and that is the most wonderful thing this planet has ever produced.

> We shall not cease from exploration
> And the end of all our exploring
> Will be to arrive where we started
> And know the place for the first time.
>
> —T. S. Eliot, "Little Gidding"

Excerpt from "Little Gidding" in *Four Quartets*, copyright 1942 by T. S. Eliot and renewed 1970 by Esme Valerie Eliot, reprinted by permission of Harcourt, Inc. and Faber & Faber, Ltd.

Appendix A

Units and Astronomical Data

The Metric System and SI Units

A system of measurement is based on the three fundamental units for length, mass, and time. By international agreement, there is a preferred subset of metric units known as the *Système International d'Unités* (SI units) that is based on the meter, kilogram, and second.

Density should be expressed in SI units as kilograms per cubic meter, but no human hand can enclose a cubic meter, so that unit does not help you easily grasp the significance of a given density. Instead, this book refers to density in grams per cubic centimeter. A gram is roughly the mass of a paperclip, a cubic centimeter is the size of a small sugar cube, and a density of 1 g/cm^3 equals the density of water, all tangible units.

Americans generally use the British system of units. In that system (now common only in the United States, Tonga, and Southern Yemen, but, ironically, no longer in Great Britain), the fundamental unit of length is the foot, composed of 12 inches. For conceptual purposes, this book expresses some quantities in both metric and English units. Instead of saying that the average person would weigh 133 N on the moon, it might help some readers to see that weight described as 30 lb. Consequently, this text often gives quantities in metric form followed by the British form in parentheses. For example the radius of the moon is 1,738 km (1,080 mi).

Log in at 4ltrpress.cengage.com/astro to view **Supplemental Tables** including a list of the fundamental SI units and their abbreviations in **Table S.1**. The metric prefixes and their abbreviations, used to describe multiples or fractions of the fundamental units, are given in **Table S.2**. Conversion factors between English and metric units are given in **Table S.3**.

Temperature Scales

In astronomy, as in most other sciences, temperatures are expressed on the Kelvin scale, although the centigrade scale is also used. The centigrade scale is also called the Celsius scale after its inventor, the Swedish astronomer Anders Celsius (1701–1744). Temperatures on the centigrade scale are measured from the freezing point of water (0°C). Temperatures on the Kelvin scale are measured from absolute zero, the temperature of an object that contains no extractable heat (0 K). In practice, no object can be as cold as absolute zero, although laboratory apparatus have reached temperatures less than 10^{-6} K. The Kelvin scale is named after the Scottish mathematical physicist William Thomson, Lord Kelvin (1824–1907). Centigrade and Kelvin degree steps are the same size.

The Fahrenheit scale fixes the freezing point of water at 32°F and the boiling point at 212°F. Named after the German physicist Gabriel Daniel Fahrenheit (1686–1736), who made the first successful mercury thermometer in 1720, the Fahrenheit scale is used routinely only in the United States.

Log in at 4ltrpress.cengage.com/astro to view **Supplemental Tables** including a list of numerical values and formulas for conversion of temperatures from one scale to another in **Table S.4**.

Astronomy Units and Constants

Astronomy, and science in general, is a way of learning about nature and understanding the universe. To test hypotheses about how nature works, scientists use observations of nature. The tables that follow contain some of the basic observations that support science's best understanding of the astronomical universe. Of course, these data are expressed in the form of numbers, not because science reduces all understanding to mere numbers, but because the struggle to understand nature is so demanding that science must use every good means available. Quantitative thinking—reasoning mathematically—is one of the most powerful techniques ever invented by the human brain. Thus these tables are not nature reduced to mere numbers but numbers that support humanity's growing understanding of the natural world around us.

Table A.1 defines units of measure commonly used by astronomers. **Table A.2** gives physical constants such as the speed of light, plus basic astronomical data such as the mass and luminosity of the sun. **Table A.3** presents the characteristics of different spectral types of

Table A.1 Units Used in Astronomy

1 angstrom (Å)	$= 10^{-8}$ cm $= 10^{-10}$ m $= 10$ nm
1 Astronomical Unit (AU)	$= 1.50 \times 10^{11}$ m $= 93.0 \times 10^{6}$ miles
1 light-year (ly)	$= 6.32 \times 10^{4}$ AU $= 9.46 \times 10^{15}$ m $= 5.88 \times 10^{12}$ miles
1 parsec (pc)	$= 2.06 \times 10^{5}$ AU $= 3.09 \times 10^{16}$ m $= 3.26$ ly
1 kiloparsec (kpc)	$= 1,000$ pc
1 megaparsec (Mpc)	$= 1,000,000$ pc

Table A.2 Astronomical Constants

Velocity of light (c)	$= 3.00 \times 10^{8}$ m/s
Gravitational constant (G)	$= 6.67 \times 10^{-11}$ m³/s²kg
Mass of H atom	$= 1.67 \times 10^{-27}$ kg
Mass of Earth ($M_{\oplus}$)	$= 5.98 \times 10^{24}$ kg
Earth equatorial radius ($R_{\oplus}$)	$= 6.38 \times 10^{6}$ m
Mass of sun ($M_{\odot}$)	$= 1.99 \times 10^{30}$ kg
Radius of sun ($R_{\odot}$)	$= 6.96 \times 10^{8}$ m
Solar luminosity ($L_{\odot}$)	$= 3.83 \times 10^{26}$ J/s
Mass of moon	$= 7.35 \times 10^{22}$ kg
Radius of moon	$= 1.74 \times 10^{3}$ km

Table A.3 Properties of Main-Sequence Stars

Spectral Type	Absolute Visual Magnitude (M_v)	Luminosity*	Temp. (K)	λ_{max} (nm)	Mass*	Radius*	Average Density (g/cm³)
O5	−5.8	500,000	40,000	72.4	40.0	17.8	0.01
B0	−4.1	20,000	28,000	100	18.0	7.4	0.1
B5	−1.1	800	15,000	190	6.4	3.8	0.2
A0	+0.7	80	9,900	290	3.2	2.5	0.3
A5	+2.0	20	8,500	340	2.1	1.7	0.6
F0	+2.6	6.3	7,400	390	1.7	1.4	1.0
F5	+3.4	2.5	6,600	440	1.3	1.2	1.1
G0	+4.4	1.3	6,000	480	1.1	1.0	1.4
G5	+5.1	0.8	5,500	520	0.9	0.9	1.6
K0	+5.9	0.4	4,900	590	0.8	0.8	1.8
K5	+7.3	0.2	4,100	700	0.7	0.7	2.4
M0	+9.0	0.1	3,500	830	0.5	0.6	2.5
M5	+11.8	0.01	2,800	1,000	0.2	0.3	10.0
M8	+16.0	0.001	2,400	1,200	0.1	0.1	63.0

*Luminosity, mass, and radius are given in terms of the sun's luminosity, mass, and radius.

main-sequence stars. Tables **A.4** and **A.5**, respectively, list the 15 nearest and the 15 brightest stars with their individual properties. The fact that these two star lists have almost no overlap is an important truth about stars that you can read more about in Chapter 6. **Table A.6** gives the physical and orbital properties of the major planets in our solar system. **Table A.7** lists the larger moons of each planet and their characteristics. The outer planets have many smaller moons that are not listed here, with more being discovered each year.

Log in at 4ltrpress.cengage.com/astro to view **Supplemental Tables** including **Table S.5**, which contains the Greek alphabet used in naming the brighter stars.

(continued p. 331)

Table A.4 **The 15 Nearest Stars**

Name	Distance (ly)	Distance (pc)	Apparent Visual Magnitude (m_v)	Absolute Magnitude (M_v)	Spectral Type
Sun			−26.7	4.8	G2
Proxima Cen	4.2	1.3	11.0	15.5	M6
α Cen A	4.4	1.3	0.0	4.4	G2
α Cen B	4.4	1.3	1.3	5.7	K5
Barnard's Star	5.9	1.8	9.5	13.2	M4
Wolf 359	7.7	2.4	13.5	16.7	M6
Lalande 21185	8.3	2.5	7.5	10.5	M2
a CMa A (Sirius A)	8.6	2.6	−1.5	1.4	A1
a CMA B (Sirius B)	8.6	2.6	8.4	11.3	white dwarf
Luyten 726-8A	8.7	2.7	12.6	15.4	M6
Luyten 726-8B	8.7	2.7	12.0	14.9	M5
Ross 154	9.7	3.0	11.0	13.6	M3
Ross 248	10.4	3.2	12.2	14.8	M6
ε Eri	10.5	3.2	3.7	6.2	K2
Luyten 789-6	10.9	3.3	12.2	14.6	M6

Data from the SIMBAD database, operated at CDS, Strasbourg, France.

Table A.5 **The 15 Brightest Stars**

Star	Name	Apparent Visual Magnitude (m_v)	Distance (pc)	Absolute Magnitude (M_v)	Spectral Type
α CMa	Sirius	−1.47	2.6	1.4	A1
α Car	Canopus	−0.72	96	−5.6	F0
α Cen	Rigil Kentaurus	−0.29	1.3	4.1	G2
α Boo	Arcturus	−0.04	11	−0.3	K2
α Lyr	Vega	0.03	7.8	0.6	A0
α Aur	Capella	0.08	13	−0.5	G8
β Ori	Rigel	0.12	240	−6.8	B8
α CMi	Procyon	0.34	3.5	2.6	F5
α Eri	Achernar	0.50	44	−5.1	B3
α Ori	Betelgeuse	0.58	130	−5.0	M2
β Cen	Hadar	0.60	160	−5.4	B1
α Aql	Altair	0.77	5.1	2.2	A7
α Cru	Acrux	0.81	98	−4.2	B0
α Tau	Aldebaran	0.85	20	−0.7	K5
α Vir	Spica	1.04	80	−3.5	B1

Data from the SIMBAD database, operated at CDS, Strasbourg, France.
For multiple star systems, the magnitude given is the combined light of all components; the spectral type is for the primary component.

You will also find **Table S.6**, the periodic chart and list of the chemical elements. One of the primary themes of this book is the history of the atoms in your body: the story of how hydrogen and helium were produced in the big bang and how the rest of the stable elements (up to uranium, atomic number 92) were produced inside stars, either slowly during giant star phases or rapidly during supernova explosions.

Table A.6 **Properties of the Planets**

PHYSICAL PROPERTIES (EARTH = ⊕)

Planet	Equatorial Radius (km)	Equatorial Radius (⊕ = 1)	Mass (⊕ = 1)	Average Density (g/cm³)	Surface Gravity (⊕ = 1)	Escape Velocity (km/s)	Sidereal Period of Rotation	Inclination of Equator to Orbit
Mercury	2,440	0.38	0.056	5.44	0.38	4.3	58.65^d	0°
Venus	6,052	0.95	0.815	5.24	0.90	10.3	243.02^d	177°
Earth	6,378	1.00	1.00	5.50	1.00	11.2	23.93^h	23.4°
Mars	3,396	0.53	0.108	3.94	0.38	5.0	24.62^h	25.2°
Jupiter	71,494	11.21	317.8	1.34	2.54	61.0	9.92^h	3.1°
Saturn	60,330	9.42	95.2	0.69	1.16	35.6	10.54^h	26.4°
Uranus	25,559	4.01	14.5	1.29	0.92	22.0	17.23^h	97.9°
Neptune	24,750	3.93	17.2	1.66	1.19	25.0	16.05^h	28.8°

h = hours; d = days

ORBITAL PROPERTIES

Planet	Semimajor Axis (AU)	Semimajor Axis (10⁶ km)	Orbital Period (years)	Orbital Period (days)	Average Orbital Velocity (km/s)	Orbital Eccentricity	Inclination to Ecliptic
Mercury	0.39	57.9	0.24	87.97	47.9	0.206	7.0°
Venus	0.72	108.2	0.62	224.68	35.0	0.007	3.4°
Earth	1.00	149.6	1.00	365.26	29.8	0.017	0° (by defn)
Mars	1.52	227.9	1.88	686.95	24.1	0.093	1.9°
Jupiter	5.20	778.3	11.9	4,334.3	13.1	0.048	1.3°
Saturn	9.58	1,433.4	29.5	10,760	9.6	0.056	2.5°
Uranus	19.23	2,876.7	84.0	30,685	6.8	0.046	0.8°
Neptune	30.10	4,503.4	164.8	60,189	5.4	0.010	1.8°

Table A.7 Principal Satellites of the Solar System

Planet	Satellite	Radius (km)	Average Distance from Planet (10³ km)	Orbital Period (days)	Orbital Eccentricity	Orbital Inclination
Earth	Moon	1,738	384.4	27.32	0.055	5.1°
Mars	Phobos	14 × 12 × 10	9.4	0.32	0.018	1.0°
	Deimos	8 × 6 × 5	23.5	1.26	0.002	2.8°
Jupiter	Amalthea	135 × 100 × 78	182	0.50	0.003	0.4°
	Io	1,820	422	1.77	0.000	0.3°
	Europa	1,565	671	3.55	0.000	0.5°
	Ganymede	2,640	1,071	7.16	0.002	0.2°
	Callisto	2,420	1,884	16.69	0.008	0.2°
	Himalia	~85	11,470	250.6	0.158	27.6°
Saturn	Janus	110 × 80 × 100	151.5	0.70	0.007	0.1°
	Mimas	196	185.5	0.94	0.020	1.5°
	Enceladus	250	238.0	1.37	0.004	0.0°
	Tethys	530	294.7	1.89	0.000	1.1°
	Dione	560	377	2.74	0.002	0.0°
	Rhea	765	527	4.52	0.001	0.4°
	Titan	2,575	1,222	15.94	0.029	0.3°
	Hyperion	205 × 130 × 110	1,484	21.28	0.104	~0.5°
	Iapetus	720	3,562	79.33	0.028	14.7°
	Phoebe	110	12,930	550.4	0.163	150°
Uranus	Miranda	242	129.9	1.41	0.017	3.4°
	Ariel	580	190.9	2.52	0.003	0°
	Umbriel	595	266.0	4.14	0.003	0°
	Titania	805	436.3	8.71	0.002	0°
	Oberon	775	583.4	13.46	0.001	0°
Neptune	Proteus	205	117.6	1.12	~0	~0°
	Triton	1,352	354.59	5.88	0.00	160°
	Nereid	170	5,588.6	360.12	0.76	27.7°

~ signifies approximately.

Observing the Sky

Observing the sky with the naked eye is as important to modern astronomy as picking up pretty pebbles is to modern geology. The sky is a natural wonder unimaginably bigger than the Grand Canyon, the Rocky Mountains, or any other site that tourists visit every year. To neglect the beauty of the sky is equivalent to geologists neglecting the beauty of the minerals they study. This supplement is meant to act as a tourist's guide to the sky. You analyzed the universe in the textbook's chapters; here you can admire it.

The brighter stars in the sky are visible even from the centers of cities with their air and light pollution. But in the countryside, only a few miles beyond the cities, the night sky is a velvety blackness strewn with thousands of glittering stars. From a wilderness location, far from the city's glare, and especially from high mountains, the night sky is spectacular.

Using Star Charts

The constellations are a fascinating cultural heritage of our planet, but they are sometimes a bit difficult to learn because of Earth's motion. The constellations above the horizon change with the time of night and the seasons.

Because Earth rotates eastward, the sky appears to rotate westward around Earth. A constellation visible in the overhead soon after sunset will appear to move westward, and in a few hours it will disappear below the horizon. Other constellations will rise in the east, so the sky changes gradually through the night.

In addition, Earth's orbital motion makes the sun appear to move eastward among the stars. Each day the sun moves about twice its own diameter, about one degree, eastward along the ecliptic. Consequently, each night at sunset, the constellations are about one degree farther toward the west.

Orion, for instance, is visible in the evening sky in January; but, as the days pass, the sun moves closer to Orion. By March, Orion is difficult to see in the western sky soon after sunset. By June, the sun is so close to Orion that the constellation sets with the sun and is invisible. Not until late July is the sun far enough past Orion for the constellation to become visible rising in the eastern sky just before dawn.

Because of the rotation and orbital motion of Earth, you need more than one star chart to map the sky. Which chart you select depends on the month and the time of night. The charts printed on the cards at the end of this book show the evening sky for each season, as viewed from the northern hemisphere at a latitude typical of the United States or central Europe.

To use the charts, select the appropriate chart and hold it overhead as shown in Figure B.1. If you face south, turn the chart until the words southern horizon are at the bottom of the chart. If you face other directions, turn the chart appropriately. Note that hours are in Standard Time; for Daylight Savings Time (spring, summer, and the first half of fall in the United States) add one hour.

Figure B.1

To use the star charts in this book, select the appropriate chart for the season and time. Hold it overhead and turn it until the direction at the bottom of the chart is the same as the direction you are facing.

DESIGN ELEMENTS

Heading background astrolabe: © Miguel S. Salmeron/Photodisc/Getty Images
Section background photo (globe sculpture): © Comstock Images/Jupiter Images
Looking Back/Ahead box image: © Faruk Ulay/iStockphoto.com
Recap box (Getting Started box in Chapter 1): © Faruk Ulay/iStockphoto.com
Quote circles (hand-drawn circles): © iStockphoto.com

Topic Summaries

Where is Earth in relation to the sun, the planets, the stars, and the galaxies?

- You live on planet Earth, which orbits our star, the sun, once a year.
- As Earth rotates once a day, you see the sun rise and set.
- The other major planets in our solar system, Mercury, Venus, Mars, Jupiter, Saturn, Uranus, and Neptune, orbit our sun in ellipses that are nearly circular.
- Our sun is just one out of the billions of stars that fill our home galaxy, the Milky Way, which is a spiral galaxy.
- The Milky Way Galaxy is just one of billions of galaxies arranged in great clusters, clouds, walls, and filaments that fill the universe.

How do astronomers describe distances?

- Astronomers use scientific notation for very large or very small numbers.
- Astronomers have also invented new units of measure such as the Astronomical Unit (AU) and the light-year (ly). One AU is the average distance from Earth to the sun. Mars, for example, orbits 1.5 AU from the sun. The light-year (ly) is the distance light can travel in one year. The nearest star is 4.2 ly from the sun.

Which objects are big relative to the others, and which are small?

- The moon is only about one-fourth the diameter of Earth, but the sun is about 110 times larger than Earth—a typical size for a star.
- Galaxies contain many billions of stars. You live in the Milky Way Galaxy, which is about 80,000 ly in diameter and contains over 100 billion stars.
- The largest things in the universe are the walls and long filaments containing many clusters of galaxies.

Are there other worlds like Earth?

- Planets as small as Earth are difficult to detect around other stars. Many stars have families of planets like our solar system, and some of those billions of planets may resemble Earth.

Key Terms

field of view The area visible in an image. Usually given as the diameter of the region.

scientific notation The system of recording very large or very small numbers by using powers of 10.

solar system The sun and its planets, asteroids, comets, and so on.

planet A non-luminous body in orbit around a star, large enough to be spherical and to have cleared its orbital zone of other objects.

star A globe of gas held together by its own gravity and supported by internal pressure, which generates energy by nuclear fusion.

Astronomical Unit (AU) Average distance from Earth to the sun; 1.5×10^8 kilometers, or 93.3 million miles.

galaxy A large system of stars, star clusters, gas, dust, and nebulae orbiting a common center of mass.

light-year (ly) Unit of distance equal to the distance light travels in 1 year.

Milky Way The hazy band of light that circles our sky, produced by the glow of our galaxy.

Milky Way Galaxy The spiral galaxy containing our sun, visible in the night sky as the Milky Way.

spiral arms Long spiral pattern of bright stars, star clusters, gas, and dust. Spiral arms extend from the center to the edge of the disk of spiral galaxies.

supercluster A cluster of galaxy clusters.

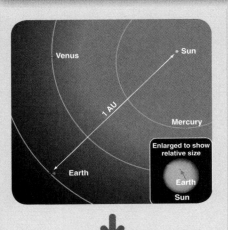

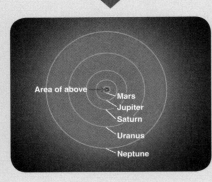

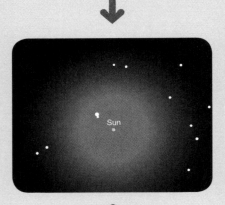

Review Questions

1. What is the largest dimension you have personal knowledge of? Have you run a mile? Hiked 10 miles? Run a marathon? Driven across the country?

2. In Figure 1.4, the division between daylight and darkness is at the right on the globe of Earth. How do you know this is the sunset line and not the sunrise line?

3. What is the difference between our solar system, our galaxy, and the universe?

4. Look at Figure 1.6. How can you tell that Mercury follows an elliptical orbit? Can you detect the elliptical shape of any other orbits in this figure or the next?

5. Which is the outermost major planet in our solar system?

6. Why are light-years more convenient than miles, kilometers, or astronomical units for measuring certain distances?

7. Why is it difficult to detect planets orbiting other stars?

8. What does the size of the star image in a photograph tell us?

9. What is the difference between the Milky Way and the Milky Way Galaxy?

10. What are the largest known structures in the universe?

11. Of the objects listed here, which would be contained inside the object shown below? Which would contain the object? Stars, planets, galaxy clusters, supercluster filaments, spiral arms.

Topic Summaries

2.1 The Stars

How do astronomers refer to stars?

- Astronomers divide the sky into 88 constellations that originated in Greek and Middle Eastern mythology. Modern constellations were added to fill in spaces.
- Modern astronomers often refer to bright stars by a name composed of the constellation name plus a Greek letter assigned according to the star's brightness within the constellation.

How can you compare the brightness of stars?

- The magnitude system is the astronomer's brightness scale. First-magnitude stars are brighter than second-magnitude stars, which are brighter than third-magnitude stars, and so on.
- The magnitude you see when you look at a star in the sky is its apparent visual magnitude, which does not take into account its distance from Earth.

2.2 The Sky and Its Motions

How does the sky appear to move as Earth rotates?

- The celestial sphere is a model of the sky. The northern and southern celestial poles are the pivots on which the sky appears to rotate. The celestial equator, an imaginary line around the sky above Earth's equator, divides the sky in half.
- The angular distance from the horizon to the north celestial pole always equals your latitude. This is an important basis for celestial navigation.
- The gravitational forces of the moon and sun act on the spinning Earth and cause it to precess like a top. Earth's axis of rotation sweeps around in a circular motion with a period of 26,000 years, and consequently the celestial poles and celestial equator move slowly against the background of the stars.

2.3 The Cycle of the Sun

What causes the seasons?

- Because Earth orbits the sun, the sun appears to move eastward along the ecliptic. The inclination of the ecliptic means the sun is in the northern hemisphere half the year, and the southern hemisphere the other half.
- In the summer, the sun is above the horizon longer and shines more directly down on the ground. Both effects cause warmer weather.

2.4 The Cycles of the Moon

Why does the moon go through phases?

- Because you see the moon by reflected sunlight, its shape appears to change as it orbits Earth over a cycle of 29.53 days.

2.5 Eclipses

What is a solar eclipse?

- A total solar eclipse occurs if a new moon passes directly between the sun and Earth and the moon's shadow sweeps over the part of the Earth's surface where you are located. If you are outside the path of totality but the penumbra sweeps over your location, you see a partial eclipse.

Key Terms

constellation One of the stellar patterns identified by name, usually of mythological gods, people, animals, or objects. Also, the region of the sky containing that star pattern.

asterism A named grouping of stars that is not one of the recognized constellations.

magnitude scale The astronomical brightness scale. The larger the number, the fainter the star.

apparent visual magnitude (m_v) A measure of the brightness of a star as seen by human eyes on Earth.

flux A measure of the flow of energy out of a surface. Usually applied to light.

celestial sphere An imaginary sphere of very large radius surrounding Earth to which the planets, stars, sun, and moon seem to be attached.

scientific model A concept that helps you think about some aspect of nature without necessarily being true.

precession The slow change in orientation of the Earth's axis of rotation. One cycle takes nearly 26,000 years.

horizon The circular boundary between the sky and Earth.

zenith The point in the sky directly above the observer.

nadir The point on the celestial sphere directly below the observer; the opposite of the zenith.

north and south celestial poles The points on the celestial sphere directly above Earth's north and south poles.

celestial equator The imaginary line around the sky directly above Earth's equator.

Key Terms

north, south, east, and west points The four cardinal directions; the points on the horizon in those exact directions.

angular distance The angle formed by lines extending from the observer to two locations in the sky.

arc minute 1/60th of a degree.

arc second 1/60th of an arc minute.

angular diameter The angle formed by lines extending from the observer to opposite edges of an object.

circumpolar constellation Any of the constellations so close to the celestial poles that they never set (or never rise) from a given location.

rotation Motion around an axis passing through the rotating body.

revolution Orbital motion about a point located outside the orbiting body.

ecliptic The apparent path of the sun around the sky.

zodiac A band centered on the ecliptic and encircling the sky.

vernal equinox The place on the celestial sphere where the sun crosses the celestial equator moving northward. Also, the beginning of spring.

summer solstice The point on the celestial sphere where the sun is at its most northerly point. Also, the beginning of summer.

autumnal equinox The point on the celestial sphere where the sun crosses the celestial equator going southward. Also, the beginning of autumn.

winter solstice The point on the celestial sphere where the sun is farthest south. Also, the beginning of winter.

perihelion The orbital point of closest approach to the sun.

aphelion The orbital point of greatest distance from the sun.

evening star Any planet visible in the sky just after sunset.

morning star Any planet visible in the sky just before sunrise.

solar eclipse The event that occurs when the moon passes directly between Earth and the sun, blocking your view of the sun.

umbra The region of a shadow that is totally shaded.

penumbra The portion of a shadow that is only partially shaded.

annular eclipse A solar eclipse in which the solar photosphere appears around the edge of the moon in a bright ring, or annulus. Features of the solar atmosphere cannot be seen during an annular eclipse.

lunar eclipse The darkening of the moon when it moves through Earth's shadow.

Saros cycle An 18-year, 11⅓-day period after which the pattern of lunar and solar eclipses repeats.

Topic Summaries

- During a total solar eclipse, the bright disk of the sun is covered, and you can see fainter features of the surrounding solar atmosphere.
- If the moon is in the farther part of its orbit during a solar eclipse, it does not appear large enough to cover the solar disk and you see only an annular eclipse.

What is a lunar eclipse?

- If a full moon passes through Earth's shadow while you are on the night side of Earth, you see the moon darken in a lunar eclipse.
- The totally eclipsed moon looks copper-red because of sunlight refracted through Earth's atmosphere.

Review Questions

1. Do people from other cultures on Earth see the same stars you see? The same constellations and asterisms?

2. In what ways is the celestial sphere a scientific model?

3. Why does the number of circumpolar constellations depend on the latitude of the observer?

4. How could you detect Earth's precession by examining star charts from ancient Egypt?

5. What phase would you see Earth in if you were on the moon when the moon is full? When the moon is at first quarter?

Topic Summaries

3.1 Astronomy before Copernicus

How did people in ancient civilizations describe Earth's place in the universe?

- Ancient philosophers accepted as a first principle that Earth was the unmoving center of the universe. Another first principle was that the heavens were perfect; and, because the circle was the only perfect geometrical form, objects in the heavens must move in uniform (constant speed) circular motion.

- The geocentric (Earth-centered) universe became part of the teachings of the great philosopher Aristotle, who argued that the sun, moon, and stars were carried around Earth on rotating crystalline spheres.

- About AD 140, Ptolemy gave mathematical form to Aristotle's model. Ptolemy kept the geocentric (Earth-centered) and uniform circular motion principles, but he added off-center circles and variable speeds to better predict the motions of the planets.

3.2 Nicolaus Copernicus

How did Copernicus revise that ancient model?

- Copernicus devised a heliocentric (sun-centered) model. He preserved the principle of uniform circular motion, but he argued that Earth rotates on its axis and circles the sun once a year. His theory was controversial because it contradicted Church teaching.

- Copernicus published his theory in his book *De Revolutionibus* in 1543, the year he died.

Why was the Copernican model gradually accepted?

- Because Copernicus kept uniform circular motion, his model did not predict the motions of the planets well, but it did offer a simple explanation of retrograde motion of planets without using large epicycles (circles on circles). He did have to include small epicycles to account for some observed planetary motions.

- The Copernican model was also more eloquent and straightforward. Venus and Mercury were treated the same as all the other planets, and the velocity of each planet was related to its distance from the sun.

3.3 Tycho Brahe, Johannes Kepler, and Planetary Motion

How did Tycho Brahe and Johannes Kepler contribute to the Copernican Revolution?

- Tycho's great contribution was to compile the most precise and detailed naked-eye observations of the planets and stars ever made, observations that were later analyzed by Kepler.

- Kepler inherited Tycho's books of observations in 1601 and used them to uncover three laws of planetary motion. He found that the planets follow ellipses with the sun at one focus, that they move faster when near the sun, and that a planet's orbital period squared is proportional to its orbital radius cubed.

Key Terms

first principle Something that seems obviously true and needs no further examination.

geocentric universe A model universe with Earth at the center, such as the Ptolemaic universe.

heliocentric universe A model of the universe with the sun at the center, such as the Copernican universe.

uniform circular motion The classical belief that the perfect heavens could move only by the combination of uniform motion along circular orbits.

parallax The apparent change in position of an object due to a change in the location of the observer. Astronomical parallax is measured in seconds of arc.

retrograde motion The apparent backward (westward) motion of planets as seen against the background of stars.

epicycle The small circle followed by a planet in the Ptolemaic theory. The center of the epicycle follows a larger circle (the deferent) around Earth.

paradigm A commonly accepted set of scientific ideas and assumptions.

ellipse A closed curve around two points, called the foci, such that the total distance from one focus to the curve and back to the other focus remains constant.

semi-major axis (a) Half of the longest diameter of an ellipse.

eccentricity (e) A number between 1 and 0 that describes the shape of an ellipse; the distance from one focus to the center of the ellipse divided by the semimajor axis.

Key Terms

empirical Description of a phenomenon without explaining why it occurs.

hypothesis A conjecture, subject to further tests, that accounts for a set of facts.

theory A system of assumptions and principles applicable to a wide range of phenomena that has been repeatedly verified.

natural law A theory that has been so well confirmed that it is almost universally accepted as correct.

mass A measure of the amount of matter making up an object.

weight The force that gravity exerts on an object.

inverse square relation A rule that the strength of an effect (such as gravity) decreases in proportion as the distance squared increases.

spring tide Ocean tide of large range that occurs at full and new moon.

neap tide Ocean tide of small range occurring at first- and third-quarter moon.

circular velocity The velocity an object needs to stay in orbit around another object.

geosynchronous satellite A satellite that orbits eastward around Earth with a period of 24 hours and remains above the same spot on Earth's surface.

center of mass The balance point of a body or system of masses. The point about which a body or system of masses rotates in the absence of external forces.

closed orbit An orbit that repeatedly returns to the same starting point.

escape velocity The initial velocity an object needs to escape from the surface of a celestial body.

open orbit An orbit that carries an object away, never to return to its starting point.

Topic Summaries

3.4 Galileo Galilei

Why was Galileo condemned by the Inquisition?

- Galileo used the newly invented telescope to observe the heavens, and he recognized the significance of what he saw there. His discoveries of the phases of Venus, the satellites of Jupiter, the mountains of the moon, and other phenomena helped undermine the Ptolemaic model.

- Galileo based his analysis on observational evidence. In 1633, he was condemned before the Inquisition for refusing to obey an order to halt his defense of Copernicus's model.

3.5 Isaac Newton, Gravity, and Orbital Motion

How did Isaac Newton change humanity's view of nature?

- Newton used the work of Kepler and Galileo to discover three laws of motion and the law of gravity. These laws made it possible to understand such phenomena as orbital motion and the tides.

- Newton's laws gave scientists a unified way to think about nature. Every effect has a cause, and science is the search for those causes.

- The 144 years from Copernicus's book *De Revolutionibus* to Newton's book *Principia* marked the beginning of modern science. From that time on, science depended on evidence to test theories and relied on the methods of mathematical analysis demonstrated by Kepler and Newton.

Review Questions

1. Why did classical Greek astronomers believe that the heavens were made up of perfect crystalline spheres moving at constant speeds?

2. How did the Ptolemaic model explain retrograde motion of the planets?

3. In what ways were the models of Ptolemy and Copernicus similar?

4. Why did the Copernican hypothesis gradually win acceptance?

5. Why is it difficult for scientists to replace an old paradigm with a new paradigm?

6. Why did Tycho Brahe expect the new star of 1572 to show parallax? Why was the lack of parallax evidence against the Ptolemaic model?

7. Explain how Kepler's laws contradict uniform circular motion.

8. What are the differences between a hypothesis, a theory, and a law?

9. Review Galileo's telescope discoveries and explain why they supported the Copernican model and contradicted the Ptolemaic model.

10. If you lived on Mars, which planets would describe retrograde loops? Which would be always seen near the sun? Which would never be visible as crescent phases?

11. Galileo was condemned by the Inquisition, but Kepler, also a supporter of Copernicus, was not. Why not?

12. Why did Newton conclude that gravitation had to be universal?

13. Explain why you might describe the orbital motion of the moon with the statement, "The moon is falling."

Topic Summaries

4.1 Radiation: Information from Space

What is light?

- Light is the visible form of electromagnetic radiation. The electromagnetic spectrum includes gamma rays, X rays, ultraviolet radiation, visible light, infrared radiation, microwaves, and radio waves.
- You can think of a photon as a bundle of waves that acts sometimes as a particle and sometimes as a wave.
- The wavelength of visible light, usually measured in nanometers (10^{-9} m), ranges from 400 nm to 700 nm. Infrared and radio photons have longer wavelengths and carry less energy. Ultraviolet, X-ray, and gamma-ray photons have shorter wavelengths and carry more energy.

4.2 Telescopes

How do telescopes work? What are their capabilities and limitations?

- Astronomers use telescopes to gather light, see fine detail, and magnify images. Two of the three powers of a telescope—light-gathering power and resolving power—depend on the telescope's diameter. Consequently, astronomers strive to build telescopes with large diameters.
- Sometimes astronomical telescopes can be linked together to form an interferometer, which has a resolution equivalent to a telescope as large in diameter as the greatest separation between the individual telescopes.
- The third power of a telescope—magnifying power—is not an inherent property of a telescope but depends on which eyepiece is used.

4.3 Observatories on Earth—Optical and Radio

How are observatories built, and how are the best locations chosen for them?

- Astronomers build optical observatories on high mountains for two reasons: the seeing is better in the calmer and dryer air, and the thinner air is more transparent, especially at infrared wavelengths.
- Radio telescopes consist of one or more large dish reflectors, which record the intensity of the radio energy coming from a spot in the sky.
- Optical and radio telescopes need to be located where light pollution and human-made radio signals do not interfere with observations of faint cosmic sources.
- Telescopes on Earth must move continuously to compensate for Earth's rotation in order to stay pointed at celestial objects.

4.4 Astronomical Instruments and Techniques

What kinds of instruments and techniques do astronomers use to record and analyze light?

- For many decades, astronomers used photographic plates and photometers, but modern electronic array detectors such as CCDs have replaced them in most applications.
- Spectrographs using prisms or a grating can form a spectrum revealing information about the composition and motion of the object being studied.

Key Terms

electromagnetic radiation Changing electric and magnetic fields that travel through space and transfer energy from one place to another; examples are light or radio waves.

wavelength The distance between successive peaks or troughs of a wave, usually represented by λ.

nanometer (nm) A unit of distance equaling one-billionth of a meter (10^{-9} m), commonly used to measure the wavelength of light.

angstrom (Å) A unit of distance commonly used to measure the wavelength of light. $1\ \text{Å} = 10^{-10}$ m.

infrared (IR) The portion of the electromagnetic spectrum with wavelengths longer than red light, ranging from 700 nm to about 1 mm, between visible light and radio waves.

ultraviolet (UV) The portion of the electromagnetic spectrum with wavelengths shorter than violet light, between visible light and X rays.

X ray Electromagnetic waves with wavelengths shorter than ultraviolet light.

gamma ray The shortest-wavelength electromagnetic waves.

photon A quantum of electromagnetic energy that carries an amount of energy that depends inversely on its wavelength.

atmospheric window Wavelength region in which our atmosphere is transparent—at visual, radio, and some infrared wavelengths.

refracting telescope A telescope that forms images by bending (refracting) light with a lens.

reflecting telescope A telescope that forms images by reflecting light with a mirror.

primary lens In a refracting telescope, the largest lens.

*remember

Review questions can be found online at 4ltrpress.cengage.com/astro.

primary mirror In a reflecting telescope, the largest mirror.

eyepiece A short-focal-length lens used to enlarge the image in a telescope. The lens nearest the eye.

focal length The focal length of a lens or mirror is the distance from that lens or mirror to the point where it focuses parallel rays of light.

chromatic aberration A distortion found in refracting telescopes because lenses focus different colors at slightly different distances. Images are consequently surrounded by color fringes.

optical telescope Telescope that gathers visible light.

radio telescope Telescope that gathers radio radiation.

light-gathering power The ability of a telescope to collect light, proportional to the area of the telescope's objective lens or mirror.

resolving power The ability of a telescope to reveal fine detail. Depends on the diameter of the telescope objective.

diffraction fringe Blurred fringe surrounding any image, caused by the wave properties of light. Because of this, no image detail smaller than the fringe can be seen.

interferometer Separated telescopes combined to produce a virtual telescope with the resolution of a much larger-diameter telescope.

seeing Atmospheric conditions on a given night. When the atmosphere is unsteady, producing blurred images, the seeing is said to be poor.

adaptive optics A computer-controlled optical system in an astronomical telescope used to partially correct for seeing.

magnifying power The ability of a telescope to make an image larger.

light pollution The illumination of the night sky by waste light from cities and outdoor lighting, which prevents the observation of faint objects.

4.5 Airborne and Space Observatories

Why do astronomers sometimes use X-ray, ultraviolet, and infrared telescopes, and why must these types of telescopes operate in the upper atmosphere or in orbit?

- X-ray, UV, and IR telescopes observe radiation from celestial objects at a wide range of temperatures and can also see through obscuring dust clouds in space.
- To observe wavelengths outside the visual window and the radio window permitted by our atmosphere, telescopes must go into the upper atmosphere or into space.
- Earth's atmosphere distorts and blurs images. Telescopes in orbit are above this seeing distortion and are limited only by diffraction fringes.

sidereal tracking [pronounced sih-dare-ee-al] The continuous movement of a telescope to keep it pointed at a star as Earth rotates.

prime focus The point at which the objective mirror forms an image in a reflecting telescope.

secondary mirror In a reflecting telescope, a mirror that directs the light from the primary mirror to a focal position.

Cassegrain focus The optical design in which the secondary mirror reflects light back down the tube through a hole in the center of the objective mirror.

Newtonian focus The optical design in which a diagonal mirror reflects light out the side of the telescope tube for easier access.

Schmidt-Cassegrain focus The optical design that uses a thin corrector plate at the entrance to the telescope tube. A popular design for small telescopes.

photographic plate The first image-recording device used with telescopes; it records the brightness of objects, but with only moderate precision.

photometer Sensitive astronomical instrument that measures the brightness of individual objects very precisely.

charge-coupled device (CCD) An electronic device consisting of a large array of light-sensitive elements used to record very faint images.

array detector Device for collecting and recording electromagnetic radiation using multiple individual detectors arrayed on the surface of a chip; for example, a CCD electronic camera.

digitized Converted to numerical data that can be read directly into a computer memory for later analysis.

false-color image A representation of graphical data with added or enhanced color to reveal detail.

spectrograph A device that separates light by wavelengths to produce a spectrum.

spectrum A range of electromagnetic radiation spread into its component wavelengths (colors); for example, a rainbow; also, representation of a spectrum as a graph showing intensity of radiation as a function of wavelength or frequency.

grating A piece of material in which numerous microscopic parallel lines are scribed. Light encountering a grating is dispersed to form a spectrum.

Topic Summaries

5.1 The Sun—Basic Characteristics

How do you know the distance, size, mass, and density of the sun?

- Observing the parallax shift of Venus in transit against the sun's disk from opposite sides of Earth was the original method for finding the distance to the sun. The diameter of the sun can be calculated from its distance and its angular size.
- Newton's laws and the motion of the planets as they orbit the sun allow the mass of the sun to be determined. Its average density can be easily calculated from its mass and diameter.

5.2 The Origin of Sunlight

How does matter produce light?

- Motion among charged particles in a solid, a liquid, or a dense gas causes the emission of blackbody radiation. Pure blackbody radiation is a continuous spectrum.
- The hotter an object is, the more energy it radiates and the shorter is its wavelength of maximum intensity, λ_{max}. This allows astronomers to estimate the temperature of the sun and other stars from their colors.

5.3 The Sun's Surface

What do astronomers see when they observe the sun?

- The photosphere is the level in the sun from which visible photons most easily escape. Its temperature is about 5,800 K.
- Energy flowing outward from the sun's interior travels as rising currents of hot gas and sinking currents of cool gas in the convective zone just below the photosphere.
- The granulation of the photosphere is produced by convection currents of gas rising from below.

5.4 Light, Matter, and Motion

How does matter interact with light to produce spectral lines?

- Electrons in an atom may occupy various permitted orbits around the nucleus but not orbits in between. The size of an electron's orbit depends on the energy stored in the electron's motion.
- An electron may be excited to a higher orbit during a collision between atoms, or it may move from one orbit to another by absorbing or emitting a photon of the proper energy.
- Shifts in the wavelengths of features in spectra of the sun and stars provide clues to the motions of their atoms. The Doppler effect, which reveals the radial velocity of the gas, the part of its velocity directed toward (blueshift) or away from (redshift) Earth.

5.5 The Sun's Atmosphere

What can you learn from the sun's spectrum?

- Because orbits of only certain energies are permitted in an atom, photons of only certain wavelengths can be absorbed or emitted. Each kind of atom has its own structure and therefore its own characteristic set of spectral lines.

Key Terms

spectral line A line in a spectrum at a specific wavelength produced by the absorption or emission of light by certain atoms.

transits of Venus Rare occasions when Venus can be seen as a tiny dot directly between Earth and the sun.

density Mass per volume.

atom The smallest unit of a chemical element, consisting of a nucleus containing protons and neutrons plus a surrounding cloud of electrons.

nucleus The central core of an atom containing protons and neutrons that carries a net positive charge.

proton A positively charged atomic particle contained in the nucleus of an atom. The nucleus of a hydrogen atom.

neutron An atomic particle with no charge and about the same mass as a proton.

electron Low-mass atomic particle carrying a negative charge.

molecule Two or more atoms bonded together.

heat Energy stored in a material as agitation among its particles.

temperature A measure of the agitation among the atoms and molecules of a material.

Kelvin temperature scale A temperature scale using Celsius degrees and based on zero being equal to absolute zero.

absolute zero The theoretical lowest possible temperature at which a material contains no extractable heat energy. Zero on the Kelvin temperature scale.

blackbody radiation Radiation emitted by a hypothetical perfect radiator. The spectrum is continuous, and the wavelength of maximum emission depends on the body's temperature.

Key Terms

wavelength of maximum intensity The wavelength at which a perfect radiator emits the maximum amount of energy. Depends only on the object's temperature.

Wien's Law A law stating that the hotter a glowing object is, the shorter will be its wavelength of maximum intensity, inversely proportional to its temperature.

Stefan-Boltzmann Law A law stating that hotter objects emit more energy than cooler objects of the same size, in proportion to the fourth power of temperature.

photosphere The bright visible surface of the sun.

sunspot Relatively dark spot on the sun that contains intense magnetic fields.

granulation The fine structure of bright grains with dark edges covering the sun's surface.

convection Circulation in a fluid driven by heat. Hot material rises and cool material sinks.

Coulomb force The electrostatic force of repulsion or attraction between charged bodies.

ion An atom that has lost or gained one or more electrons.

ionization The process in which atoms lose or gain electrons.

binding energy The energy needed to pull an electron away from its atom.

quantum mechanics The study of the behavior of atoms and atomic particles.

permitted orbit One of the unique orbits that an electron may occupy in an atom.

isotopes Atoms that have the same number of protons but a different number of neutrons.

energy level One of a number of states an electron may occupy in an atom, depending on its binding energy.

Topic Summaries

- If light from a blackbody such as the sun's photosphere passes through a low-density gas such as the sun's atmosphere on its way to your spectrograph, the gas can absorb photons of certain wavelengths, producing an absorption spectrum.
- The sun's spectrum can tell you its chemical composition through the presence of spectral lines of a certain element. However, you must proceed with care because the strengths of lines also depend on the temperature of the gas.
- If you look at a low-density gas that is excited to emit photons, you see an emission spectrum.
- The solar atmosphere consists of two layers of hot, low-density gas, the chromosphere and the corona.
- The chromosphere is most easily visible during total solar eclipses. Its pink color is caused by the Balmer emission lines in its spectrum.
- Filtergrams of the chromosphere reveal spicules and filaments.
- The corona is the sun's outermost atmospheric layer. It is composed of a very-low-density, very hot gas extending far from the visible sun. Astronomers have evidence that its high temperature—2,000,000 K or more—is maintained by effects of the magnetic carpet.
- Parts of the corona give rise to the solar wind.

5.6 Solar Activity

Why does the sun have a cycle of activity, and how does that affect Earth?

- Sunspots seem dark because they are slightly cooler than the rest of the photosphere. The average sunspot is about twice the size of Earth and contains magnetic fields a few thousand times stronger than Earth's.
- Solar astronomers can study the motion, density, and temperature of gases inside the sun through helioseismology.
- Astronomers can measure magnetic fields on the sun by measuring the splitting of some spectral lines caused by the Zeeman effect.
- The average number of sunspots varies over a period of about 11 years and appears to be related to a magnetic cycle. Alternate sunspot cycles have reversed magnetic polarity, which is explained by the Babcock model.
- The sunspot cycle does not repeat exactly, and the Maunder minimum seems to have been a time when solar activity was very low and Earth's climate was slightly colder.
- The sun rotates differentially with regions far from the equator rotating slower than equatorial regions.
- Spectroscopic observations of other stars reveal that many have spots and magnetic fields that follow long-term cycles like the sun's.
- Prominences occur in the chromosphere; their arched shapes show that they are formed of ionized gas trapped in the magnetic field.
- Flares are sudden eruptions of X-ray, ultraviolet, and visible radiation plus high-energy atomic particles produced when magnetic fields on the sun interact and reconnect. Flares are important because they can have dramatic effects on Earth, such as communications blackouts and auroras.
- Spacecraft images show long streamers extending from the corona out into space. CMEs can produce auroras and other phenomena if they strike Earth.

Review Questions

1. How was the distance to the sun first determined?

2. How is the mass of the sun determined?

3. Define density. How is the density of the sun determined?

4. Why do hot stars look bluer than cool stars?

5. Why does the amount of blackbody radiation emitted depend on the temperature of the object?

6. Why can't you see deeper into the sun than the photosphere?

7. What evidence can you give that granulation is caused by convection?

8. Why is the binding energy of an electron related to the size of its orbit?

9. Why do different atoms have different lines in their spectra?

10. Describe two ways an atom can become excited.

11. What is the difference between an isotope and an ion?

12. Explain why ionized calcium can form absorption lines, but ionized hydrogen cannot.

13. What kind of spectrum does a neon sign produce?

14. How can the Doppler effect explain wavelength shifts in both light and sound?

15. If a nebula contains mostly hydrogen excited to emit photons, what kind of spectrum would you expect it to produce?

16. Explain why the absence of spectral lines of a given element in the solar spectrum would not necessarily mean that element is absent from the sun.

17. How can astronomers detect structure in the sun's chromosphere?

18. What evidence can you give that the corona has a very high temperature?

19. What heats the chromosphere and corona to high temperatures?

20. How are astronomers able to explore the layers of the sun below the photosphere?

21. What evidence can you give that sunspots are magnetic?

22. How does the Babcock model explain the sunspot cycle?

23. What does the shape of a prominence reveal?

24. How can solar flares affect Earth?

25. The upper two images below show two solar phenomena. What are they, and how are they related? How do they differ?

26. The lower image was recorded in the extreme ultraviolet by the SOHO spacecraft. Explain the features you see.

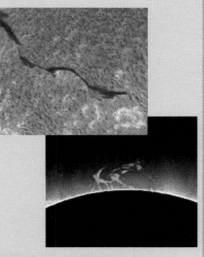

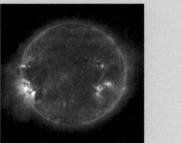

Key Terms

excited atom An atom in which an electron has moved from a lower to a higher energy level.

ground state The lowest permitted electron energy level in an atom.

quantum leap Jumps of electrons from one orbit or energy state to another.

Doppler effect The change in the wavelength of radiation due to relative radial motion of source and observer.

blueshift A Doppler shift toward shorter wavelengths caused by a velocity of approach.

redshift A Doppler shift toward longer wavelengths caused by a velocity of recession.

radial velocity (V_r) That component of an object's velocity directed away from or toward Earth.

chromosphere Bright gases just above the photosphere of the sun.

corona The faint outer atmosphere of the sun, composed of low-density, high temperature gas.

continuous spectrum A spectrum in which there are no absorption or emission lines.

absorption spectrum (dark-line spectrum) A spectrum that contains absorption lines.

absorption line A dark line in a spectrum produced by the absence of photons absorbed by atoms or molecules.

emission spectrum (bright-line spectrum) A spectrum produced by photons emitted by an excited gas.

emission line A bright line in a spectrum caused by the emission of photons from atoms.

Kirchhoff's Laws A set of laws that describe the absorption and emission of light by matter.

Key Terms

transition The movement of an electron from one atomic energy level to another.

Lyman series Spectral lines in the ultraviolet spectrum of hydrogen produced by transitions whose lowest energy level is the ground state.

Balmer series A series of spectral lines produced by hydrogen in the near-ultraviolet and visible parts of the spectrum. The three longest-wavelength Balmer lines are visible to the human eye.

Paschen series Spectral lines in the infrared spectrum of hydrogen produced by transitions whose lowest energy level is the third.

filtergram A photograph (usually of the sun) taken in the light of a specific region of the spectrum—for example, an H$_\alpha$ filtergram.

filament A solar eruption, seen from above, silhouetted against the bright photosphere.

spicule A small, flamelike projection in the chromosphere of the sun.

coronagraph A telescope designed to capture images of faint objects such as the corona of the sun that are near relatively bright objects.

magnetic carpet The network of small magnetic loops that covers the solar surface.

solar wind Rapidly moving atoms and ions that escape from the solar corona and blow outward through the solar system.

helioseismology The study of the interior of the sun by the analysis of its modes of vibration.

Maunder butterfly diagram A graph showing the latitude of sunspots versus time, first plotted by W. W. Maunder in 1904.

Zeeman effect The splitting of spectral lines into multiple components when the atoms are in a magnetic field.

Maunder minimum A period between 1645 and 1715 of less numerous sunspots and other solar activity.

active region A magnetic region on the solar surface that includes sunspots, prominences, flares, and the like.

differential rotation The rotation of a body in which different parts of the body have different periods of rotation. This is true of the sun, the Jovian planets, and the disk of the galaxy.

dynamo effect The process by which a rotating, convecting body of conducting matter, such as Earth's core, can generate a magnetic field.

convective zone The region inside a star where energy is carried outward as rising hot gas and sinking cool gas.

Babcock model A model of the sun's magnetic cycle in which the differential rotation of the sun winds up and tangles the solar magnetic field. This is thought to be responsible for the sunspot cycle.

prominence Eruption on the solar surface. Visible during total solar eclipses.

flare A violent eruption on the sun's surface.

reconnection On the sun, the merging of magnetic fields to release energy in the form of flares.

aurora The glowing light display that results when a planet's magnetic field guides charged particles toward the north and south magnetic poles, where they strike the upper atmosphere and excite atoms to emit photons.

coronal mass ejection (CME) Matter ejected from the sun's corona in powerful surges guided by magnetic fields.

coronal hole An area of the solar surface that is dark at X-ray wavelengths, thought to be associated with divergent magnetic fields and the source of the solar wind.

Topic Summaries

6.1 Star Distances

How far away are the stars?

- Distance is critical in astronomy. Finding the luminosities, diameters, and masses of stars requires first finding their distances.
- Astronomers can measure the distance to nearby stars by observing their parallaxes. Stellar distances are commonly expressed in parsecs. One parsec is 206,265 AU—the distance to an imaginary star whose parallax is 1 arc second.
- Stars farther away than about 170 pc have parallaxes too small to measure from ground-based observatories.

6.2 Apparent Brightness, Intrinsic Brightness, and Luminosity

How much energy do stars make?

- Once you know the distance to a star, you can find its intrinsic brightness, which can be expressed as its absolute magnitude. The absolute magnitude of a star equals the apparent magnitude it would have if it were 10 pc away.
- The luminosity of a star, which can be found from its absolute magnitude, is a measure of the total energy radiated by the star. Luminosity can be expressed in watts, but luminosities of stars are often given in units of the luminosity of the sun.

6.3 Star Temperatures

How can you tell a star's temperature using its spectrum?

- The hydrogen Balmer lines are weak in cool stars because atoms are not excited out of the ground state. In hot stars, the Balmer lines are weak because atoms are excited to higher orbits or are ionized. Only at intermediate temperatures are the Balmer lines strong.
- The strengths of many spectral lines in a star's spectrum can be used to tell you its temperature. Stars are classified in the temperature spectral sequence: O, B, A, F, G, K, M.
- Long after the spectral sequence was created, astronomers discovered L and T objects with temperatures even cooler than the M stars.

6.4 Star Sizes

How big are stars?

- The H–R diagram is a plot of luminosity versus surface temperature. It is an important graph in astronomy because it sorts the stars into categories by size.
- Roughly 90 percent of normal stars, including the sun, fall on the main sequence, with the hotter main-sequence stars being more luminous. The giants and supergiants, however, are much larger and lie above the main sequence in the diagram. Some of the white dwarfs are hot stars, but they fall below the main sequence because they are so small.
- The large size of the giants and supergiants means their atmospheres have low densities. Giant stars, luminosity class III, have narrow spectral lines, and supergiants, class I, have extremely narrow lines. Class V main-sequence stars have relatively broad spectral lines.

Key Terms

stellar parallax (p) The small apparent shift in position of a nearby star relative to distant background objects due to Earth's orbital motion.

parsec (pc) The distance to a hypothetical star whose parallax is 1 second of arc. 1 pc = 206,265 AU = 3.26 ly.

intrinsic brightness A measure of the amount of light a star produces.

flux A measure of the flow of energy out of a surface. Usually applied to light.

absolute visual magnitude (M_v) Intrinsic brightness of a star. The apparent visual magnitude the star would have if it were 10 pc away.

luminosity (L) The total amount of energy a star radiates per second at all wavelengths.

spectral class A star's position in the temperature classification system O, B, A, F, G, K, M, based on the appearance of the star's spectrum.

spectral sequence The arrangement of spectral classes (O, B, A, F, G, K, M) ranging from hot to cool.

Hertzsprung–Russell (H–R) diagram A plot of the intrinsic brightness versus the surface temperature of stars. It separates the effects of temperature and surface area on stellar luminosity and is commonly plotted as absolute magnitude versus spectral type but also as luminosity versus surface temperature or color.

main sequence The region of the H–R diagram running from upper left to lower right, which includes roughly 90 percent of all stars generating energy by nuclear fusion.

giant Large, cool, highly luminous star in the upper right of the H–R diagram, typically 10 to 100 times the diameter of the sun.

supergiant Exceptionally luminous star whose diameter is 100 to 1,000 times that of the sun.

red dwarf A faint, cool, low-mass, main-sequence star.

white dwarf Dying star at the lower left of the H–R diagram that has collapsed to the size of Earth and is slowly cooling off.

luminosity class A category of stars of similar luminosity, determined by the widths of lines in their spectra.

spectroscopic parallax The method of determining a star's distance by comparing its apparent magnitude with its absolute magnitude as estimated from its spectrum.

binary stars Pairs of stars that orbit around their common center of mass.

visual binary system A binary star system in which the two stars are separately visible in the telescope.

spectroscopic binary system A star system in which the stars are too close together to be visible separately. We see a single point of light, and only by taking a spectrum can we determine that there are two stars.

eclipsing binary system A binary star system in which the stars cross in front of each other as seen from Earth.

light curve A graph of brightness versus time commonly used in analyzing variable stars and eclipsing binaries.

mass–luminosity relation The more massive a main-sequence star is, the more luminous it is.

- Stars so far away that their parallaxes are too small to measure can have their distances estimated by the technique of spectroscopic parallax.

6.5 Star Masses—Binary Stars

How much matter do stars contain?

- The only direct way you can find the mass of a star is by studying binary stars. When two stars orbit a common center of mass, astronomers find their masses by observing the period and sizes of their orbits.

- Few binary star systems are easy to analyze; most are spectroscopic binaries in which the component stars are known only by the alternating Doppler shifts of their spectral lines.

- Eclipsing binary star systems allow measurement not just of the component stars' masses but also independent checks on their temperatures and diameters.

6.6 Typical Stars

What is the typical star like?

- Given the mass and diameter of a star, you can find its average density. On the main sequence, the stars are about as dense as the sun, but the giants and supergiants are very-low-density stars. Some are much thinner than air. The white dwarfs, lying below the main sequence, are tremendously dense.

- The mass–luminosity relation says that the more massive a main sequence star is, the more luminous it is. Giants and supergiants do not follow this relation in general, and white dwarfs not at all.

- A survey of stars in the neighborhood of the sun shows that giants and supergiants are rare and that red dwarfs, stars at the low-luminosity end of the main sequence, are the most common type but are also faint and hard to find.

1. Why are Earth-based parallax measurements limited to the nearest stars?

2. What does luminosity measure that is different from what absolute visual magnitude measures?

3. Why are hydrogen Balmer lines strong in the spectra of medium-temperature stars and weak in the spectra of hot and cool stars?

4. Why does the luminosity of a star depend on both its radius and its temperature?

5. How can you be sure that giant stars really are larger than main-sequence stars?

6. Why do astronomers conclude that white dwarfs must be very small?

7. What observations would you make to classify a star according to its luminosity? Why does that method work?

8. Why does the orbital period of a binary star depend on its mass?

9. Why don't astronomers know the inclination of a spectroscopic binary? How do they know the inclination of an eclipsing binary?

10. If all of the stars in the photo to the left are members of the same star cluster, then they all have about the same distance. Then why are three of the brightest much redder than the rest? What kind of star are they?

Topic Summaries

7.1 Stellar Structure

What are the insides of stars like?

- For a star to be stable, it must maintain hydrostatic equilibrium and the deep layers must be hotter and denser and have higher pressure than the less-deep layers.
- Energy in a star must flow from the hot interior to the cool exterior by conduction, convection, or radiation. Conduction is not usually important inside stars.
- Much of what astronomers know about stars is based on detailed mathematical models of the interior of stars.

7.2 Nuclear Fusion in the Sun and Stars

How do the stars make energy?

- Stars generate energy near their centers, where conditions allow nuclear fusion reactions to combine hydrogen nuclei to make helium nuclei.
- Energy generated in a star's core flows outward either as photons moving through a radiative zone or as rising currents of hot gas and sinking currents of cool gas in a convective zone.
- Stars of the sun's mass or less make energy via the proton–proton chain; more massive main-sequence stars fuse hydrogen into helium in the CNO cycle.
- Observations of too few neutrinos coming from the sun's core are now explained by the changing of neutrinos back and forth among three different types called "flavors." The neutrinos confirm that the sun makes its energy by hydrogen fusion.
- The relationship between pressure and temperature, called the pressure–temperature thermostat, ensures that stars generate just enough energy to be stable.

7.3 Main-Sequence Stars

What determines the properties of main-sequence stars?

- The luminosity of a star depends on its mass. More massive stars have more weight to support, and their pressure–temperature thermostats must make more energy. That makes them more luminous.
- The main sequence has a minimum mass because objects less massive than 0.08 solar masses cannot get hot enough to begin hydrogen fusion.

How long can a star survive?

- The more massive a star is, the faster it uses up its hydrogen fuel. A 25-solar-mass star will exhaust its hydrogen and die in only about 7 million years.

7.4 The Birth of Stars

How are stars born?

- Stars form from the gas and dust of the interstellar medium.
- Astronomers know there is an interstellar medium because they can see emission nebulae, reflection nebulae, and dark nebulae. They can also detect interstellar absorption lines in the spectra of distant stars.

Key Terms

conservation of mass One of the basic laws of stellar structure: The total mass of the star must equal the sum of the shell masses.

conservation of energy One of the basic laws of stellar structure: The total luminosity must equal the sum of energy generated in all of the layers.

hydrostatic equilibrium One of the basic laws of stellar structure: The weight on each layer must be balanced by the pressure in that layer.

energy transport One of the basic laws of stellar structure: Energy must move from hot to cool regions by conduction, radiation, or convection.

opacity The resistance of a gas to the passage of radiation.

stellar model A table of numbers representing the conditions in various layers within a star.

nuclear forces The two forces of nature that only affect the particles in the nuclei of atoms.

nuclear fission Reactions that break the nuclei of atoms into fragments.

nuclear fusion Reactions that join the nuclei of atoms to form more massive nuclei.

Coulomb barrier The electrostatic force of repulsion between bodies of like charge, commonly applied to atomic particles.

proton–proton chain A series of three nuclear reactions that builds a helium atom by adding together protons. The main energy source in the sun.

deuterium An isotope of hydrogen in which the nucleus contains a proton and a neutron.

neutrino A neutral, nearly massless atomic particle that travels at or nearly at the speed of light.

CNO (carbon–nitrogen–oxygen) cycle A series of nuclear reactions that use carbon as a catalyst to combine four hydrogen nuclei to make one helium nucleus plus energy, effective in stars more massive than the sun.

brown dwarf A stellar object with such low mass that it cannot raise its central temperature high enough to sustain hydrogen fusion.

zero-age main sequence (ZAMS) The location in the H–R diagram where stars first reach stability as hydrogen-burning stars.

interstellar medium (ISM) The gas and dust distributed between the stars.

interstellar dust Microscopic solid grains in the interstellar medium.

nebula A relatively dense cloud of interstellar gas and dust.

interstellar reddening The process in which dust scatters blue light out of starlight and makes the stars look redder.

molecular cloud A dense interstellar gas cloud in which atoms are able to link together to form molecules such as H_2 and CO.

shock wave A sudden change in pressure that travels as an intense sound wave.

emission nebula A cloud of glowing gas excited by ultraviolet radiation from hot stars.

HII region A region of ionized hydrogen around a hot star.

reflection nebula A nebula produced by starlight reflecting off dust particles in the interstellar medium.

dark nebula A cloud of gas and dust seen silhouetted against a brighter nebula.

star cluster A group of stars that formed together and orbit a common center of mass.

stellar association A group of stars that formed together but are not gravitationally bound to one another.

protostar A collapsing cloud of gas and dust destined to become a star.

Bok globule Small, dark cloud only about 1 ly in diameter that contains 10 to 1,000 solar masses of gas and dust, thought to be related to star formation.

Herbig–Haro object A small nebula that varies irregularly in brightness, evidently associated with star formation.

T Tauri star A young star surrounded by gas and dust, understood to be contracting toward the main sequence.

- The dust in the interstellar medium makes distant stars look fainter and redder than expected.
- Molecular clouds are sites of star formation. Such clouds can be triggered to collapse by collision with a shock wave that compresses and fragments the gas cloud, producing a cluster of protostars.

How do you know that theories of star formation are correct?

- A contracting protostar is cooler than a main-sequence star. The dust in the cocoon absorbs the protostar's light and re-radiates it as infrared radiation.
- As a protostar grows hot enough to begin hydrogen fusion at its core, it settles onto the main sequence.
- Very young star clusters contain large numbers of T Tauri stars.
- Some protostars have been found emitting jets of gas. Where these flows strike existing clouds of gas, astronomers can see Herbig–Haro objects. The jets are evidently focused by disks of gas and dust around the protostars.
- The Great Nebula in Orion is an active region of star formation. The bright stars in the center of the nebula formed within the last few million years, and infrared telescopes detect protostars buried inside the molecular cloud behind the visible nebula.

1. What opposing forces are balanced in a stable star?
2. Why does the CNO cycle require a higher temperature than the proton–proton chain?
3. Why is there a mass–luminosity relation for stars on the main sequence?
4. Why does a star's life expectancy depend on its mass?
5. How are Herbig–Haro objects related to star formation? T Tauri stars?

Topic Summaries

8.1 Giant Stars

What happens to a star when it uses up the last of the hydrogen fuel in its core?

- As a star uses up the last of its hydrogen, the nuclear reactions die down, the core contracts, heats up, and forms a hydrogen-fusion shell. Energy flowing from the hydrogen-fusion shell swells the star into a cool giant or a supergiant.
- Eventually contraction of the star's core ignites helium, first in the core and later in a shell. Massive stars can eventually fuse carbon and other elements.
- H–R diagrams demonstrate how stars in a cluster begin their evolution at the same time but evolve at different rates, depending on their masses.

8.2 The Deaths of Lower-Main-Sequence Stars

How will the sun and stars smaller than the sun die?

- Red dwarfs less massive than 0.4 solar mass have very little hydrogen left when they die. They cannot become giant stars and will remain on the main sequence for a great amount of time.
- Medium-mass stars between about 0.4 and 4 solar masses become cool giants and fuse helium but will never be hot enough in their cores to fuse carbon.
- Some medium-mass stars, up to about 4 solar masses, produce planetary nebulae and then become white dwarfs.

8.3 The Evolution of Binary Systems

What happens if an evolving star is in a binary star system?

- Mass can transfer in a binary system through accretion disks around receiving stars. Hot accretion disks can emit light and X rays.
- Mass transferred onto a white dwarf can build and erupt as a nova explosion. As long as mass transfers form new layers of fuel, eruptions can occur repeatedly.

8.4 The Deaths of Massive Stars

How do massive stars die?

- The massive stars on the upper main sequence fuse nuclear fuels up to iron, then the iron core collapses and triggers a type II supernova.
- A type Ia supernova can occur when mass transferred onto a white dwarf increases its mass above the Chandrasekhar limit, collapses, and immediately fuses its remaining nuclear fuel. A type Ib supernova occurs when a massive star in a binary system loses its outer layers of hydrogen before exploding.

8.5 Neutron Stars

What are neutron stars, and how do you know they actually exist?

- Theory predicts that a collapsing core cannot support itself as a white dwarf if its mass is greater than the Chandrasekhar limit. If its mass lies between 1.4 solar masses and about 3 solar masses, it can halt its contraction and form a neutron star.
- A neutron star is supported by degenerate neutron pressure. Theory predicts that a neutron star should spin very fast, have a surface temperature of millions of degrees K, and have a powerful magnetic field.

Key Terms

nova From the Latin, meaning "new," a sudden and temporary brightening of a star making it appear as a new star in the sky, evidently caused by an explosion of nuclear fuel on the surface of a white dwarf.

supernova A "new star" in the sky that is roughly 4,000 times more luminous than a normal nova and longer lasting, evidently the result of an explosion of a star.

giant star Large, cool, highly luminous star in the upper right of the H–R diagram, typically 10 to 100 times the diameter of the sun.

supergiant star Exceptionally luminous star whose diameter is 100 to 1,000 times that of the sun.

horizontal branch The location in the H–R diagram of giant stars fusing helium.

planetary nebula An expanding shell of gas ejected from a medium-mass star during the latter stages of its evolution.

open cluster A cluster of 100 to 1,000 stars with an open, transparent appearance, usually relatively young and located in the disk of the galaxy. The stars are not tightly grouped.

globular cluster A star cluster containing 100,000 to 1 million stars in a sphere about 75 ly in diameter, generally old, metal-poor, and found in the spherical component of the galaxy.

turnoff point The point in an H–R diagram at which a cluster's stars turn off of the main sequence and move toward the red-giant region, revealing the approximate age of the cluster.

degenerate matter Extremely high-density matter in which pressure no longer depends on temperature due to quantum mechanical effects.

compact object One of the three final states of stellar evolution, which generates no nuclear energy and is much smaller and denser than a normal star.

Chandrasekhar limit The maximum mass of a white dwarf, about 1.4 solar masses. A white dwarf of greater mass cannot support itself and will collapse.

Roche lobe The volume of space a star controls gravitationally within a binary system.

angular momentum A measure of the tendency of a rotating body to continue rotating. Mathematically, the product of mass, velocity, and radius.

accretion disk The rotating disk that forms in some situations as matter is drawn gravitationally toward a central body.

type I supernova A supernova whose spectrum contains no hydrogen lines.

type II supernova A supernova explosion caused by the collapse of a massive star.

synchrotron radiation Radiation emitted when high-speed electrons move through a magnetic field.

supernova remnant The expanding shell of gas and dust marking the site of a supernova explosion.

neutron star A small, highly dense star, with radius about 10 km, composed almost entirely of tightly packed neutrons.

pulsar A source of short, precisely timed radio bursts, understood to be spinning neutron stars.

lighthouse model The explanation of a pulsar as a spinning neutron star sweeping beams of electromagnetic radiation around the sky.

general theory of relativity Einstein's theory that describes gravity as due to curvature of space-time.

Topic Summaries

- Pulsars were discovered in 1967.
- Astronomers detect pulsars if their beams sweep across Earth as they spin.
- A neutron star slows as it radiates its energy into space, though millisecond pulsars appear to be old and fast.
- The pulsars found in binary systems allow astronomers to estimate the masses of neutron stars.
- Planets have been found orbiting at least one neutron star. They may be the remains of a companion star that was mostly devoured by the neutron star.

8.6 Black Holes

What are black holes, and how do you know they really exist?

- Black holes occur when the collapsing core of a supernova greater than 3 solar masses contracts to such a small size that no radiation can escape.
- There are two relativistic effects on an object falling into a black hole. Compared to the observer, the object's clock would slow and light would be redshifted. The object would heat and deform from powerful tidal forces.
- Binary star systems where mass flows into a compact object with mass greater than 3 solar masses and emits X rays indicate possible black holes.
- Black holes and neutron stars at the center of accretion disks can eject powerful beams of radiation and gas. Such beams have been detected.
- Gamma-ray bursts appear to be related to violent events involving neutron stars and black holes. Merging binary compact objects or shifts in neutron star crusts may produce short bursts.

gravitational radiation Expanding waves in a gravitational field that transport energy through space at the speed of light, as predicted by general relativity.

millisecond pulsar A pulsar with a pulse period of only a few milliseconds.

singularity An object of zero radius and infinite density.

black hole A mass that has collapsed to such a small volume that its gravity prevents the escape of all radiation. Also, the volume of space from which radiation may not escape.

event horizon The boundary of the region of a black hole from which no radiation may escape. No event that occurs within the event horizon is visible to a distant observer.

Schwarzschild radius (R_S) The radius of the event horizon around a black hole.

time dilation The slowing of moving clocks or clocks in strong gravitational fields.

gravitational redshift The lengthening of the wavelength of a photon as it escapes from a gravitational field.

gamma-ray burst A sudden, powerful burst of gamma rays.

hypernova Produced when a very massive star collapses into a black hole. Thought to be a possible source of gamma-ray bursts.

*remember

Review questions can be found online at 4ltrpress.cengage.com/astro.

Topic Summaries

9.1 The Discovery of the Galaxy

How do you know you live in a galaxy?

- The hazy band of the Milky Way is our wheel-shaped galaxy seen from within, but its size and shape are not obvious. William and Caroline Herschel counted stars over the sky to show that it seemed to be shaped like a disk with the sun near the center, but they could not see very far into space because of interstellar gas and dust.

- In the early 20th century, Harlow Shapley calibrated Cepheid variable stars to find the distance to globular clusters and demonstrated that our galaxy is much larger than the part you can see easily and that the sun is not at the center.

- Modern observations suggest that our galaxy contains a disk component about 80,000 ly in diameter and that the sun is two-thirds of the way from the center to the visible edge. The central bulge around the center and an extensive halo containing old stars and little gas and dust make up the spherical component.

- The mass of the galaxy can be found from its rotation curve. Kepler's third law tells astronomers the galaxy contains over 100 billion solar masses, and the rising rotation curve at great distance from the center shows that the halo must contain much more mass. Because that material is not emitting detectable electromagnetic radiation, astronomers call it dark matter.

9.2 Spiral Arms and Star Formation

How do you know ours is a spiral galaxy, and what are the spiral arms?

- The most massive stars live such short lives that they don't have time to move from their place of birth. Because they are concentrated along the spiral arms, astronomers conclude that the spiral arms are sites of star formation.

- Spiral arms can be traced through the sun's neighborhood by using spiral tracers such as O and B stars; but to extend the map over the entire galaxy, astronomers must use infrared and radio observations to see through the gas and dust.

- The spiral density wave theory suggests that the spiral arms are regions of compression that move around the disk. When an orbiting gas cloud overtakes the compression wave, the gas cloud forms stars. Another process, self-sustaining star formation, may act to modify the arms with branches and spurs as the birth of massive stars triggers the formation of more stars by compressing neighboring gas clouds.

9.3 The Origin and History of the Milky Way

How did our galaxy form and evolve?

- The oldest open star clusters indicate that the disk of our galaxy is only about 9 billion years old. The oldest globular clusters appear to be at least 13 billion years old, so our galaxy must have begun forming at least 13 billion years ago.

- Stellar populations are an important clue to the formation of our galaxy. The first stars to form in our galaxy, termed population II stars, were poor in elements heavier than helium—elements that astronomers call metals. As generations of stars manufactured metals and spread them back into the interstellar medium, the metal abundance of more recent generations increased. Population I stars, including the sun, are richer in metals than population II stars.

Key Terms

Cepheid variable stars Variable stars with pulsation periods of 1 to 60 days whose period of variation is related to their luminosity.

instability strip The region of the H–R diagram in which stars are unstable to pulsation. A star evolving through this strip becomes a variable star.

period–luminosity relation The relation between period of pulsation and intrinsic brightness among Cepheid variable stars.

proper motion The rate at which a star moves across the sky, measured in seconds of arc per year.

calibrate To make observations of reference objects, checks on instrument performance, calculations of units conversions, and so on, needed to completely understand measurements of unknown quantities.

disk component All material confined to the plane of the galaxy.

kiloparsec (kpc) A unit of distance equal to 1,000 pc or 3,260 ly.

spiral arms Long spiral pattern of bright stars, star clusters, gas, and dust. Spiral arms extend from the center to the edge of the disk of spiral galaxies.

spherical component The part of the galaxy including all matter in a spherical distribution around the center (the halo and central bulge).

halo The spherical region of a spiral galaxy, containing a thin scattering of stars, star clusters, and small amounts of gas.

central bulge The spherical cloud of stars that lies at the center of spiral galaxies.

rotation curve A graph of orbital velocity versus radius in the disk of a galaxy.

Key Terms

dark halo The low-density extension of the halo of our galaxy, believed to be composed of dark matter.

dark matter Nonluminous matter that is detected only by its gravitational influence.

spiral tracer Object used to map the spiral arms—for example, O and B associations, open clusters, clouds of ionized hydrogen, and some types of variable stars.

density wave theory Theory proposed to account for spiral arms as compressions of the interstellar medium in the disk of the galaxy.

self-sustaining star formation The process by which the birth of stars compresses the surrounding gas clouds and triggers the formation of more stars.

population I star Stars rich in atoms heavier than helium, nearly always relatively young stars found in the disk of the galaxy.

population II star Stars poor in atoms heavier than helium, nearly always relatively old stars found in the halo, globular clusters, or the central bulge.

metals In astronomical usage, all atoms heavier than helium.

monolithic collapse model An early hypothesis that says that the galaxy formed from the collapse of a single large cloud of turbulent gas over 13 billion years ago.

Sagittarius A* The powerful radio source located at the core of the Milky Way Galaxy.

Topic Summaries

- Because the halo is made up of population II stars and the disk is made up of population I stars, astronomers conclude that the halo formed first and the disk later. The monolithic collapse model, that the galaxy formed from a single, roughly spherical cloud of turbulent gas and gradually flatted into a disk, has been amended to include mergers with other galaxies and later infalling gas contributing to the disk.

9.4 The Nucleus

What lies at the very center of our galaxy?

- The nucleus of the galaxy is not visible at visual wavelengths, but radio, infrared, and X-ray radiation can penetrate the gas and dust in space. These wavelengths reveal crowded central stars and warmed dust.

- The very center of the Milky Way Galaxy is marked by a radio source, Sagittarius A*. The source must be less than a few astronomical units in diameter, but the motion of stars around the center shows that it must contain roughly 4.3 million solar masses. A supermassive black hole is the only object that could contain so much mass in such a small space.

Review Questions

1. Why isn't it possible to locate the center of our galaxy from the appearance of the Milky Way?

2. Why is there a period–luminosity relation for Cepheid variable stars?

3. How can astronomers use Cepheid variable stars to find distances?

4. Why didn't astronomers before Shapley realize how large the galaxy is?

5. How can astronomers find the mass of the galaxy?

6. What evidence is there that our galaxy has an extended halo of dark matter?

7. Why do spiral tracers have to be short lived?

8. What evidence is there that the density wave theory is not fully adequate to explain spiral arms in our galaxy?

9. How do astronomers know how old our galaxy is?

10. Why do astronomers conclude that stars with metal-poor spectra are older than stars with metal-rich spectra?

11. How do the orbits of stars around the Milky Way Galaxy help explain its origin?

12. What evidence contradicts the monolithic collapse model for the origin of our galaxy?

13. What is the evidence that the center of our galaxy contains a massive black hole?

14. Why must astronomers use infrared telescopes to observe the motions of stars around Sgr A*?

15. Why does the galaxy shown at the top below have so much dust in its disk? How big do you suppose the halo of this galaxy really is?

16. Why are the spiral arms in the galaxy at the bottom below blue? What color would the halo be if it were bright enough to see in this photo?

Topic Summaries

10.1 The Family of Galaxies

What types of galaxies exist?

- Galaxies can be divided into three classes, with subclasses specifying details of the galaxy's shape.
- Elliptical galaxies contain little gas and dust and cannot make new stars. Consequently, they lack hot, blue stars and have a relatively red color.
- Spiral galaxies contain more gas and dust and support active star formation. The massive, hot, and blue stars give these galaxies a blue color.
- A spiral galaxy's halo and nuclear bulge usually lack gas and dust and contain little star formation and so have a red color.
- Irregular galaxies have no obvious shape but contain gas and dust and support star formation.

10.2 Measuring the Properties of Galaxies

How do astronomers measure the distances to galaxies, and how does that allow the sizes, luminosities, and masses of galaxies to be determined?

- Astronomers find the distance to galaxies using objects of known luminosity, such as Cepheid variable stars and Type 1a supernovae explosions.
- Astronomers can estimate the distance to a galaxy by dividing its apparent rate of recession (its radial velocity) by the Hubble constant. Apparent rate of recession is related to distance by the Hubble law.
- Astronomers measure the masses of galaxies in several ways. The most precise is the rotation curve method, but it can be applied only to nearby galaxies.
- Observations show that galaxies contain 10 to 100 times more dark matter than visible matter. Signs of the dark matter are more obvious in outer halos.
- Stars near the centers of galaxies are orbiting at high velocities, suggesting the presence of supermassive black holes in the centers of most galaxies.

10.3 The Evolution of Galaxies

Why are there different kinds of galaxies, and how do galaxies evolve?

- Rich clusters of galaxies contain fewer spirals than do poor clusters of galaxies, and that is evidence that galaxies evolve by collisions and mergers.
- When galaxies collide, tides twist and distort their shapes, and the resulting compression of gas clouds triggers star formation.
- There is clear evidence that our own Milky Way Galaxy is devouring some of the small galaxies that orbit nearby and that it has consumed other small galaxies in the past.
- Shells of stars, counter-rotating parts of galaxies, streams of stars in the halos of galaxies, and multiple nuclei are evidence that galaxies can merge.
- The merger of two larger galaxies can scramble star orbits and drive bursts of star formation to use up gas and dust.
- Spiral galaxies have thin, delicate disks and appear not to have suffered mergers with large galaxies.
- A galaxy moving through the gas in a cluster of galaxies can be stripped of its own gas and dust and become an S0 galaxy.

Key Terms

elliptical galaxy A galaxy that is round or elliptical in outline and contains little gas and dust, no disk or spiral arms, and few hot, bright stars.

spiral galaxy A galaxy with an obvious disk component containing gas; dust; hot, bright stars; and spiral arms.

barred spiral galaxy A spiral galaxy with an elongated nucleus resembling a bar from which the arms originate.

irregular galaxy A galaxy with a chaotic appearance, large clouds of gas and dust, and both population I and II stars, but without spiral arms.

Large Magellanic Cloud The larger of two irregular galaxies visible in the southern sky passing near our Milky Way galaxy.

Small Magellanic Cloud The smaller of two irregular galaxies visible in the southern sky passing near our Milky Way galaxy.

megaparsec (Mpc) A unit of distance equal to 1,000,000 pc.

standard candle Object of known brightness that astronomers use to find distance—for example, Cepheid variable stars and supernovae.

distance ladder The calibration used to build a distance scale reaching from the size of Earth to the most distant visible galaxies.

look-back time The amount by which you look into the past when you look at a distant galaxy, a time equal to the distance to the galaxy in light-years.

Hubble law The linear relation between the distances to galaxies and the apparent velocity of recession.

Key Terms

Hubble constant A measure of the rate of expansion of the universe, the average value of the apparent velocity of recession divided by distance, about 70 km/s/Mpc.

rotation curve method A method of determining a galaxy's mass by observing the orbital velocity and orbital radius of stars in the galaxy.

rich galaxy cluster A cluster containing a thousand or more galaxies, usually mostly ellipticals, scattered over a volume only a few Mpc in diameter.

poor galaxy cluster An irregularly shaped cluster that contains fewer than 1,000 galaxies, many of which are spiral, and no giant ellipticals.

Local Group The small cluster of a few dozen galaxies that contains our Milky Way Galaxy.

starburst galaxy A galaxy undergoing a rapid burst of star formation.

active galaxy A galaxy whose center emits large amounts of excess energy, often in the form of radio emission. Active galaxies have massive black holes in their centers into which matter is flowing.

radio galaxy A galaxy that is a strong source of radio signals.

active galactic nuclei (AGN) The centers of active galaxies that are emitting large amounts of excess energy.

tidal tail A long streamer of stars, gas, and dust torn from a galaxy during its close interaction with another passing galaxy.

galactic cannibalism The theory that large galaxies absorb smaller galaxies.

ring galaxy A galaxy that resembles a ring around a bright nucleus, resulting from a head-on collision between two galaxies.

Topic Summaries

- At great distances, astronomers see that galaxies were smaller, more irregular, and closer together. Also, there were more spirals and fewer ellipticals long ago.

10.4 Active Galaxies and Quasars

What is the energy source for active galaxies, what can trigger the activity, and what does that reveal about the history of galaxies?

- Seyfert galaxies are spirals with small, highly luminous cores. Spectra of Seyfert galaxy nuclei show that they contain highly excited, rapidly moving gas.
- Because the brightness of AGN can change noticeably in only a few minutes or hours, they must be smaller than a few AU in diameter.
- The lobes in double-lobed radio galaxies appear to be inflated by jets ejected from the nuclei of the galaxies.
- Quasars have very high redshifts, evidence of great distance. To be visible from so far, quasars must be ultraluminous and provide a window into time where galaxies were just forming.
- The spectra of hazy objects near quasars and the spectra of quasar fuzz show that quasars are the active cores of very distant galaxies.
- Evidence is strong that AGN contain supermassive black holes into which matter is flowing through hot accretion disks. This can eject jets in opposite directions that push into the intergalactic medium and inflate radio lobes like balloons.
- The mass of a central black hole can be found by observing the velocity of stars orbiting the black hole or the rotational speed of its accretion disk.
- According to the unified model, what you see depends on the tilt of the accretion disk. Seen face-on, the core produces broad spectral lines. Seen edge-on, the core is hidden and the spectrum contains narrow lines.
- Many galaxies appear to contain dormant supermassive black holes at their centers. Only when a supermassive black hole is fed does it erupt.
- Matter flowing into a hot accretion disk around a black hole can eject high-energy jets of gas and radiation perpendicular to the disk.
- Interactions between galaxies can throw matter into the center, feed the black hole, and trigger eruptions.
- Supermassive black holes powering AGN and quasars cannot have been formed by dying stars but may have formed as the nuclear bulges of their host galaxies began to form and contract.

*remember

Review questions can be found online at 4ltrpress.cengage.com/astro.

Seyfert galaxy An otherwise normal spiral galaxy with an unusually bright, small core that fluctuates in brightness.

double-lobed radio source A galaxy that emits radio energy from two regions (lobes) located on opposite sides of the galaxy.

quasar Small, powerful source of energy in the active core of a very distant galaxy.

unified model An attempt to explain the different types of active galactic nuclei using a single model viewed from different directions.

blazar A type of active galaxy nucleus that is especially variable and has few or no spectral emission lines.

Topic Summaries

11.1 Introduction to the Universe

Does the universe have an edge and a center?

- Astronomers conclude that it is impossible for the universe to have an edge because it introduces logical inconsistencies. If the universe has no edge, then it cannot have a center.

- The darkness of the night sky leads to the conclusion that the universe is not infinitely old. If it were infinite in extent and age, then every spot on the sky would glow as brightly as the surface of a star. This problem, commonly known as Olbers's paradox, leads to the conclusion that the universe had a beginning.

11.2 The Big Bang Theory

How do you know that the universe began with a big bang?

- Edwin Hubble's 1929 discovery that the redshift of a galaxy is proportional to its distance is known as the Hubble law and indicates that the universe is expanding. Tracing this expansion backward in time brings you to an initial high-density, high-temperature state commonly called the big bang.

- The galaxies do not recede from a single point. They recede from each other as space expands between them.

- The CMB is blackbody radiation with a temperature of about 2.73 K uniformly spread over the entire sky. It is the light from the big bang released from matter at the time of recombination and now redshifted by a factor of 1,100.

- The background radiation is clear evidence that the universe began with a big bang.

- During the first 3 minutes of the big bang, nuclear fusion converted some of the hydrogen into helium but was unable to make many other heavy atoms because no stable nuclei exist with atomic weights of 5 or 8. Today hydrogen and helium are common in the universe, but heavier atoms are rare.

11.3 Space and Time, Matter and Energy

How does the universe expand?

- In its major features, the universe is isotropic and homogeneous. It looks the same in all directions and in all locations.

- Isotropy and homogeneity together lead to the cosmological principle, the idea that there are no special places in the universe. Except for local differences, every place is the same.

- General relativity explains that cosmic redshifts are caused by the stretching of photon wavelengths as they travel through expanding space-time.

- Model universes with flat (uncurved) space-time are infinite and can expand forever. A flat universe would have an average density equal to what is called the critical density. Modern observations indicate that the universe is flat.

- The amounts of deuterium and lithium-7 show that normal baryonic matter can make up only about 4 percent of the critical density. Dark matter must be nonbaryonic and makes up less than 30 percent of the critical density. For the universe to be flat, there must be another major component aside from baryonic matter and dark matter.

Key Terms

cosmology The study of the nature, origin, and evolution of the universe.

Olbers's paradox The conflict between theory and evidence regarding the darkness of the night sky.

observable universe The part of the universe that you can see from your location in space and in time.

big bang The high-density, high-temperature state from which the expanding universe of galaxies began.

Hubble time The age of the universe, equivalent to 1 divided by the Hubble constant. The Hubble time is the age of the universe if it has expanded since the big bang at a constant rate.

cosmic microwave background radiation (CMB) Radiation from the hot matter of the universe soon after the big bang. The large redshift makes it appear to come from a blackbody with a temperature of 2.7 K.

antimatter Matter composed of antiparticles, which upon colliding with a matching particle of normal matter annihilate and convert the mass of both particles into energy. The antiproton is the antiparticle of the proton, and the positron is the antiparticle of the electron.

recombination The stage, within 400,000 years of the big bang, when the gas became transparent to radiation.

dark age The period of time after the glow of the big bang faded into the infrared and before the birth of the first stars, during which the universe expanded in darkness.

reionization The stage in the early history of the universe when ultraviolet photons from the first stars ionized the gas filling space.

isotropy The observation that, in its general properties, the universe looks the same in every direction.

homogeneity The observation that, on the large scale, matter is uniformly spread through the universe.

cosmological principle The assumption that any observer in any galaxy sees the same general features of the universe.

open universe A model of the universe in which space-time is curved in such a way that the universe is infinite.

closed universe A model of the universe in which space-time is curved to meet itself and the universe is finite.

flat universe A model of the universe in which space-time is not curved.

critical density The average density of the universe needed to make its curvature flat.

nonbaryonic matter Proposed dark matter made up of particles other than protons and neutrons (baryons).

cold dark matter Dark matter that is made of slow-moving particles.

flatness problem In cosmology, the peculiar circumstance that the early universe must have contained almost exactly the right amount of matter to make space-time flat.

horizon problem In cosmology, the circumstance that the primordial background radiation seems much more isotropic than can be explained by the standard big bang theory.

inflationary big bang A version of the big bang theory, derived from grand unified theories of particle physics, that includes a rapid expansion when the universe was very young.

11.4 Modern Cosmology

How has the universe evolved, and what will be its fate?

- The inflationary theory proposes that the universe expanded dramatically a tiny fraction of a second after the big bang.
- Inflation solves the flatness problem because the sudden inflation forced the universe to become flat, just as a spot on an inflating balloon becomes flatter as the balloon inflates.
- Inflation solves the horizon problem because the part of the universe that is now observable was so small before inflation that energy could move and equalize the temperature everywhere.
- Observations of type Ia supernovae reveal that the expansion of the universe is speeding up. This is thought to be due to "dark energy."
- The nature of dark energy is unknown. It may be described by Einstein's cosmological constant, or it may change with time, in which case astronomers refer to it as quintessence.
- The observed value of the Hubble constant implies that the universe is 13.7 billion years old. The future fate of the universe depends on the nature of dark energy. If dark energy increases in strength with time, the universe could end in a big rip.
- The sudden inflation of the universe could have magnified tiny quantum mechanical fluctuations in space-time. These very large-scale but weak differences in gravity worked with dark matter to draw together baryonic (ordinary) matter and create the large-scale structure.
- Precise observations of irregularities in the CMB and in the large-scale structure of the universe confirm that the universe is flat and contains about 4 percent baryonic matter, 23 percent dark matter, and 73 percent dark energy.

Review Questions

1. How can Earth be located at the center of the observable universe if you accept the Copernican principle?
2. Why couldn't atomic nuclei exist when the universe was less than 2 minutes old?
3. Why must the universe have been very uniform during its first 50,000 years?
4. What evidence shows that the expansion of the universe is accelerating?

cosmological constant A constant in Einstein's equations of space and time that represents a force of repulsion.

quintessence A possible form of dark energy that can change in strength as the universe ages.

big rip The fate of the universe if dark energy increases with time and galaxies, stars, and even atoms are eventually ripped apart by the accelerating expansion of the universe.

supercluster A cluster of galaxy clusters.

large-scale structure The distribution of clusters and superclusters of galaxies in filaments and walls enclosing voids.

Topic Summaries

12.1 The Great Chain of Origins

What theory explains the origin of the solar system?

- Modern astronomy reveals that the matter in our solar system was formed in the big bang, and the atoms heavier than helium were cooked up in successive generations of stars. The sun and planets evidently formed from a cloud of gas in the interstellar medium.
- Hot disks of gas and dust have been detected around young stars and are understood to be where planets could form.
- The solar nebula hypothesis proposes that the planets formed in a disk of gas and dust around the protostar that became the sun.

12.2 A Survey of the Planets

What are the observed properties of the solar system that must be explained by the theory of its origin?

- The solar system is disk shaped. The orbital revolutions of the planets, as well as most of their axial rotations and most of the orbits of their moons, share a common direction of motion.
- The planets are divided into two types: Terrestrial planets and Jovian planets.
- The Jovian worlds have ring systems and large families of moons. The Terrestrial planets have no rings and few moons.

12.3 Space Debris: Planet Building Blocks

What are asteroids and comets and what are their connections to meteors and meteorites?

- Asteroids are irregular in shape and heavily cratered by collisions. Most asteroids lie in a belt between Mars and Jupiter, although some follow orbits that cross into the inner solar system.
- Comets are icy bodies that fall into the inner solar system along long elliptical orbits.
- A comet gas tail is ionized gas carried away by the solar wind. A comet dust tail is solid debris released from the nucleus and blown outward by the pressure of sunlight.
- Meteoroids that fall into Earth's atmosphere are vaporized by friction and become visible as meteors. Meteoroids that reach the ground are called meteorites.
- Evidence suggests that most meteorites are fragments of asteroids. The vast majority of meteors appear to be bits of debris from comets.

12.4 The Story of Planet Formation

How do planets form?

- The age of a rocky body can be found by radioactive dating. The oldest objects in our solar system are some meteorites that have ages of 4.6 billion years. This is taken to be the age of the solar system.
- Planets begin growing by accreting solid material. But once a planet approaches about 15 Earth masses, it can begin growing by gravitational collapse as it pulls in gas from the solar nebula.

Key Terms

solar nebula theory The theory that the planets formed from a spinning disk of material around the forming sun.

asteroid Small, rocky world. Most asteroids orbit between Mars and Jupiter in the asteroid belt.

comet One of the small, icy bodies that orbit the sun and produce tails of gas and dust when they approach the sun.

Terrestrial planet An Earthlike planet—small, dense, rocky.

Jovian planet Jupiter-like planet with a large diameter and low density.

Galilean satellite The four largest satellites of Jupiter, named after their discoverer, Galileo.

volatile Easily evaporated.

Kuiper belt The collection of icy planetesimals orbiting in a region from just beyond Neptune out to 50 AU or more.

meteor A small bit of matter heated by friction to incandescent vapor as it falls into Earth's atmosphere.

meteoroid A meteor in space before it enters Earth's atmosphere.

meteorite A meteor that survives its passage through the atmosphere and strikes the ground.

carbonaceous chondrite Stony meteorite that contains small glassy spheres called chondrules and volatiles. These chondrites may be the least-altered remains of the solar nebula still present in the solar system.

meteor shower A display of meteors that appear to come from one point in the sky, understood to be cometary debris.

half-life The time required for half of the radioactive atoms in a sample to decay.

Key Concepts

uncompressed density The density a planet would have if its gravity did not compress it.

ice line Boundary beyond which water vapor could freeze to form ice.

condensation sequence The sequence in which different materials condense from the solar nebula depending on distance from the sun.

planetesimal One of the small bodies that formed from the solar nebula and eventually grew into protoplanets.

condensation The growth of a particle by addition of material from surrounding gas, atom by atom.

accretion The sticking together of solid particles to produce a larger particle.

protoplanet Massive object, destined to become a planet, resulting from the coalescence of planetesimals in the solar nebula.

gravitational collapse The process by which a forming body such as a planet gravitationally captures gas rapidly from the surrounding nebula.

differentiation The separation of planetary material inside a planet according to density.

outgassing The release of gases from a planet's interior.

heavy bombardment The intense cratering that occurred sometime during the first 0.5 billion years in the history of the solar system.

NEO (Near Earth Object) A small solar system body (asteroid or comet) with an orbit near enough to Earth that it poses some threat of eventual collision.

evolutionary theory An explanation of a phenomenon involving slow, steady processes of the sort seen happening in the present day.

catastrophic theory An explanation of a phenomenon involving special, sudden, perhaps violent, events.

Topic Summaries

- The condensation sequence explains that the Terrestrial planets formed in the inner solar system where only denser minerals could condense to form solids, while the Jovian planets formed farther out where ices could condense.
- The Terrestrial planets may have formed slowly from the accretion of planetesimals of similar composition and then differentiated later when radioactive decay heated the planet's interiors.
- Earth's first atmosphere was outgassed.
- Disks of gas and dust may not last long enough to form Jovian planets by core accretion and then by gravitational collapse. Some models suggest Jovian planets can form rapidly by gravitational collapse without core accretion.
- The asteroids formed as rocky planetesimals between Mars and Jupiter, but Jupiter prevented them from forming a planet.
- Comets are believed to have formed as icy planetesimals in the outer solar system; comets falling in from the Oort cloud become long-period comets.
- Other icy bodies in the outer solar system make up the Kuiper belt. Objects from the Kuiper belt can become short-period comets.
- All of the old surfaces in the solar system were heavily cratered by an early bombardment of the debris that filled the solar system when it was young.

12.5 Planets Orbiting Other Stars

What do astronomers know about other solar systems?

- Debris disks appear to be produced by dust released by collisions among objects in other planetary systems and may contain planets.
- Planets orbiting other stars have been detected because they create small Doppler shifts in the stars' spectra. Planets have also been detected as they cross in front of their star and dim the star's light.
- Nearly all extrasolar planets found so far are massive, Jovian worlds. Lower-mass Terrestrial planets are harder to detect but are presumably common.

Review Questions

1. What produced the helium now present in the sun's atmosphere? In Jupiter's atmosphere? In the sun's core?

2. According to the solar nebula theory, why is the sun's equator nearly in the plane of Earth's orbit?

3. Why does the solar nebula theory predict that planetary systems are common?

4. What is the difference between condensation and accretion?

5. What evidence can you cite that planets orbit other stars?

heat of formation In planetology, the heat released by infalling matter during the formation of a planetary body.

Oort cloud The hypothetical source of comets, a swarm of icy bodies understood to lie in a spherical shell extending to 100,000 AU from the sun.

debris disk A disk of dust found by infrared observations around some stars. The dust is debris from collisions among asteroids, comets, and Kuiper belt objects.

extrasolar planet A planet orbiting a star other than the sun.

Topic Summaries

13.1 A Travel Guide to the Terrestrial Planets

How can comparison help you understand the Terrestrial planets?

- The Terrestrial worlds differ mainly in size, but they all have low-density rocky crusts, mantles of dense rock, and metallic cores.

- Comparative planetology alerts you to expect that cratered surfaces are old, that heat flowing out of a planet drives geological activity, and that the nature of a planet's atmosphere depends on both the size of the planet and its temperature.

- Earth's moon illustrates important principles such as cratering and flooding of basins by lava.

13.2 Earth: The Active Planet

What are the main features of Earth when you view it as a planet?

- Earth has passed through three stages as it has evolved: (1) differentiation, forming a liquid metallic core generating a magnetic field; (2) cratering and basin formation; and (3) continued slow surface evolution.

- Earth is dominated by plate tectonics that breaks the crust into moving plates. This process is driven by heat flowing upward from the interior.

- Earth's primary atmosphere was probably mostly carbon dioxide, but that gas dissolves in seawater, and plant life has created oxygen, so the primary atmosphere has been replaced by a secondary atmosphere.

13.3 The Moon

How does size determine the geological activity and evolution of a planetary body?

- Earth's moon formed in a molten state and partly differentiated, but it contains little metal and has a low density.

- Because it is a small world, Earth's moon has lost most of its internal heat and is no longer geologically active. Its old highlands are heavily cratered, but the lowland maria are filled by smooth lava flows that formed soon after the end of the heavy bombardment.

- The large-impact hypothesis suggests the moon formed when an impact between Earth and a very large planetesimal ejected debris that formed a disk around Earth. The moon formed from that disk.

13.4 Mercury

How does Mercury compare to Earth?

- Mercury is smaller than Earth but larger than Earth's moon. It is airless and has an old, heavily cratered surface.

- Mercury has a much higher density than Earth's moon and must have a large metallic core. It may have suffered a major impact when it was young that drove off some of the lower-density crust and mantle rock and left Mercury with a metallic core that is larger than the condensation sequence would imply.

- Mercury has long curving ridges formed by compression of the crust when its large metallic core solidified and contracted.

Key Terms

comparative planetology Understanding planets by searching for and analyzing contrasts and similarities among them.

mantle The layer of dense rock and metal oxides that lies between the molten core and Earth's surface or a similar layer in another planet.

***P* wave** A type of seismic wave involving compression and decompression of the material through which it passes.

***S* wave** A type of seismic wave involving lateral motion of the material through which it passes.

primary atmosphere A planet's first atmosphere.

secondary atmosphere A planet's atmosphere that replaces the primary atmosphere, for example by outgassing, impact of volatile-bearing planetesimals, or biological activity.

greenhouse effect The process by which a carbon dioxide atmosphere traps heat and raises the temperature of a planetary surface.

global warming The gradual increase in the surface temperature of Earth caused by human modifications to Earth's atmosphere.

plate tectonics The constant destruction and renewal of Earth's surface by the motion of sections of crust.

rift valley A long, straight, deep valley produced by the separation of crustal plates.

midocean rise One of the undersea mountain ranges that push up from the seafloor in the center of the oceans.

basalt Dark igneous rock, characteristic of solidified lava.

subduction zone A region of a planetary crust where a tectonic plate slides downward.

Key Terms

folded mountain range A long range of mountains formed by the compression of a planet's crust.

maria (mare) One of the lunar lowlands filled by successive flows of dark lava, from the Latin for "sea."

albedo The ratio of the amount of light reflected from an object to the amount of light received by the object. Albedo equals 0 for perfectly black and 1 for perfectly white.

ejecta Pulverized rock scattered by meteorite impacts on a planetary surface.

anorthosite Aluminium- and calcium-rich silicate rock found in the lunar highlands.

breccia Rock composed of fragments of older rocks bonded together.

large-impact hypothesis Hypothesis that the moon formed from debris ejected during a collision between Earth and a large planetesimal.

magma ocean The exterior of the newborn moon, a shell of molten rock hundreds of kilometers deep.

multiringed basin Large impact feature (crater) containing two or more concentric rims formed by fracturing of the planetary crust.

late heavy bombardment The sudden temporary increase in the cratering rate in our solar system that occurred about 4 billion years ago.

micrometeorite Meteorite of microscopic size.

coronae On Venus, large, round geological faults in the crust caused by the intrusion of magma below the crust.

runaway greenhouse A greenhouse effect so dramatic that it amplifies itself, becoming stronger with time.

permafrost Permanently frozen soil.

shield volcano Wide, low-profile volcanic cone produced by highly liquid lava.

Topic Summaries

13.5 Venus

How do distance from the sun and size of a planet affect the properties of its atmosphere?

- Although Venus is almost as big as Earth, it has a thick, cloudy atmosphere of carbon dioxide that hides the surface from sight. The surface can be studied by radar mapping.
- The carbon dioxide atmosphere drives an intense greenhouse effect and makes Venus's surface hot enough to melt lead.
- Venus is slightly closer to the sun than Earth and was too warm for liquid water oceans to persist and dissolve carbon dioxide from the atmosphere, so Venus suffered a runaway greenhouse effect.
- The hot crust of Venus is dominated by volcanism but not by plate tectonics.

13.6 Mars

What is the evidence that surface conditions on Mars were originally more Earthlike than they are at present?

- Mars is about half the diameter of Earth; it has a thin atmosphere and has lost much of its internal heat.
- The loss of atmospheric gases depends on the size of a planet and its temperature. Mars is cold, but it is small and has a low escape velocity, and many of its lighter gases have leaked away.
- Some water may have leaked away because UV radiation from the sun broke it into hydrogen and oxygen, but some water is frozen in the polar caps and in the soil. Orbiters have also measured large amounts of water frozen below the surface.
- Rovers have found clear signs that liquid water flowed over the surface in at least some places and therefore evidence that the Martian climate was different in the past. The northern lowlands may even have held an ocean once.
- Outflow channels and valley networks are visible from orbit, but water cannot exist as a liquid on Mars now because of its low temperature and low atmospheric pressure.
- The southern hemisphere of Mars is old cratered terrain, but some large volcanoes lie in the north. The sizes of these volcanoes indicate that the crust does not have horizontal motions and plate tectonics.
- Some volcanism may still occur on Mars, but because the planet is small it has cooled and is not very active geologically.
- The two moons of Mars are probably captured asteroids.

Review Questions

1. What are the common stages in the development of Terrestrial planets?
2. Why would you expect planets to have differentiated?
3. Why doesn't Earth have as many craters as its moon or Venus?
4. What kind of erosion is now active on Earth's moon?

outflow channel Geological features on Mars and Earth caused by flows of vast amounts of water released suddenly.

valley network A system of dry drainage channels on Mars that resembles the beds of rivers and tributary streams on Earth.

Topic Summaries

14.1 A Travel Guide to the Outer Planets

What are the properties of the Jovian planets?

- The Jovian planets—Jupiter, Saturn, Uranus, and Neptune—are large, low-density worlds rich in hydrogen and helium.
- The atmospheres of the Jovian planets are marked by cloud belts parallel to their equators.
- Jupiter and Saturn, usually called "gas giants," are composed mostly of liquid hydrogen and might instead be called "liquid giants." Uranus and Neptune contain water in liquid and solid form and therefore are sometimes called "ice giants."
- All of the Jovian worlds have large systems of satellites and rings that have had complex histories.

14.2 Jupiter

- Jupiter's atmosphere contains three layers of clouds formed of hydrogen-rich molecules such as ammonia and water.
- The clouds are in bands parallel to the equator called zones and belts. Zones are high-pressure regions of rising gas, and belts are lower-pressure areas of sinking gas.
- Spots in Jupiter's atmosphere are circulating weather patterns.
- Models indicate that Jupiter has a core of heavy elements and a deep mantle of liquid metallic hydrogen in which the planet's magnetic field is generated.
- The magnetic field around Jupiter traps high-energy particles from the sun to form intense radiation belts.
- Jupiter must be very hot inside because heat is flowing out of it.
- Jupiter's ring is composed of dark particles that strongly forward-scatter light, which means the particles are very small. They are probably composed of dust resulting from meteorite impacts on moons.
- Jupiter's ring, like all of the rings in the solar system, lies inside the planet's Roche limit, where moons would be torn apart or unable to form.

How do you know that some moons have been geologically active?

- Grooves on Ganymede, smooth ice and cracks on Europa, and active volcanoes on Io show that tidal heating has made these moons active.

14.3 Saturn

- Saturn is less dense than water and contains a small core and less metallic hydrogen than Jupiter.
- The cloud layers on Saturn occur at the same temperature as those on Jupiter, but, because Saturn is further from the sun and colder, the cloud layers are deeper in the hydrogen atmosphere below a layer of methane haze.

How are planetary rings formed and maintained?

- Saturn's rings are composed of icy particles ranging in size from boulders to dust. In some regions the ice is purer than in other regions.
- Grooves and other features in the rings can be produced by resonances with moons or by waves that propagate through the rings.
- Narrow rings and sharp ring edges can be confined by shepherd satellites.

Key Terms

oblateness The flattening of a spherical body, usually caused by rotation.

liquid metallic hydrogen A form of liquid hydrogen that is a good electrical conductor, inferred to exist in the interiors of Jupiter and Saturn.

magnetosphere The volume of space around a planet within which the motion of charged particles is dominated by the planetary magnetic field rather than the solar wind.

belt–zone circulation The atmospheric circulation typical of Jovian planets in which dark belts and bright zones encircle the planet parallel to its equator.

forward scattering The optical property of finely divided particles to preferentially direct light in the original direction of the light's travel.

Roche limit The minimum distance between a planet and a satellite that can hold itself together by its own gravity. If a satellite's orbit brings it within its planet's Roche limit, tidal forces will pull the satellite apart.

tidal heating The heating of a planet or satellite because of friction caused by tides.

shepherd satellite A satellite that, by its gravitational field, confines particles to a planetary ring.

ovoid The oval features found on Miranda, a satellite of Uranus.

occultation The passage of a larger body in front of a smaller body.

dwarf planet A body that orbits the sun, is not a satellite of a planet, is massive enough to pull itself into a spherical shape but not massive enough to clear out other bodies in and near its orbit. For example, Pluto, Eris, and Ceres.

plutino One of the icy Kuiper belt objects that, like Pluto, is caught in a 3:2 orbital resonance with Neptune.

1. How can Jupiter have a liquid interior and not have a liquid surface?

2. How does the dynamo effect account for the magnetic fields of Jupiter and the other Jovian planets?

3. Why are the belts and zones on Saturn less distinct than those on Jupiter?

4. Explain why geological activity on Jupiter's moons varies with distance from the planet.

5. What makes Saturn's F ring and the rings of Uranus and Neptune so narrow?

6. What is the evidence that Enceladus is geologically active?

7. Why is the atmospheric activity of Uranus unlike that of Jupiter, Saturn, and Neptune?

8. What are the seasons on Uranus like?

9. Why are Uranus and Neptune greenish-blue and blue, respectively?

10. What evidence is there that Neptune's moon Triton has been geologically active recently?

11. What evidence indicates that catastrophic impacts have occurred in the solar system's past?

12. Two images of Uranus show it as it would look to the eye and through a red filter that enhances methane clouds in the northern hemisphere. Why didn't Voyager 2 photograph the northern hemisphere? What do these features tell you about atmospheric circulation on Uranus?

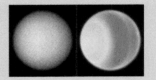

- The rings are short lived. They cannot be as old as their planet and must be replenished now and then with material from meteorites and comets colliding with moons.

- Titan has a cold, cloudy nitrogen and methane atmosphere. Sunlight entering Titan's atmosphere can convert methane into complex carbon-rich molecules to form haze and particles that settle out to coat the surface with dark organic goo. Methane lakes may have been detected on Titan's surface in radar images made by the Cassini probe.

- Enceladus has a light surface with some uncratered regions. Geysers of water and ice spray from the south polar region and provide ice particles to the E ring.

14.4 Uranus

How do the two smaller Jovian worlds—Uranus and Neptune—differ from their more massive siblings?

- Uranus is much less massive than Jupiter, and its internal pressure cannot produce liquid hydrogen. It has a heavy-element core and a mantle of slushy or solid ice and rock below a hydrogen-rich atmosphere.

- The atmosphere of Uranus is almost featureless at visual wavelengths with a pale green-blue color.

- The larger moons of Uranus are icy and heavily cratered, with some signs of past geological activity.

- The rings of Uranus are narrow hoops confined by shepherd satellites. The particles appear to be ice with traces of methane darkened by radiation.

- Uranus rotates "on its side," with its axis nearly in the plane of its orbit, perhaps because of a major impact during its early history.

14.5 Neptune

- Neptune is an ice giant like Uranus with little or no liquid hydrogen in its interior.

- The atmosphere of Neptune, marked by faint patterns of belt–zone circulation, is rich in hydrogen and colored blue by traces of methane.

- Neptune's satellite system is odd in that Nereid has an extremely elliptical orbit and Triton orbits backwards.

- Triton is icy with a thin atmosphere and frosty polar caps. Smooth areas suggest past geological activity, and dark smudges mark the location of active nitrogen geysers.

- The rings of Neptune are made of dark icy particles in narrow hoops. Neptune's rings contain arcs produced by the gravitational influence of one or more moons.

14.6 Pluto

- Pluto is a small, icy world with three moons, one of which is relatively large. The moons are in orbits highly inclined to Pluto's orbit around the sun. Pluto has a thin atmosphere.

Why do astronomers classify Pluto as a "dwarf planet"?

- Pluto is related to other icy objects in the Kuiper belt, some of which are larger than Pluto.

- Pluto does not meet all the IAU's criteria to be classified as a planet: though spherical, it hasn't cleared its orbital zone of smaller objects, making it a dwarf planet.

Topic Summaries

15.1 The Nature of Life

What is life?

- The process of life extracts energy from the environment, maintains the organism, and modifies the surroundings to promote the organism's survival.
- Living things have a physical basis—the arrangement of matter and energy that makes life possible. Life on Earth is based on carbon chemistry.
- Living things must also have controlling units of information, which can be passed to each new generation.
- Genetic information for life on Earth is stored in long carbon-chain molecules such as DNA.
- The DNA molecule stores information in the form of chemical bases linked together like the rungs of a ladder. Copied by the RNA molecule, the patterns of bases act as recipes for the manufacture of proteins and enzymes that are respectively the main structural and control components of the life process.
- When a cell divides, the DNA molecules split lengthwise and duplicate themselves so that each of the new cells can receive a copy of the genetic information.
- Errors in duplication or damage to the DNA molecule can produce mutants, organisms that contain new DNA information and have new properties. Variation in genetic codes can become widespread among individuals in a species.
- Natural selection determines which of these variations are best suited to survive, and the species evolves to fit its environment.

15.2 The Origin of Life

How did life originate on Earth?

- The oldest fossils on Earth are at least 3.4 billion years old. Those fossils provide evidence that life began in the oceans.
- Fossil evidence indicates that life began on Earth as simple organisms, such as bacteria, and evolved into more complex creatures.
- The Miller experiment shows that the chemical building blocks of life form naturally under a wide range of circumstances.
- Biologists hypothesize that chemical evolution concentrated simple molecules into a diversity of larger stable molecules, but those molecules did not reproduce copies of themselves.
- Biological evolution began when molecules started reproducing.
- Not until about 0.5 billion years ago did life forms become large and complex, during what is called the Cambrian explosion.
- Life emerged from the oceans only about 0.4 billion years ago, and human intelligence developed over the last 3 million (0.003 billion) years.

Could life begin on other worlds?

- Life as it is known on Earth requires liquid water and thus a specific range of temperatures.
- No other planet in our solar system appears to harbor life at present. Most are too hot or too cold, although life might have begun on Mars before it became too cold and dry.
- Liquid water exists, and therefore Earthlike life is at least possible, under the surfaces of Jupiter's moon Europa, Saturn's moon Enceladus, and possibly Jupiter's

Key Terms

biological evolution The process of mutation, variation, and natural selection by which life adjusts itself to its changing environment.

mutant Offspring born with DNA that is altered relative to parental DNA.

natural selection The process by which the best genetic traits are preserved and accumulated, allowing the fittest organisms and species to survive and proliferate.

DNA (deoxyribonucleic acid) The long carbon-chain molecule that records information to govern the biological activity of the organism. DNA carries the genetic data passed to offspring.

amino acid Carbon-chain molecule that is the building block of protein.

protein Complex molecule composed of amino acid units.

enzyme Special protein that controls processes in an organism.

RNA (ribonucleic acid) Long carbon-chain molecules that use the information stored in DNA to manufacture complex molecules necessary to the organism.

chromosome A body within a living cell that contains genetic information responsible for the determination and transmission of hereditary traits.

gene A unit of DNA—or sometimes RNA—information responsible for controlling an inherited physiological trait.

stromatolite A layered formation caused by mats of algae or bacteria combined with sediments.

Miller experiment An experiment that attempted to reproduce early Earth conditions and showed how easily amino acids and other organic compounds can form.

primordial soup The rich solution of organic molecules in Earth's first oceans.

chemical evolution The chemical process that led to the growth of complex molecules on primitive Earth. This did not involve the reproduction of exact molecules.

multicellular An organism composed of more than one cell.

Cambrian explosion A geologically brief period about 540 million years ago during which fossil evidence indicates Earth life became complex and diverse. Cambrian rocks contain the oldest easily identifiable fossils.

habitable zone A region around a star within which planets have temperatures that permit the existence of liquid water on their surfaces.

water hole The interval of the radio spectrum between the 21-cm hydrogen emission line and the 18-cm OH emission line; these are likely wavelengths to use in the search for extraterrestrial life.

SETI The Search for Extra-Terrestrial Intelligence.

Drake equation The equation that estimates the total number of communicative civilizations in our galaxy.

moons Ganymede and Callisto. Saturn's moon Titan has abundant organic compounds but does not have liquid water.

- Because the origin of life and its evolution into intelligent creatures took so long on Earth, scientists do not consider short-lived stars such as middle- and upper-main-sequence stars as likely hosts for life-bearing planets.
- Main-sequence G and K stars are thought to be likely candidates to host planets with life. The fainter M stars are also possibilities.
- The habitable zone around a star may be larger than scientists had expected, given the wide variety of living things now found in extreme environments on Earth and the possibility of tidal heating of moons orbiting large planets.

15.3 Communication with Distant Civilizations

Can Earthlings communicate with civilizations on other worlds?

- Because of distance, speed, and fuel, travel between the stars seems almost impossible for humans or for aliens who might visit Earth.
- Radio communication may be possible, but a real conversation would be difficult because of very long travel times for radio signals.
- Broadcasting a radio beacon of pulses would distinguish the signal from naturally occurring radio emission and identify the source as a technological civilization. The signal can be anticoded in hope it would be easy for another civilization to decode.
- The best part of the radio spectrum for communication is called the water hole, the wavelength range from the 21-cm spectral line of hydrogen to the 18-cm line of OH. Even so, millions of radio wavelengths need to be tested to fully survey the water hole for a given target star.
- Sophisticated searches are now underway to detect radio transmissions from civilizations on other worlds, but such SETI programs are hampered by limited computer power and radio noise pollution from human civilization.
- The number of civilizations in our galaxy that are at a technological level and able to communicate while humans are listening is limited by the lifetimes of their and our civilizations.

1. If life is based on information, what is that information?

2. What would happen to a life form if the information handed down to offspring was always the same? How would that endanger the future of the life form?

3. How does the DNA molecule produce a copy of itself?

4. Give an example of natural selection acting on new DNA patterns to select the most advantageous characteristics.

5. What evidence do scientists have that life on Earth began in the sea?

6. Why do scientists think that liquid water is necessary for the origin of life?

7. What is the difference between chemical evolution and biological evolution?

8. What was the significance of the Miller experiment?

9. How does intelligence make a creature more likely to survive?

10. Why are upper-main-sequence (high-luminosity) stars unlikely sites for intelligent civilizations?

11. Why is it reasonable to suspect that travel between stars is nearly impossible?

12. How does the stability of technological civilizations affect the probability that Earth can communicate with them?

13. What is the water hole, and why would it be a good "place" to look for other civilizations?

14. The star cluster NGC 2264 contains cool red giants and main-sequence stars from hot blue stars all the way down to red dwarfs. Discuss the likelihood that planets orbiting any of these stars might be home to life. Don't neglect to estimate the age of the cluster.

celestialprofile

Mercury

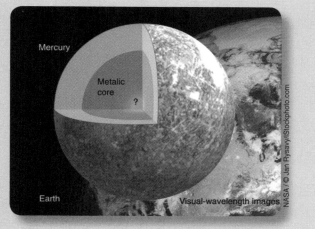

Mercury is slightly more than one-third the diameter of Earth. Its high density must mean it has a large iron core. The amount of heat it retains is unknown.

Mercury

Motion:

Average distance from Sun	0.387 AU (5.79 × 10^7 km)
Eccentricity of orbit	0.206
Inclination of orbit to ecliptic	7.0°
Orbital period	88.0 d (0.241 y)
Period of rotation (sidereal)	58.6 d (direct)
Inclination of equator to orbit	0°

Characteristics:

Equatorial diameter	4.88 × 10^3 km (0.383 $D_\oplus$)
Mass	3.31 × 10^{23} kg (0.0558 $M_\oplus$)
Average density	5.44 g/cm^3
Uncompressed density	5.4 g/cm^3
Surface gravity	0.38 Earth gravity
Surface temperature	−170 to 430°C (−275 to 805°F)
Average albedo	0.1
Oblateness	0

Venus

Venus is only 5 percent smaller than Earth. Its atmosphere is perpetually cloudy, and its surface is hot enough to melt lead. There is evidence that it has a hot core about the size of Earth's.

Venus

Motion:

Average distance from Sun	0.723 AU (1.08 × 10^8 km)
Eccentricity of orbit	0.007
Inclination of orbit to ecliptic	3.4°
Orbital period	0.615 y (224.68 d)
Period of rotation (sidereal)	243.02 d (retrograde)
Inclination of equator to orbit	177°

Characteristics:

Equatorial diameter	1.21 × 10^4 km (0.949 $D_\oplus$)
Mass	4.87 × 10^{24} kg (0.815 $M_\oplus$)
Average density	5.24 g/cm^3
Uncompressed density	4.2 g/cm^3
Surface gravity	0.90 Earth gravity
Surface temperature	470°C (880°F)
Average albedo	0.76 (cloud tops)
Oblateness	0

Earth

Mars

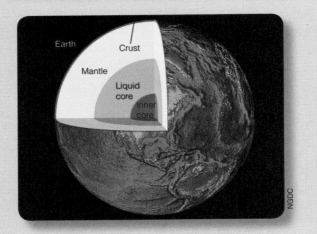

Earth's surface is marked by high continents and low seafloors, but the crust is only 10 to 60 km thick. Below that are a deep mantle and an iron core.

Mars is only half the diameter of Earth and probably retains some internal heat, but the size and composition of its core are not well known.

Earth

Motion:

Average distance from Sun	1.00 AU (1.50×10^8 km)
Eccentricity of orbit	0.017
Inclination of orbit to ecliptic	0° [by definition]
Orbital period	1.000 y (365.26 d)
Period of rotation (solar)	24.00 h
Period of rotation (sidereal)	23.93 h
Inclination of equator to orbit	23.4°

Characteristics:

Equatorial diameter	1.28×10^4 km
Mass	5.97×10^{24} kg
Average density	5.50 g/cm^3
Uncompressed density	4.1 g/cm^3
Surface gravity	1.00 Earth gravity
Surface temperature	−90 to 60°C (−130 to 140°F)
Average albedo	0.39
Oblateness	0.0034

Mars

Motion:

Average distance from Sun	1.52 AU (2.28×10^8 km)
Eccentricity of orbit	0.093
Inclination of orbit to ecliptic	1.9°
Orbital period	1.881 y (686.95 d)
Period of rotation (sidereal)	24.62 h (direct)
Inclination of equator to orbit	25.2°

Characteristics:

Equatorial diameter	6.79×10^3 km (0.532 $D_{\oplus}$)
Mass	6.42×10^{24} kg (0.108 $M_{\oplus}$)
Average density	3.94 g/cm^3
Uncompressed density	3.3 g/cm^3
Surface gravity	0.38 Earth gravity
Surface temperature	−140 to 15°C (−220 to 60°F)
Average albedo	0.16
Oblateness	0.009

celestialprofile

Jupiter

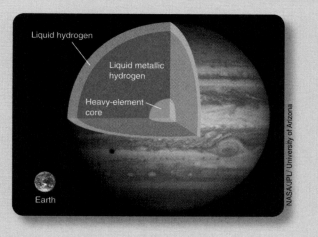

Jupiter is mostly a liquid hydrogen planet with a small core of heavy elements, not much bigger than Earth.

NASA/JPL/University of Arizona

Saturn

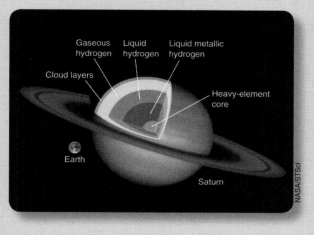

Density, oblateness, and gravity measurements made by planetary probes allow planetary astronomers to model Saturn's interior.

NASA/STScI

Jupiter

Motion:

Average distance from Sun	5.20 AU
Eccentricity of orbit	0.048
Inclination of orbit to ecliptic	1.3°
Orbital period	11.9 y
Period of rotation (sidereal)	9.92 h (direct)
Inclination of equator to orbit	3.1°

Characteristics:

Equatorial diameter	1.43×10^5 km (11.2 $D_{\oplus}$)
Mass	1.90×10^{27} kg (318 $M_{\oplus}$)
Average density	1.34 g/cm³
Gravity at base of clouds	2.5 Earth gravities
Temperature at cloud tops	145 K (−200°F)
Albedo	0.51
Oblateness	0.064

Saturn

Motion:

Average distance from Sun	9.58 AU
Eccentricity of orbit	0.056
Inclination of orbit to ecliptic	2.5°
Orbital period	29.5 y
Period of rotation (sidereal)	10.54 h (direct)
Inclination of equator to orbit	26.4°

Characteristics:

Equatorial diameter	1.21×10^5 km (9.4 $D_{\oplus}$)
Mass	5.68×10^{26} kg (95.2 $M_{\oplus}$)
Average density	0.69 g/cm³
Gravity at base of clouds	1.2 Earth gravities
Temperature at cloud tops	95 K (−290°F)
Albedo	0.61
Oblateness	0.102

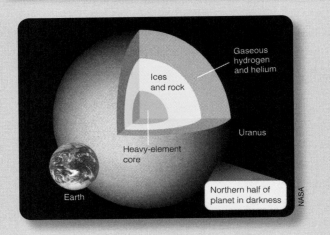

Uranus rotates on its side, and, when Voyager 2 flew past in 1986, the planet's south pole was pointed almost directly at the sun.

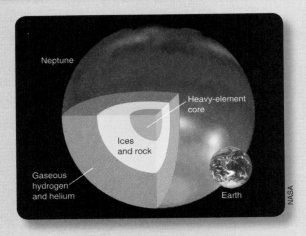

Neptune's axis was tipped slightly away from the sun when the Hubble Space Telescope recorded the image used to make this diagram. The interior is much like Uranus's, but there is more heat flowing outward from Neptune.

Uranus

Motion:

Average distance from Sun	19.23 AU
Eccentricity of orbit	0.046
Inclination of orbit to ecliptic	0.8°
Orbital period	84.0 y
Period of rotation (sidereal)	17.23 h (retrograde)
Inclination of equator to orbit	97.9°

Characteristics:

Equatorial diameter	5.11×10^4 km (4.01 $D_\oplus$)
Mass	8.69×10^{25} kg (14.5 $M_\oplus$)
Average density	1.29 g/cm³
Gravity at cloud tops	0.9 Earth gravities
Temperature above cloud tops	55 K (−360°F)
Albedo	0.35
Oblateness	0.023

Neptune

Motion:

Average distance from Sun	30.1 AU
Eccentricity of orbit	0.010
Inclination of orbit to ecliptic	1.8°
Orbital period	164.8 y
Period of rotation (sidereal)	16.05 h (direct)
Inclination of equator to orbit	28.8°

Characteristics:

Equatorial diameter	4.95×10^4 km (3.93 $D_\oplus$)
Mass	1.03×10^{26} kg (17.2 $M_\oplus$)
Average density	1.66 g/cm³
Gravity at cloud tops	1.2 Earth gravities
Temperature above cloud tops	55 K (−360°F)
Albedo	0.35
Oblateness	0.017

celestialprofile

Sun

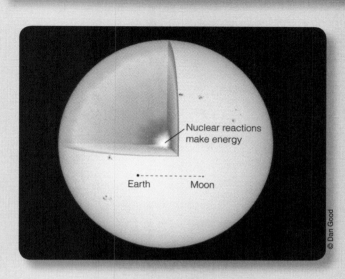

Nuclear reactions make energy

Earth · · · · · Moon

© Dan Good

This visible image of the sun shows a few sunspots and is cut away to show the location of energy generation at the sun's center. The Earth–moon system is shown for scale.

The Sun

From Earth:

Average distance from Earth	1.00 AU (1.50×10^8 km)
Average angular diameter	0.53° (32 arc minutes)
Period of rotation	25.4 days at equator
Apparent visual magnitude	−26.74

Characteristics:

Radius	6.96×10^5 km
Mass	1.99×10^{30} kg
Average density	1.41 g/cm³
Escape velocity at surface	618 km/s
Luminosity	3.83×10^{26} J/s
Surface temperature	5,780 K
Central temperature	15×10^6 K
Spectral type	G2 V
Absolute visual magnitude	+4.83

Moon

The moon

Visual-wavelength images

Earth

NASA

Earth's moon is about one-fourth the diameter of Earth. Its low density indicates that it contains little iron, but the size of its core and the amount of remaining heat are uncertain.

The Moon

Motion:

Average distance from Earth	3.84×10^5 km
Eccentricity of orbit	0.055
Inclination of orbit to ecliptic	5.1°
Orbital period (sidereal)	27.3 d
Orbital period (synodic)	29.5 d
Period of rotation (sidereal)	27.3 d (synchronous)
Inclination of equator to orbit	6.7°

Characteristics:

Equatorial diameter	3.48×10^3 km ($0.273\ D_{\oplus}$)
Mass	7.35×10^{22} kg ($0.0123\ M_{\oplus}$)
Average density	3.36 g/cm³
Uncompressed density	3.3 g/cm³
Surface gravity	0.17 Earth gravity
Surface temperature	−170 to 130°C (−275 to 265°F)
Average albedo	0.06
Oblateness	0

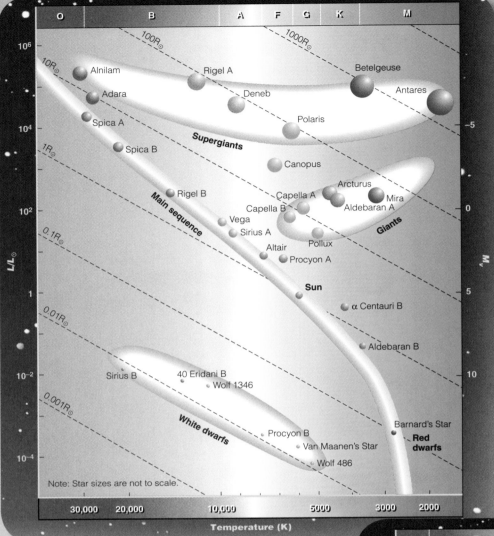

To get a better feel for how our sun relates to other stars, examine these two H–R diagrams.

This H–R diagram shows the luminosity and temperature of many well-known stars, including our sun. Now you can compare the sun with other stars you can see in the sky. The sizes of the stars on both H–R diagrams are not to scale. Using the size of the sun, sketch a few of the largest stars in these diagrams to scale.

This H–R diagram shows the masses of stars along the main sequence, as well as the masses of some giants and supergiants. Note that the sun's mass is defined as 1.

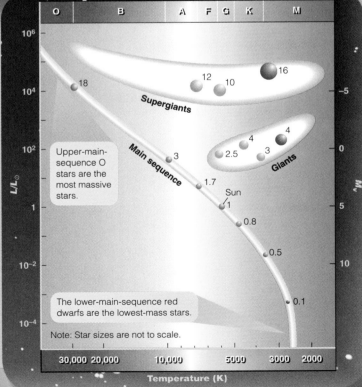

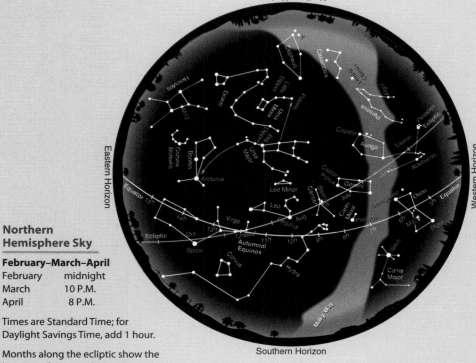

Northern Hemisphere Sky

February–March–April

February   midnight
March   10 P.M.
April   8 P.M.

Times are Standard Time; for Daylight Savings Time, add 1 hour.

Months along the ecliptic show the location of the sun during the year.

Numbers along the celestial equator show right ascension.

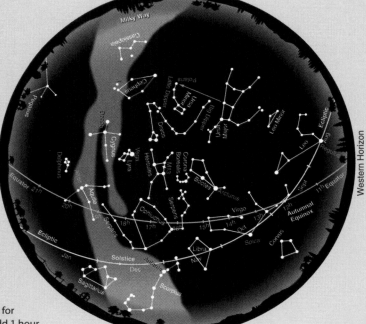

Northern Hemisphere Sky

May–June–July

May   midnight
June   10 P.M.
July   8 P.M.

Times are Standard Time; for Daylight Savings Time, add 1 hour.

Months along the ecliptic show the location of the sun during the year.

Numbers along the celestial equator show right ascension.

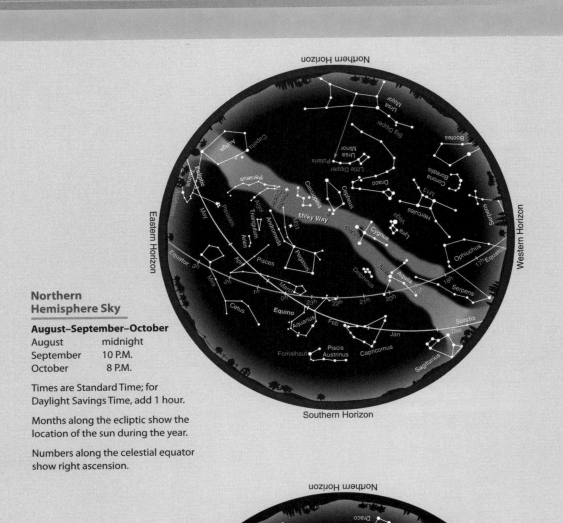

Northern Horizon

Eastern Horizon

Western Horizon

Southern Horizon

Northern Hemisphere Sky

August–September–October

August	midnight
September	10 P.M.
October	8 P.M.

Times are Standard Time; for Daylight Savings Time, add 1 hour.

Months along the ecliptic show the location of the sun during the year.

Numbers along the celestial equator show right ascension.

Northern Horizon

Eastern Horizon

Western Horizon

Southern Horizon

Northern Hemisphere Sky

November–December–January

November	midnight
December	10 P.M.
January	8 P.M.

Times are Standard Time; for Daylight Savings Time, add 1 hour.

Months along the ecliptic show the location of the sun during the year.

Numbers along the celestial equator show right ascension.